Structural Health Monitoring
10APWSHM

The 10th Asia-Pacific Workshop on Structural Health Monitoring (10APWSHM), 8th to 10th December 2024, Sendai, Japan.

Editors
S. Xue, K. Ikago, L. Xie and M. Cao
Tongji University, Tohoku University, Tohoku Institute of Technology

Peer review statement

All papers published in this volume of "Materials Research Proceedings" have been peer reviewed. The process of peer review was initiated and overseen by the above proceedings editors. All reviews were conducted by expert referees in accordance to Materials Research Forum LLC high standards.

Published under License by **Materials Research Forum LLC**
Millersville, PA 17551, USA

Published as part of the proceedings series
Materials Research Proceedings
Volume 50 (2025)

ISSN 2474-3941 (Print)
ISSN 2474-395X (Online)

ISBN 978-1-64490-350-6 (print)
ISBN 978-1-64490-351-3 (eBook)

This book contains information obtained from authentic and highly regarded sources. Reasonable efforts have been made to publish reliable data and information, but the author and publisher cannot assume responsibility for the validity of all materials or the consequences of their use. The authors and publishers have attempted to trace the copyright holders of all material reproduced in this publication and apologize to copyright holders if permission to publish in this form has not been obtained. If any copyright material has not been acknowledged please write and let us know so we may rectify in any future reprint.

Distributed worldwide by

Materials Research Forum LLC
105 Springdale Lane
Millersville, PA 17551
USA
https://mrforum.com

Manufactured in the United State of America
10 9 8 7 6 5 4 3 2 1

Table of Contents

Preface

The 10th Asia-Pacific Workshop on Structural Health Monitoring (10APWSHM) was held December 2024 in Sendai, Japan. As global attention continues to focus on structural safety and sustainability, Structural Health Monitoring (SHM) technology has become increasingly important in civil, maritime, and aerospace applications. This workshop aims to promote the latest advances in SHM technology and provide a platform for academic and industrial exchange.

Structural Health Monitoring is a multidisciplinary field, and its successful development and implementation require close collaboration between academics, researchers, original equipment manufacturers (OEMs), and end-users. Both fundamental research and practical applications are essential for developing SHM systems that meet stringent requirements for sensitivity, reliability, and confidence. Continuing the tradition of previous workshops, this event will focus on emerging technologies and practical applications, particularly in early warning systems, loss prevention, and structural safety assurance.

The organizing committee looks forward to welcoming experts and scholars from Asia, Australia, North America, and Europe. All submitted papers will undergo rigorous peer review to ensure technical originality and significance. This work will advance the development and application of SHM technology globally, contributing to the construction of safer and more reliable infrastructure.

We gratefully acknowledge the support of our sponsors:
- USAF Asian Office for Aerospace Research and Development (USAF, AOARD)
- The Obayashi Foundation
- Sendai Tourism, Convention and International Association
- Tohoku Institute of Technology
- The International Research Institute of Disaster Science, Tohoku University
- Tongji University
- Aerospace (Open Access Journal in MDPI)

The organizing committee especially appreciates the significant contributions made by previous workshops to the SHM field. From Melbourne (2002) to Tokyo, Shenzhen, Hobart, Hongkong, and Cairns, each workshop has advanced the field. This tradition of excellence will continue in Sendai, making new contributions to the development of structural health monitoring technology.

We anticipate that participants will establish new collaborations, inspire innovative research directions, and gain valuable insights into the latest trends and challenges in structural health monitoring. The committee is committed to providing a high-quality platform for academic exchange, promoting the development and implementation of SHM technology, and ultimately contributing to safer and more resilient structures worldwide.

Prof S. Xue	Prof K. Ikago	Prof W.K. Chiu	Prof A. Ogawa	Dr N. Rajic	Dr S. Takeda
Co-chair,	Co-chair,	Co-chair,	Co-chair,	Co-chair,	Co-chair,
Tohoku Institute of Technology, Japan and Tongji University, China	Tohoku University, Japan	Monash University, Australia	Keio University, Japan	Defence Science & Technology Group, Australia	Japan Aerospace Exploration Agency, Japan

Sponsors

The Organising Committee would like to express our sincere
appreciation to the following sponsors:

公益財団法人
大林
財団

未来のエスキースを描く。

Sendai Tourism, Convention and
International Association

TOHOKU INSTITUTE
OF TECHNOLOGY
東北工業大学

IRIDeS
International Research Institute
of Disaster Science
災害科学国際研究所

aerospace
an Open Access Journal by MDPI

10APWSHM 2024 COMMITTEES

Organising Committee
Workshop Chairs

Prof S. Xue	**Prof K. Ikago**	**Prof W.K. Chiu**	**Prof A. Ogawa**	**Dr N. Rajic**	**Dr S. Takeda**
Tohoku Institute of Technology, Japan and Tongji University, China	Tohoku University, Japan	Monash University, Australia	Keio University, Japan	Defence Science & Technology Group, Australia	Japan Aerospace Exploration Agency, Japan

Local Organising Committee

M. Cao, Tohoku Institute of Technology, Japan
N. Funaki, Tohoku Institute of Technology, Japan
K. Ikago, Tohoku University, Japan
Y. Xia, Tongji University, China
L. Xie, Tongji University, China
T. Hida, Ibaraki University, Japan
S. Xue, Tohoku Institute of Technology, Japan and Tongji University, China

International Organising Committee

D. Adams (Vanderbilt University, USA)
S. Alampalli (New York State Department of Transportation, USA)
C. Boller (Universität des Saarlandes, Germany)
M. Buderath (Airbus, Germany)
F.K. Chang (Stanford University, USA)
M. Derriso (USAF, AFRL, USA)
K.Q. Ding (China Special Equipment Inspection and Research Institute, China)
S.S. Ding (CRRC Qingdao Sifang Co Ltd, China)
V. Giurgiutiu (University of South Carolina, USA)
A. Guemes (Universidad Politecnica de Madrid, Spain)
P. Huthwaite (Imperial College, UK)
H. Li (Harbin Institute of Technology, China)
C. Lissenden (Penn State University, USA)
Y.C. Koay (VicRoads, Australia)
C.G. Koh (National University of Singapore, Singapore)
D. Le (Texas Technical University, USA)
V. Le Cam (IFSTTAR, Nantes, France)
Y. Lei (Xiamen University, China)
K. Loh (University of California, San Diego, USA)
J.Lynch (University of Michigan, USA)
A. Mal (University of California, Los Angeles, USA)
H. Muruyama (The University of Tokyo, Japan)
T. Nagayama (The University of Tokyo, Japan)
W. Ostachowicz (Polish Academy of Sciences, Poland)
B. Prosser (NASA Langley Research Center, USA)
L. Richards (NASA Armstrong Flight Research Center, USA)
N. Salowitz (University of Wisconsin, USA)
L. Salvino (US Office of Naval Research, USA)
H. Sohn (KAIST, Korea)
H. Speckmann (Airbus Operations GmbH, Germany)
D. Stargel (USAF, AFOSR, USA)
Z. Su (Hong Kong Polytechnic University, Hong Kong)
M. Todd (University of California, San Diego, USA)
H. Widyastuti (Sepuluh Nopember Institute of Technology, Indonesia)
S.T. Xue, (Tohoku Institute of Technology, Sendai, Japan)
F.G. Yuan (North Carolina State University, USA)
S.F. Yuan (Nanjing University of Aeronautics and Astronautics, China)

Keynote speakers

Prof. Xilin lv
Tongji University

Prof. Wingkong Chin
Monash University

Dr. Shin-ichi Takeda
Japan Aerospace Exploration Agency

Prof. Liyu Xie
Tongji University

Prof. Fuh-Gwo Yuan
North Carolina State University

Structural Health Monitoring: 10APWSHM
Materials Research Proceedings 50 (2025) 1-7

Materials Research Forum LLC
https://doi.org/10.21741/9781644903513-1

A SHM damage diagnosis model evolution mechanism for individual aircraft structure

Hutao Jing[1,a] and Shenfang Yuan[1,b] *

[1]Research Center of Structural Health Monitoring and Prognosis, State Key Laboratory of Mechanics and Control for Aerospace Structures, Nanjing University of Aeronautics and Astronautics, 29 Yudao Street, Nanjing 210016, China

[a]jinghutao@nuaa.edu.cn, [b]ysf@nuaa.edu.cn

Keywords: Guided Wave, Structural Health Monitoring, Individual Aircraft Structure, Hidden Markov Model, Model Evolution Mechanism

Abstract. The concept of aircraft health management is evolving from conventional fleet-based management to individual aircraft-based management. Accurate damage diagnosis with guided wave (GW)-based structural health monitoring (SHM) is of great significance for individual aircraft in service. However, both the damage propagation and monitoring of individual aircraft structures are affected by various uncertainties, such as time-varying environmental and operational conditions, different flight missions, and different damage morphologies. Consequently, employing a prior trained damage diagnosis model inevitably introduces errors, thereby limiting the engineering applicability of SHM. To achieve more reliable damage diagnosis for in-service aircraft structures, this paper proposes a whole lifetime data-based damage diagnosis model evolution mechanism. Multi-source data from the design, service, and maintenance stages are used to continuously evolve the probabilistic damage diagnosis model, enabling it to track the specific damage propagation. The proposed method is validated through fatigue crack monitoring experiments of a typical aircraft load-carrying structure. The results demonstrate that it can significantly improve the damage diagnosis performance for in-service aircraft structures under the influence of uncertainties.

Introduction

The health and life management of aircraft is evolving form traditional fleet-based management to a new concept of individual aircraft management [1]. Structural health monitoring (SHM) techniques can realize online monitoring of structural damage, which is crucial for ensuring structural safety, prolonging service life, and reducing maintenance costs [2,3]. The application of SHM has the potential to achieve condition-based maintenance (CBM) for individual aircraft [4].

Among SHM techniques, active guided wave (GW)-based SHM is considered one of the most promising methods due to its ability to cover a wide range and its sensitivity to small damage [5,6]. Various damage diagnosis algorithms have been developed till now, ranging from conventional damage indices to guided wave imaging, as well as the latest machine learning and probabilistic statistical learning methods [7,8]. However, existing damage diagnosis model training typically relies on prior data obtained from ground design and tests. An individual aircraft is subject to diverse uncertainties throughout its whole service time, such as time-varying environmental and operational conditions, different flight missions and damage morphologies [9]. In this case, models trained offline struggle to ensure their reliability over the long service life of individual aircraft, which presents challenges for the engineering application of SHM.

To deal with this issue, some researches have focused on training damage diagnosis models online that better fit individual aircraft by utilizing in-service data instead of prior data. Yang et al. [10] trained a linear relationship between GW signal features and crack length based on experimental data from batch specimens. After that a Bayesian approach was used to update the parameters with a small amount of crack length measurement data. Yuan et al. [11] proposed an online updated Gaussian process model-based crack diagnosis method. They improved the accuracy of online crack diagnosis through multiple updates.

In this paper, to achieve more accurate online damage diagnosis for individual aircraft structures, a model evolution mechanism is proposed and combined with a hidden Markov model (HMM) in probabilistic statistical learning. The proposed method is validated through fatigue crack monitoring experiments of a typical aircraft load-carrying structure. The validation result demonstrates that it can significantly improve the damage diagnosis performance for in-service aircraft structures under the influence of uncertainties.

Dynamic evolutionary GW-HMM-based damage diagnosis

Basics of the active GW-based SHM

The principle of the active GW-based SHM is to use piezoelectric transducers (PZTs) permanently integrated on the structure to excite and acquire GW signals, as shown in Fig. 1. The occurrence of structural damage may lead to changes in the GW propagating on the structure. A damage-related feature can be extracted by comparing the GW baseline signal and the online monitoring signal. Damage diagnosis can be further achieved by interpreting the damage feature.

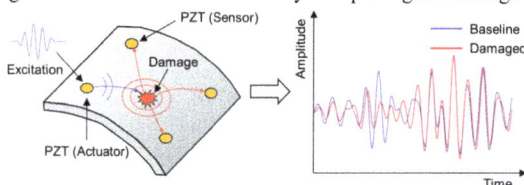

Fig.1 Principle of active GW-based SHM

Offline GW-HMM-based damage diagnosis

HMM can model the non-stationary time series like the GW feature sequence for damage diagnosis. It contains two stochastic processes, including the stochastic transition between hidden states and stochastic generation of observation sequences at each hidden state. Fig. 2 depicts the basic principle of conventional offline GW-HMM-based damage diagnosis.

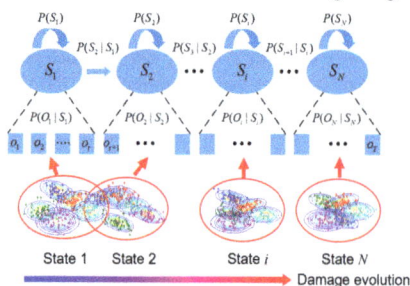

Fig.2 Principle of offline GW-HMM-based damage diagnosis

A GW-HMM can be expressed by a compact notation $\delta=[\pi, A, B]$. Its parameters include S, which indicates different hidden states. The total number of hidden states in the model is denoted as N. π is an initial state distribution vector. A is the state-transition probability matrix. B is the state-dependent observation density. $O=\{o_1, o_2, ..., o_T\}$ represents the observation sequence of GW damage features obtained from a total of T observations.

Structural Health Monitoring: 10APWSHM
Materials Research Proceedings 50 (2025) 1-7

Materials Research Forum LLC
https://doi.org/10.21741/9781644903513-1

For the task of damage diagnosis, the parameters of GW-HMM need to be adjusted offline to maximize the probability $P(O|\delta)$, which is the probability of an observation sequence O generated by the model δ, as shown in Eq. 1:

$$\delta_{opt} = \text{argmax } P(O|\delta). \tag{1}$$

where δ_{opt} denotes the optimal GW-HMM model parameters.

The Baum-Welch algorithm is used for the offline training of GW-HMM, which is a variant of the expectation-maximization (EM) algorithm. Its basic idea is to maximize the log-likelihood of $P(O|\delta)$ of the offline training data iteratively. Before the iteration, parameters of GW-HMM need to be initialized. The training will stop if the convergence criterion shown in Eq. 2 is satisfied or the maximum iteration number n_{max} is reached:

$$\log(P(O|\delta_{n+1})) - \log(P(O|\delta_n)) < \varepsilon. \tag{2}$$

where ε represents the convergence threshold.

After obtaining the optimal parameters of the GW-HMM through offline training, probability features representing damage information can be extracted from the model. By calibrating a mapping relationship between the features and the actual damage sizes of the offline training data, damage diagnosis can be achieved.

Dynamic evolutionary GW-HMM-based damage diagnosis

When an individual aircraft begins its service, the structures can be considered in a healthy state. The online GW monitoring data obtained during this period is used to dynamically evolve the offline-trained GW-HMM into an online health GW-HMM model. Corresponding probability feature extraction and mapping relationship calibration also need to be evolved synchronously. This is because the online GW monitoring data is believed to better represent the specific service conditions and material properties of the individual aircraft, eliminating the need to rely on average monitoring data from the fleet.

When the service time of individual aircraft increases, the probability of structural damage initiation significantly rises. At this point, in addition to online GW monitoring, regular offline inspections are typically conducted to confirm the presence of structural damage and its actual size. By combining offline inspection data with online GW monitoring data, further dynamic evolution of the GW-HMM can be achieved. Specifically, if damage propagation reaches a new stage, new hidden states can be introduced into the GW-HMM to represent it. Corresponding probability feature extraction and mapping relationship calibration will also be evolved to adapt to the latest structural damage state. Moreover, the actual damage sizes obtained from inspections can be used to correct the mapping relationship, improving the accuracy of damage diagnosis. The above dynamic evolution process will be repeatedly carried out each time damage propagation in the individual aircraft structure enters a new stage or after the implementation of offline inspections.

Fig. 3 shows the entire process of the proposed dynamic evolutionary GW-HMM-based damage diagnosis method.

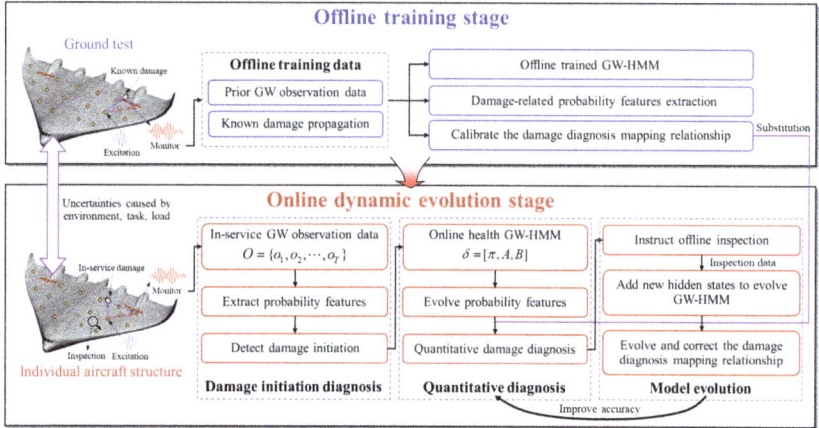

Fig.3 Principle of dynamic evolutionary GW-HMM-based damage diagnosis

Experimental validation

Validation setup

To prove the effectiveness of the proposed method, a validation experiment is performed on a kind of landing gear beam specimen under variable amplitude fatigue loading. This structure is made of aluminum alloy and designed referring to the critical region of a real aircraft's landing gear, as shown in Fig. 4. In this beam, there are two designed notches, where stress concentration exists. The stress concentration makes the structure prone to cracking after a period of fatigue loading. In addition, at one of the notches, a pre-crack is made by wire cutting with a length of 2 mm and 1 mm in depth. A total of six PZTs are integrated on the structure to perform active GW monitoring, which are PZT1 to PZT6.

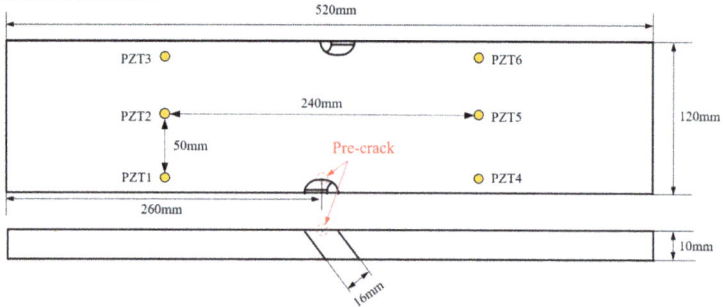

Fig.4 Landing gear beam specimen

The fatigue test setup is shown in Fig. 5. A material test system MTS809 is used to apply fatigue load. The dynamic load is a 10Hz variable amplitude spectrum with peak value V_{max}=98.9kN and valley value V_{min}=12kN. The active GW SHM system developed by the authors' group is employed to excite and collect GW signals. A 3-cycle tone-burst signal with a center frequency of 100kHz is used as the excitation signal.

Fig.5 Fatigue test setup

When the structure is healthy, a set of GW signals is collected initially as the baseline signal. Then during the fatigue loading GW signals are continuously excited and collected at an interval of 5s. Totally seven specimens are tested in this study, labeled as S1-S7.

Experimental results

In this paper, two kinds of damage indices, denoted as DI_1 and DI_2 are adopted as damage features to form the GW-HMM observation to provide enough information for accurate damage diagnosis. They are mean square deviation and spectrum magnitude difference, respectively. The damage indices are calculated by using the S0 mode direct wave, as shown in Fig. 6.

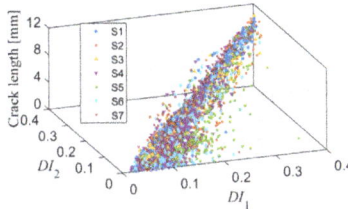

Fig.6 Damage indices versus crack length of batch specimens

From Fig. 6 differences can be found in the damage indices of batch specimens with the same crack length, which is caused by random fatigue loading and dispersion among batch specimens.

Method validation results

In this paper, a leave-one-out validation method is adopted to validate the effectiveness of the proposed method. All the data obtained from specimens S1-S6 are used for model training and S7 is chosen for online damage diagnosis.

Fig. 7(a) shows the offline training results of the GW-HMM using batch specimens as prior training data. Based on this, three crack length alarm thresholds of 2mm, 4mm and 8mm are set for the landing gear beam structure in this paper. When GW-based SHM diagnoses that the crack propagation reaches the alarm threshold, an inspection is triggered to obtain the actual crack length. Therefore, the GW-HMM after each dynamic evolution is shown in Fig. 7(b), 7(c) and 7(d), respectively. As the fatigue crack grows, the number of hidden states that represent the evolution of the structural damage state gradually increases.

Materials Research Forum LLC
https://doi.org/10.21741/9781644903513-1

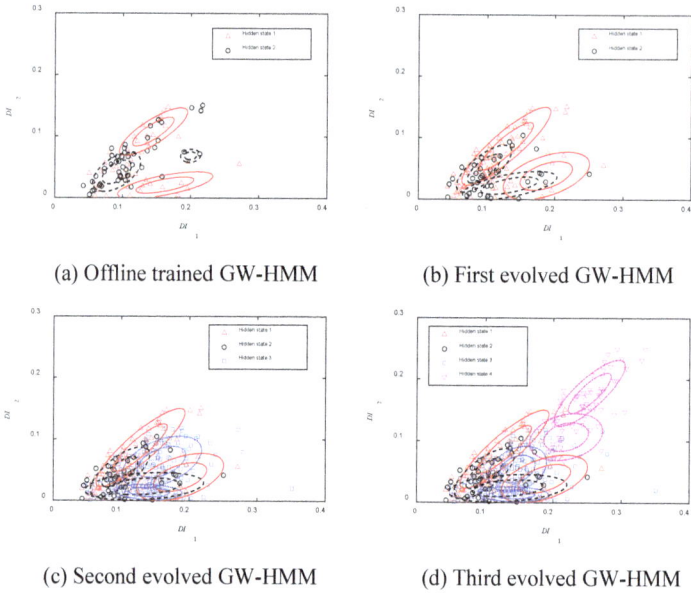

(a) Offline trained GW-HMM

(b) First evolved GW-HMM

(c) Second evolved GW-HMM

(d) Third evolved GW-HMM

Fig.7 Offline trained GW-HMM and its dynamic evolution

After GW-HMM training, probability features representing structural damage are extracted and then a mapping relationship for damage diagnosis is calibrated. Fig. 8 shows the damage diagnosis result of specimen S7 and how it changes with the dynamic evolution of the model. It can be found that the dynamic evolution of the model effectively reduces the diagnosis error during long-term service. The maximum error decreases from 2.1mm by the offline GW-HMM to 1.2mm.

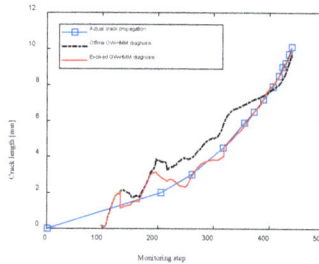

Fig.8 Crack length diagnosis result for specimen S7

Conclusion

This paper proposes a whole lifetime data-based model evolution mechanism to achieve more accurate online damage diagnosis for individual aircraft structures. Through dynamic evolution for three times, the proposed method has achieved satisfactory online fatigue crack diagnosis results on landing gear beam structures. For a total of seven specimens, the diagnosis maximum error is reduced from 2.1mm to 1.2mm.

Acknowledgment

This study is supported by National Natural Science Foundation of China (Grant No. 51921003 and 52275153), Frontier Technologies R&D Program of Jiangsu (Grant No. BF2024068), Fundamental Research Funds for the Central Universities (Grant No. NI2023001), the Fund of Prospective Layout of Scientific Research for Nanjing University of Aeronautics and Astronautics and the State Key Laboratory of Mechanics and Control for Aerospace Structures (Nanjing University of Aeronautics and astronautics) (Grant No. MCAS-S-0423G04).

References

[1] M. Kordestani, M.E. Orchard, K. Khorasani and M. Saif, An overview of the state of the art in aircraft prognostic and health management strategies, IEEE T. Instrum. Meas. 72 (2023) 1-15. https://doi.org/10.1109/TIM.2023.3236342

[2] V. Giurgiutiu, Structural health monitoring: with piezoelectric wafer active sensors, second ed., Academic Press, Burlington, 2014. https://doi.org/10.1016/B978-0-12-418691-0.00007-1

[3] A. Güemes, A. Fernandez-Lopez, A.R. Pozo and J. Sierra-Pérez, Structural health monitoring for advanced composite structures: a review, J. Compos. Sci. 4 (2020) 13. https://doi.org/10.3390/jcs4010013

[4] R. Gorgin, Y. Luo and Z.J. Wu, Environmental and operational conditions effects on Lamb wave based structural health monitoring systems: A review, Ultrasonics 105 (2020) 106114. https://doi.org/10.1016/j.ultras.2020.106114

[5] M. Mitra and S. Gopalakrishnan, Guided wave based structural health monitoring: A review, Smart Mater. Struct. 25(5) (2016) 053001. https://doi.org/10.1088/0964-1726/25/5/053001

[6] P. Cawley, Guided waves in long range nondestructive testing and structural health monitoring: Principles, history of applications and prospects, NDT & E Int. (2023) 103026. https://doi.org/10.1016/j.ndteint.2023.103026

[7] L. Qiu, S.F. Yuan, Q. Bao, H.F. Mei and Y.Q. Ren, Crack propagation monitoring in a full-scale aircraft fatigue test based on guided wave-Gaussian mixture model, Smart Mater. Struct. 25(5) (2016) 055048. https://doi.org/10.1088/0964-1726/25/5/055048

[8] S. Sawant, S. Patil, J. Thalapil, S. Banerjee and S. Tallur, Temperature variation compensated damage classification and localisation in ultrasonic guided wave SHM using self-learnt features and Gaussian mixture models, Smart Mater. Struct. 31(5) (2022) 055008. https://doi.org/10.1088/1361-665X/ac5ce3

[9] S.R. Yeratapally, M.G. Glavicic, C. Argyrakis and M.D. Sangid, Bayesian uncertainty quantification and propagation for validation of a microstructure sensitive model for prediction of fatigue crack initiation, Reliab. Eng. Syst. Saf. 164 (2017) 110-123. https://doi.org/10.1016/j.ress.2017.03.006

[10] J.S. Yang, J.J. He, X.F. Guan, D.J. Wang, H.P. Chen, W.F. Zhang and Y.M. Liu, A probabilistic crack size quantification method using in-situ Lamb wave test and Bayesian updating, Mech. Syst. Signal Process. 78 (2016) 118-133. https://doi.org/10.1016/j.ymssp.2015.06.017

[11] S.F. Yuan, H. Wang and J. Chen, A pzt based on-line updated guided wave-gaussian process method for crack evaluation, IEEE Sens. J. 20(15) (2019) 8204-8212. https://doi.org/10.1109/JSEN.2019.2960408

Structural Health Monitoring: 10APWSHM
Materials Research Proceedings 50 (2025) 8-14

Materials Research Forum LLC
https://doi.org/10.21741/9781644903513-2

Research on framework and condition control for SHM reliability evaluation

XU Qiuhui[1,a], YUAN Shenfang[1,b]* and CHEN Jian[1,c]

[1] Research Center of Structural Health Monitoring and Prognosis, State Key Lab of Mechanics and Control for Aerospace Structures, Nanjing University of Aeronautics and Astronautics, Nanjing 210016, P. R. China

[a]qhx@nuaa.edu.cn, [b]ysf@nuaa.edu.cn, [c]cj1108@nuaa.edu.cn

Keywords: Structural Health Monitoring, Reliability Evaluation, Evaluation Condition Control, Guided Wave-Based Monitoring, Crack Detection and Sizing

Abstract. The structural health monitoring (SHM) reliability evaluation is a key aspect that needs to be urgently addressed to promote the broader application of SHM. However, existing limited studies are mainly based on the non-destructive testing/evaluation (NDT/E) reliability metrics in a straightforward way without a systematic analysis of where these metrics originated from, especially the evaluation conditions which are very important to apply these metrics. In fact, both NDT/E and SHM belong to the instrument and measurement area. Therefore, this paper first performs a systematic analysis of the whole framework of instrument reliability evaluation and condition control. Based on these analyses, considering the special online application scenario of SHM, the overall framework of SHM reliability evaluation and criteria for evaluation condition control are proposed. Finally, the proposed method is demonstrated on crack monitoring reliability evaluation by guided wave-SHM on aircraft structures.

1. Introduction

Structural health monitoring (SHM) is an important technology that uses permanently installed sensors and diagnostic algorithms to assess the real-time condition of structures, particularly for detecting damage. This technology helps improve structural reliability and reduce maintenance costs. Over the years, SHM has been applied in various engineering fields, especially in aerospace[1][2]. However, there is still a significant gap between research and the practical application of SHM methods, largely due to the lack of effective performance assessment techniques.

Although discussions on SHM reliability evaluation have been ongoing for over two decades, research in this area remains limited. Some researchers have attempted to directly apply traditional NDT/E reliability methods to SHM. Like NDT/E, these methods can generally be classified into two main approaches: hit/miss analysis and the "$\hat{a}$ vs. a" analysis[3]. The hit/miss approach is suited for systems that output a binary result—either detecting damage or not. For example, to evaluate the reliability of guided wave-based SHM (GW-SHM) for impact damage in carbon fiber-reinforced composite structures, Aliabadi and Yue used the hit/miss analysis to calculate key reliability metrics such as probability of detection (POD) and probability of false alarm (PFA)[4]. These metrics are commonly used in NDT/E reliability assessments. The "$\hat{a}$ vs. a" analysis applies if the system provides a continuous signal response correlated with damage size. For the reliability evaluation of GW-SHM methods in crack monitoring, Chang and Janapati conducted experiments on 30 aluminum specimens, collecting damage index data "$\hat{a}$" for various cut crack lengths "a" and used the "$\hat{a}$ vs. a" data to assess POD for GW-SHM[5]. While applying NDT/E reliability methods to SHM is a straightforward approach, aircraft SHM differs from NDT in that it is primarily applied online, with SHM sensors and monitoring systems operating alongside aircraft structures under complex service conditions[6].

Structural Health Monitoring: 10APWSHM   Materials Research Forum LLC
Materials Research Proceedings 50 (2025) 8-14   https://doi.org/10.21741/9781644903513-2

Regarding SHM reliability evaluation, since SHM methods are applied usually online in service, there exist a large number of uncertainties. Different uncertainties dominate in different application scenarios. Under these diverse situations, how to evaluate SHM reliability and present the user with useful reliability information still needs to be deeply researched. However, among the existing discussions, many concepts are mixed with each other, and limited metrics developed so far are only used for a specific object or application scenario. No uniformly applicable SHM reliability metric has been proposed.

Therefore, this paper first performs an analysis of the framework of instrument reliability evaluation and condition control. Both NDT/E and SHM belong to the instrument and measurement area, supported by established ISO standards like ISO 5725 [7][8] and ISO/IEC Guide 98[9], which define reliability concepts and methods for managing various uncertainties. Based on these analyses, considering the special online application scenario of SHM, the framework of SHM reliability evaluation and criteria for evaluation condition control are proposed. Finally, the proposed method is demonstrated on crack monitoring reliability evaluation by GW-SHM on aircraft structures.

The paper is organized as follows: Section 2 provides a systematic analysis of the whole framework of instrument reliability evaluation, and emphasis is paid to the evaluation condition control. On this basis, the overall framework of SHM reliability evaluation and criteria for evaluation condition control are proposed in Section 3. Section 4 demonstrates the application case of the Dual-RE. Finally, the conclusions are given in Section 5.

2. Reliability evaluation of instrument

Analysis is provided regarding instrument accuracy and uncertainty metrics, and their evaluation condition control.

2.1 Instrument reliability evaluation framework

According to ISO5725, the framework of instrument reliability evaluation includes three core parts: (1) test accuracy; (2) test uncertainty; (3) test conditions [7][8]. About the framework, the following key points need to be emphasized:

(1) A complete test result from an instrument system generally includes two parts: the estimated value of the measured quantity and the uncertainty of the test. Therefore, the evaluation of test accuracy and test uncertainty together constitute the overall framework of instrument reliability evaluation. To evaluate the accuracy of the estimated test value, met-rics such as error, sub-precision metrics, and overall precision metrics are generally used [39]. To evaluate test uncertainty, metrics such as standard deviation, confidence probability, and confidence intervals (CI) are usually used.

(2) Clarifying test conditions is essential for instrument reliability evaluation. Accord-ing to ISO5725, the evaluation is generally carried out under repeatability conditions, inter-mediate conditions, and reproducibility conditions. There may be significant differences in the accuracy and uncertainty metrics under different conditions.

Currently, most of the existing SHM reliability evaluation studies did not notice these aspects and focus more on monitoring uncertainty and paid little attention to the evaluation condition control.

2.2 Test accuracy and test uncertainty

ISO 5725 employs two terms to describe the test accuracy: trueness and precision. Trueness indicates how close the estimated values from multiple tests are to the true value, while precision indicates the degree of closeness between the estimated values in multiple tests. The test trueness is usually measured by error. Error is the difference between the estimated test value and the true or accepted reference value; a large error is associated with low trueness; a small error is associated with high trueness. Error is an overall characterization of the test trueness and is usually expressed in two forms: absolute error and relative error. The test precision can be characterized using sub-precision metrics, such as linearity, hysteresis, sensitivity error, repeatability, stability, etc. It can also be characterized by overall precision metrics, commonly represented by standard deviation or by a synthetic of sub-precision metrics[7].

When employing an instrument system, even when testing the same object under identical conditions and following the same procedure, the estimated value may vary across multiple tests. This is the phenomenon referred to as test uncertainty. Test uncertainty denotes the level of uncertainty surrounding the estimated test results, specifically, the dispersion of these results. To evaluate test uncertainty, metrics include standard deviation, confidence probability, and CI. When the estimated test values appear densely near the test mean, the standard deviation is small; Conversely, the standard deviation is relatively large. The CI is determined by confidence probability, referring to the interval of possible ranges of the estimated test values. The CI is usually centered on the mean of the estimated values, indicating that the estimated value will fall within the interval with a certain confidence probability[9].

2.3 Test conditions control

In addition to the test object, test accuracy and uncertainty are also affected by many factors, including (1) operator; (2) testing system; (3) instrument calibration; (4) test environment; (5) test interval. ISO5725 specifies three conditions including two extreme conditions—repeatability and reproducibility conditions—as well as intermediate conditions[7].

Under repeatability conditions, factors (1) to (5) are controlled as much as possible without affecting the estimated test values. These tests are conducted by a specific operator using a specific instrument in a specific environment, at short intervals, on a fixed measured quantity. Repeatability conditions present the test scenario with the fewest influencing factors, resulting in the least dispersion in the estimated test values. This setup primarily reflects the effect of the test system's own uncertainties on the test estimates of fixed measurement objects, indicating the test reliability under the influence of the test system's inherent uncertainties.

Under reproducibility conditions, influencing factors (1) to (5) are not controlled and can significantly impact the estimated test values. This means that the estimated test value is affected not only by the tested object itself but also by various external factors, such as operator differences, instrument variations, calibration methods, and environmental conditions. These external factors can sometimes have a greater impact than the tested object itself, leading to significant changes in the estimated test values. Reproducibility conditions thus provide a reflection of test reliability in real-world engineering applications.

In addition to these two extreme test conditions, there are also intermediate conditions where one or more of the influencing factors (1) to (5) change. These intermediate conditions are primarily used to evaluate test reliability under the influence of certain important factors in practical applications.

3. SHM reliability evaluation framework

In addition to following the instrument reliability evaluation framework, SHM reliability evaluation must also take service conditions into account. However, it is clearly impossible to account for all uncertainties present during service. Therefore, it is essential to focus on the dominant uncertainties. To address this, a condition control-based Dual-RE for SHM is proposed, which operates under two different condition control scenarios to address user concerns about the reliability of the SHM sensor and system, as well as the impact of dominant uncertainties during service:

(1) Approximate Repeatability Condition: This scenario evaluates the reliability of the SHM sensor and monitoring system themselves.

(2) Intermediate Condition: This scenario assesses SHM reliability under the influence of

Structural Health Monitoring: 10APWSHM
Materials Research Proceedings 50 (2025) 8-14

Materials Research Forum LLC
https://doi.org/10.21741/9781644903513-2

significant service factors.

The overall framework for SHM reliability evaluation is illustrated in Fig. 1. It includes two key components corresponding to the different controlled conditions: Integrated Sensor-Based SHM Reliability Evaluation (IS-SHM-RE), which aims to control as many monitoring influence factors as possible under approximate repeatability conditions; and Critical Service Condition-Based SHM Reliability Evaluation (CSC-SHM-RE), which introduces typical critical service factors while controlling other factors as much as possible under intermediate conditions.

Regarding different SHM monitoring levels, such as damage detection, localization, and sizing, different metrics should be used. For example, to evaluate the accuracy and uncertainty of damage detection, the POD curve, lower bound of CI, and $a_{90/95}$ can be used. For damage sizing reliability evaluation, maximum absolute error and root mean square error are used to evaluate the accuracy, while standard deviation, confidence probability, and CI can be used to characterize the uncertainty.

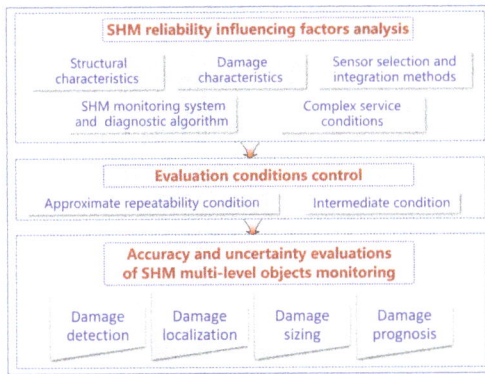

Fig. 1. The overall framework of SHM reliability evaluation.

4. Demonstration of Dual-RE for GW-SHM

GW-SHM is one of the important SHM methods. In this section, the crack on a kind of typical aircraft lug structure is monitored by GW-SHM. Reliabilities including the accuracy and uncertainty of the GW-SHM crack detection and sizing are evaluated by using Dual-RE paradigm.

4.1 Demonstration setup

The designed attachment lug is made of LY12 aluminum alloy with 5 mm thickness, with the PZT smart sensor layer co-cured[10][11]. All the lug structures are manufactured by using the same material, design style, dimensions, and manufacturing process.

The GW-SHM monitoring system, developed by the authors' group is adopted [12]. Two typical damage indexes (DIs), are utilized to quantify the crack [13]. The GW-Gaussian process (GW-GP) diagnosis algorithm is designed in the SHM system software to describe the relationship between DI and damage size[14].

4.2 IS-SHM-RE of the GW-GP SHM method

For IS-SHM-RE, the approximate repeatability conditions are controlled. The monitoring environment is controlled to be room temperature and pressure. The electrical discharge machining method is used to create fixed morphology and specific-size crack damage.

(1) IS-SHM-RE for damage detection

To evaluate the reliability of the crack detection, the POD curve, lower bound CI of POD, and $a_{90/95}$ are used. In POD analysis, the threshold a_{th}^{IS} is set to minimize the PFA while ensuring a high POD, balancing between PFA and POD. Under approximate repeatability conditions, multiple diagnostics are performed in a healthy state. Here, $\hat{a}_{th}^{IS}$ is set to 0.3 mm during IS-SHM-RE, resulting in a PFA of 0 and ensuring a high POD.

The linear regression model is employed to fit the relationship between $\hat{a}^{IS}$ and a^{IS}, shown in Fig. 2(a). The metric $a_{90/95}$ is 0.8 mm, meaning the GW-GP SHM can detect damage with at least 90% probability and 95% confidence when the crack length is 0.8 mm or larger.

(2) IS-SHM-RE for damage sizing
To evaluate the accuracy of SHM sizing, maximum absolute error e_{MAX}^{IS} and root mean square error e_{RMSE}^{IS} are calculated, with e_{MAX}^{IS} is 0.8 mm and e_{RMSE}^{IS} is 0.3 mm.

To assess the uncertainty of SHM damage sizing, the standard deviation s^{IS} and the 95% CI are utilized. As shown in Fig. 2(b), SHM sizing results exhibit dispersion at different crack lengths since the results are from different specimens. Linear regression is adopted. The regression results indicate that s^{IS} is 0.3 mm, with the 95% CI shown as a dotted line in Fig. 2(b), indicating a 95% probability that SHM diagnostic results fall within this interval.

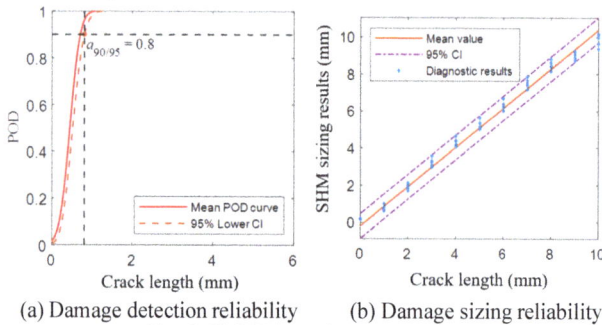

(a) Damage detection reliability   (b) Damage sizing reliability
Fig. 2. IS-SHM-RE for GW-GP SHM.

4.3 CSC-SHM-RE of the GW-GP SHM method
In this section, the same GW-GP SHM method is utilized to diagnose real fatigue cracks under load spectrum conditions. In this evaluation, 2 important critical uncertainty factors are introduced, namely load and damage. The monitoring is performed under the service load spectrum and the damage is a real fatigue crack which has its uncertainty of angle, depth, width, and morphology.

(1) CSC-SHM-RE for damage detection
Similar to the IS-SHM-RE, a PFA and POD trade-off method is used to determine the threshold $\hat{a}_{th}^{CSC}$. The threshold is decided in the initial period of monitoring when no obvious crack growth is observed. In this study, $\hat{a}_{th}^{CSC}$ is set to 0.9 mm.

The relationship between the growing crack length diagnosed by the SHM and the actual growing crack length is shown in Fig. 3(a), fitted by a linear regression model. Then, the POD curve is plotted using "$\hat{a}$ vs a" analysis, as shown in Fig. 3(a). The $a_{90/95}$ is 1.2 mm.

(2) CSC-SHM-RE for damage sizing
For evaluating the accuracy of SHM sizing, maximum absolute error e_{MAX}^{CSC}, and root mean square error e_{RMSE}^{CSC} metrics are used. For CSC-SHM-RE under the influence of service load and damage, $e_{MAX}^{CSC} = 1.9$ mm and $e_{RMSE}^{CSC} = 0.6$ mm.

To assess the uncertainty of SHM sizing, the standard deviation s^{CSC} and 95% CI are calculated. A linear regression of SHM sizing results and crack length is conducted as shown in Fig. 3(b). The s^{CSC} is 0.4 mm, with the 95% CI shown as a dotted line in Fig. 3(b). Compared to IS-SHM-RE in

Structural Health Monitoring: 10APWSHM
Materials Research Proceedings 50 (2025) 8-14

Materials Research Forum LLC
https://doi.org/10.21741/9781644903513-2

Fig. 2(b), CSC-SHM-RE shows larger SHM sizing dispersion under the service loads and damage.

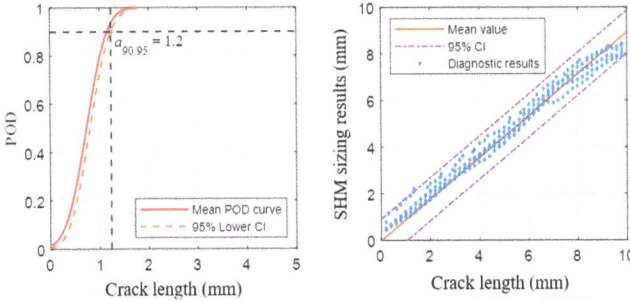

(a) Damage detection reliability   (b) Damage sizing reliability
Fig. 3. CSC-SHM-RE for GW-GP SHM.

4.4 Discussions

For IS-SHM-RE, $a_{90/95}$ is 0.8 mm, while for CSC-SHM-RE under the influence of service load and damage, $a_{90/95}$ is 1.2 mm. The IS-SHM-RE of error metrics e_{MAX} and e_{RMSE} for GW-GP SHM sizing are 0.8 mm and 0.3 mm, while 1.9 mm and 0.6 mm during CSC-SHM-RE. The comparison of Fig. 2 (IS-SHM-RE) and Fig. 3 (CSC-SHM-RE) indicates that IS-SHM-RE exhibits lower uncertainty in crack sizing. The results demonstrate that IS-SHM-RE achieves higher accuracy and lower uncertainty under strictly controlled conditions, specifically approximate repeatability conditions. In contrast, the CSC-SHM-RE results show larger errors and uncertainties due to the effects of service loads and damage in intermediate conditions. Both IS-SHM-RE and CSC-SHM-RE in the Dual-RE framework are important: IS-SHM-RE is essential for evaluating the reliability of integrated SHM sensors and systems, while CSC-SHM-RE is critical for assessing SHM reliability under typical service conditions. Therefore, clearly defining evaluation conditions is crucial, as different conditions will lead to different evaluation outcomes.

5. Conclusion

A condition control-based Dual-RE for SHM is proposed in this paper and demonstrated. The main conclusions are summarized as:

(1) According to ISO5725, the instrument reliability evaluation should include three core parts: (a) test accuracy; (b) test uncertainty; (c) test conditions. When evaluating SHM reliability, it is also crucial to include these three core aspects: monitoring accuracy, monitoring uncertainty, and evaluation conditions.

(2) Clearly specifying and controlling the evaluation conditions are essential, as different conditions certainly lead to different results. The demonstration performed in this paper shows this.

(3) Trying to cover all possible service uncertainties is apparently impossible due to the numerous uncertainties that exist in different applications. The proposed Dual-RE method provides a standard framework for assessing SHM reliability. According to different application scenarios, different key intermediate conditions can be selected and controlled to perform various CSC-SHM-RE procedures to address the user's concerns.

Acknowledgment

The authors are grateful for the support from National Natural Science Foundation of China [Grant No. 51921003, No. 52275153]; Frontier Technologies R&D Program of Jiangsu [Grant No. BF2024068]; Fundamental Research Funds for the Central Universities [Grant No. NI2023001]; The Fund of Prospective Layout of Scientific Research for Nanjing University of Aeronautics and

Structural Health Monitoring: 10APWSHM   Materials Research Forum LLC
Materials Research Proceedings 50 (2025) 8-14   https://doi.org/10.21741/9781644903513-2

Astronautics; Research Fund of State Key Laboratory of Mechanics and Control for Aerospace Structures (Nanjing University of Aeronautics and Astronautics) [Grant No. MCAS-I-0423G01]. The Postgraduate Research & Practice Innovation Program of Jiangsu Province [Grant No. KYCX22_0347].

References

[1] S. Yuan, Y. Ren, L. Qiu, H. Mei. A multi-response-based wireless impact monitoring network for aircraft composite structures. IEEE T. Ind. Electron. 63(12) (2016) 7712-7722. https://doi.org/10.1109/TIE.2016.2598529

[2] S. Yuan, H. Jing, Y. Wang, J. Zhang. A whole service time SHM damage quantification model hierarchical evolution mechanism. Mech. Syst. Signal Proc. 209 (2024) 111064. https://doi.org/10.1016/j.ymssp.2023.111064

[3] M. HDBK, "Nondestructive evaluation system reliability assessment," Dep. Def. Handb., vol. 7, 2009.

[4] N. Yue, M.H. Aliabadi, Hierarchical approach for uncertainty quantification and reliability assessment of guided wave-based structural health monitoring. Struct. Health Monit. 20 (5) (2021) 2274-2299. https://doi.org/10.1177/1475921720940642

[5] V. Janapati, F. Kopsaftopoulos, F. Li, S.J. Lee, F.K. Chang, Damage detection sensitivity characterization of acousto-ultrasound-based structural health monitoring techniques. Struct. Health Monit. 15 (2) (2016) 143-161. https://doi.org/10.1177/1475921715627490

[6] F. Falcetelli, N. Yue, R. Di Sante, D. Zarouchas, Probability of detection, localization, and sizing: The evolution of reliability metrics in Structural Health Monitoring. Struct. Health Monit. 21 (6) (2022) 2990-3017. https://doi.org/10.1177/14759217211060780

[7] ISO 5725-1, "Accuracy (trueness and precision) of measurement methods and results-Part 1: General principles and definitions- 2nd ed.", Int. Stand. Organ. 2023.

[8] ISO 5725-2, "Accuracy (trueness and precision) of measurement methods and results-Part 2: Basic method for the determination of repeatability and reproducibility of a standard measurement method- 2nd ed.", Int. Stand. Organ. 2019.

[9] ISO/IEC Guide 98-3, "Uncertainty of measurement-Part 3: Guide to the expression of uncertainty in measurement", Int. Stand. Organ./Int. Electro Tech. Comm. 2008.

[10] Y. Ren, Q. Xu, S. Yuan, Improving accuracy of damage quantification based on two-level consistency control of PZT layers. Chin. J. Aeronaut. 36 (3) (2023) 241-253. https://doi.org/10.1016/j.cja.2022.09.021

[11] Y. Ren, S. Zhang, S. Yuan, L. Qiu, In-situ integration and performance verification of large-scale PZT network for composite aerospace structure. Smart Mater. Struct. 32 (5) (2023) 055010. https://doi.org/10.1088/1361-665X/acc436

[12] L. Qiu, S. Yuan, On development of a multi-channel PZT array scanning system and its evaluating application on UAV wing box. Sens. Actuator A-Phys. 151 (2) (2009) 220-230. https://doi.org/10.1016/j.sna.2009.02.032

[13] S. Torkamani, S. Roy, M. E. Barkey, E. Sazonov, S. Burkett, S. Kotru, A novel damage index for damage identification using guided waves with application in laminated composites. Smart Mater. Struct, 23 (9) (2014) 095015. https://doi.org/10.1088/0964-1726/23/9/095015

[14] S. Yuan, H. Wang, J. Chen, A PZT based on-line updated guided wave-gaussian process method for crack evaluation. IEEE Sens. J. 20 (15) 2019 8204-8212. https://doi.org/10.1109/JSEN.2019.2960408

Structural Health Monitoring: 10APWSHM
Materials Research Proceedings 50 (2025) 15-23

Materials Research Forum LLC
https://doi.org/10.21741/9781644903513-3

Structural deformation monitoring under complex boundary constraints using boundary parameter-optimized iFEM

Tianyu Dong[1,a] and Shenfang Yuan[1,b*]

[1] Research Center of Structural Health Monitoring and Prognosis, State Key Lab of Mechanics and Control for Aerospace Structures, Nanjing University of Aeronautics and Astronautics, Nanjing 210016, P. R. China

[a]dongty@nuaa.edu.cn, [b]ysf@nuaa.edu.cn

Keywords: Deformation Monitoring, Deployable Structure, Complex Boundary Constraints, Inverse Finite Element Method, Boundary Parameter Optimization

Abstract. Spacecraft in orbit are gradually developing in the direction of large-scale, complex, and distributed. These aircraft structures will undergo complex thermal deformation because of time-varying and distributed thermal excitation in harsh operating environments. Real-time and accurate structural deformation monitoring is important to ensure the spacecraft's performance in orbit. The inverse Finite Element Method (iFEM) is the most promising strain-based deformation reconstruction algorithm for the independent of the material properties and external load information. However, iFEM needs to obtain the strain field of the structure and accurately define the geometry dimension and boundary displacements. For large deployable aerospace structures, the special characteristics of the unfolding mechanism lead to complex boundary constraints, leading to low shape-sensing accuracy of iFEM if characterized inaccurately. This paper proposed a shape-sensing method that combined iFEM with boundary parameters optimization to deal with the limitation. Based on the genetic algorithm, the parameters of boundary constraints in iFEM are optimized to achieve an accurate representation of complex boundary constraints and high-precision deformation reconstruction. The numerical experiment of a sub-panel structure with complex boundary constraints of a deployable Space-borne antenna was carried out to validate the effectiveness of the proposed method.

1. Introduction

Space-borne deployable antennas are the fastest-growing and most potential antenna type, for superior performance benefits like high gain and long-range capability [1]. Due to the physical constraints of the launch vehicle's launch and loading platforms, the antenna array is typically divided into several sub-panels, which are folded and collected around the satellite before use, and then unfolded outside the satellite cabin via elastic hinges or other extension mechanisms when the satellite enters a predetermined orbit [2]. When the satellite is in orbit, the Space-borne active phased array antenna is in a cyclic hot and cold alternating environment, which will cause thermal deformation of antenna structures. To ensure satellite performance and meet the needs of aerospace engineering applications, real-time deformation monitoring of space-borne deployable antennas is of great significance. However, the different thermal deformations of each antenna sub-panel, as well as the thermal coupling problem between antenna sub-panels caused by the unfolding mechanism, make the boundary conditions of sub-panels more complicated. As a result, the structural deformation of the deployable antenna becomes difficult to predict, which poses a great challenge to the deformation monitoring technology.

The inverse Finite Element Method (iFEM) proposed by Tessler and Spangler et al. in 2005 [3] has shown the most promise in recent years. This method can achieve deformation monitoring without the knowledge of load information and material parameters of the structure. It only needs the surface strains as the algorithm input, and the structural geometry and boundary displacements

Structural Health Monitoring: 10APWSHM   Materials Research Forum LLC
Materials Research Proceedings 50 (2025) 15-23   https://doi.org/10.21741/9781644903513-3

of the structure to establish an accurate iFEM model. iFEM has been numerically and experimentally verified for deformation monitoring of various aerospace structures [4,5,6,7,8]. In recent years, iFEM has been applied to shape sensing of large deformations [9,10], as well as damage detection[11,12], crack extension monitoring [13], and delamination damage identification[14]. However, these studies generally simplify real aerospace structures into shell structures or cantilever beam structures under ideal boundary conditions with determined displacement value, such as fixed and hinged boundary conditions [15,16]. The complexity of structural geometry and boundary constraints in practical applications is rarely considered. For real aerospace structures subject to complex boundary conditions, the boundary displacements are difficult to accurately obtain, thus affecting the deformation reconstruction accuracy of iFEM. Some research has been conducted by Oboe et al. [17] and Colombo et al. [18] on the approach based on a weighted superposition of iFEM elementary models with different ideal boundary conditions. The current solution is The iFEM elementary models used for weighted superposition must employ the same inverse elements and discretization. The weighting coefficients for each iFEM elementary model are optimized using the least squares principle to minimize the error between the reconstructed displacement field and the reference displacement field of the weighted iFEM model. However, the method based on the weighted superposition of iFEM elementary models has its limitations. The selection of the iFEM elementary model will influence the reconstruction outcomes. Consequently, it is not well-suited for structures with complex and unpredictable boundary conditions.

Based on the above analysis, this paper proposes the boundary parameter-optimized iFEM. The elastic boundary conditions with undetermined parameters are used to define the complex boundary constraints. After assembling the local reconstruction equations of all inverse elements to obtain the global reconstruction equation system of the structure, the elastic boundary constraint equations are further assembled to provide boundary constraint conditions for solving the displacement. The genetic algorithm is used to optimize the parameters of boundary constraints. The rest of the article is arranged as follows. In Section 2, the proposed method is introduced. In Section 3, the proposed method is numerically validated with a composite plate with spring-supported boundary conditions. The last section is the conclusion.

2. The Boundary Parameter-Optimized iFEM

The inverse finite element method discretizes the structure into multiple inverse elements, such as the four-node inverse shell elements (iQS4). For each inverse element, a weighted error function $\Phi(\mathbf{a}^e)$ is defined based on the least squares principle to obtain the transformation relationship between the measured strains and the displacement of the nodes:

$$\Phi(\mathbf{a}^e) = \left\| \mathbf{e}_m\left(x_i, y_i, \mathbf{a}^e\right) - \mathbf{e}_m^\varepsilon \right\|^2 + \left\| 2h\boldsymbol{\kappa}\left(x_i, y_i, \mathbf{a}^e\right) - 2h\boldsymbol{\kappa}^\varepsilon \right\|^2 + 10^{-4} \left\| \mathbf{e}_s\left(x_i, y_i, \mathbf{a}^e\right) \right\|^2 , \tag{1}$$

where (x_i, y_i) is the coordinates of the *i-th* strain measuring point, $\mathbf{e}_m\left(x_i, y_i, \mathbf{a}^e\right)$, $\boldsymbol{\kappa}\left(x_i, y_i, \mathbf{a}^e\right)$ and $\mathbf{e}_s\left(x_i, y_i, \mathbf{a}^e\right)$ are the analytical plane strain, analytical bending curvature, and analytical transverse shear strain of iQS4 at the strain measuring point. $\mathbf{e}_m^\varepsilon$ and $\boldsymbol{\kappa}^\varepsilon$ are the measured plane strain and the measured bending curvature at the strain measuring point. The minimum value of the error function can be obtained by taking the partial derivative of the node displacement:

$$\frac{\partial \Phi(\mathbf{a}^e)}{\partial \mathbf{a}^e} = 0 \tag{2}$$

Eq. 2 can be further written in the form of local reconstruction equations:

Structural Health Monitoring: 10APWSHM
Materials Research Proceedings 50 (2025) 15-23

Materials Research Forum LLC
https://doi.org/10.21741/9781644903513-3

$$\mathbf{k}^e \mathbf{a}^e = \mathbf{f}^e \tag{3}$$

where $\mathbf{k}^e$ is the coefficient matrix of the local reconstruction equations, with size 24×24. $\mathbf{f}^e$ is the coefficient vector of the unit local reconstruction equations, the size of which is 24×1. $\mathbf{a}^e$ is the element node displacement vector to be solved.

After calculating the local reconstruction equations of all iQS4, the global reconstruction equations of the structure can be obtained through coordinate transformation and element assembly. The global reconstruction equations are as follows:

$$\mathbf{KU} = \mathbf{F} \tag{4}$$

in which the coefficient matrices $\mathbf{K}$ of the global reconstruction equations are assembled from the coefficient matrices $\mathbf{k}^e$ of the local reconstruction equations of all iQS4, and coefficient vectors $\mathbf{F}$ of the global reconstruction equation are assembled from the coefficient vectors $\mathbf{f}^e$ of the local reconstruction equations of all iQS4.

To solve the element node displacement vectors $\mathbf{U}$, constraint conditions need to be applied to the global reconstruction equations. For aerospace structures with complex boundary constraints, ideal boundary conditions fail to accurately represent the true characteristics of these constraints, which leads to low reconstruction accuracy of iFEM. The proposed boundary parameter-optimized iFEM uses elastic boundary conditions with undetermined parameters to define the complex boundary constraints. The elastic boundary constraint equations of the contained element nodes can be written in the following form:

$$\mathbf{k}_{elastic} \mathbf{a}^e_{constrained} = \mathbf{f}_{elastic} \tag{5}$$

where the undetermined boundary parameters include coefficient matrix $\mathbf{k}_{elastic}$ and coefficient vector $\mathbf{f}_{elastic}$. For the element nodes with boundary constraint positions, the boundary constraint conditions are obtained by establishing the elastic constraint equation that the element nodes are subjected to, without the need to know the displacement values of the element nodes as in conventional iFEM that uses ideal fixed or simply supported boundary conditions.

Based on intelligent algorithms such as the genetic algorithm (GA) to optimize elastic boundary parameters, the elastic boundary conditions can be used to approximate any complex boundary constraints. The optimal elastic boundary parameters can be considered to be very close to the actual complex boundary constraints of the structure. The objective function is defined as the maximum relative error $\%e_{max}$ of the reconstructed displacement and the measured displacement:

$$\%e_{max} = 100 \times \frac{\left| \mathbf{w}^{iFEM} - \mathbf{w}^{test} \right|_{max}}{w^{test}_{max}} \tag{6}$$

where $\mathbf{w}^{test}$ is the vector composed of the measured displacement, w^{test}_{max} is the maximum measured displacement amplitude, and is the vector composed of the reconstructed displacement of the corresponding position. The fitness function of GA is:

$$F[\%e_{max}] = \frac{1}{\%e_{max}} \tag{7}$$

After the process of select operation, cross operation, and variation operation, and after generations of evolution, the algorithm converges to the best chromosome, which is the optimal or sub-optimal solution to the problem. By substituting the optimized elastic boundary parameters $k_{elastic}$ and $f_{elastic}$ into the global reconstruction equations to obtain the constrained global reconstruction equations:

$$\mathbf{K}^R \mathbf{U} = \mathbf{F}^R .\tag{8}$$

$\left(K^R\right)^{-1} F^R$ is the displacement vector consisting of all inverse element nodes and can further be interpolated to obtain the full-field displacements.

3. Numerical validation

3.1 FEM model of a honeycomb plate with spring-supported boundary conditions

To simulate the thermal deformation of a real deployable Space-borne antenna on-orbit, a finite element model of a composite honeycomb plate structure is established based on Abaqus. A given distributed thermal excitation is applied to one side surface of the honeycomb plate for thermal deformation analysis. This model represents a typical composite sandwich structure, with dimensions of 1500 mm × 3125 mm × 20 mm. The structure has aluminum blocks embedded in the honeycomb at the four corners for installation, reflecting the actual boundary conditions of a deployable space-borne antenna structure. The boundary constraints in the FEM model are simulated as the aluminum blocks grounded through springs, as shown in Fig. 1, which are divided into Spring-up marked in purple, and Spring-down marked in orange.

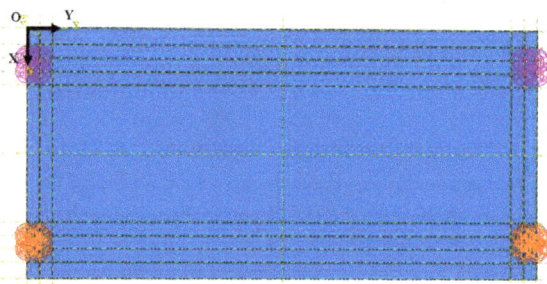

Fig. 1 FEM model of the honeycomb plate structure

Considering that the boundary constraints along the X-axis direction in practical applications are approximately fixed, the stiffness along the X-axis direction is set to a relatively high value, which is an order of magnitude higher than the two other directions. The setting values of spring stiffness along the X-axis, Y-axis, and Z-axis are shown in Table 1. By applying a specific experimental temperature field on the surface of the honeycomb plate, the simulated strain field and the simulated displacement field under the elastic boundary constraints can be obtained. The FEM displacement along the Z-axis is shown in Fig. 2.

Table 1 Spring stiffness Settings for elastic boundaries [N/mm]

X-axis Down	X-axis UP	Y-axis Down	Y-axis Up	Z-axis Down	Z-axis Up
100000	100000	14555	11086	4709.1	4000

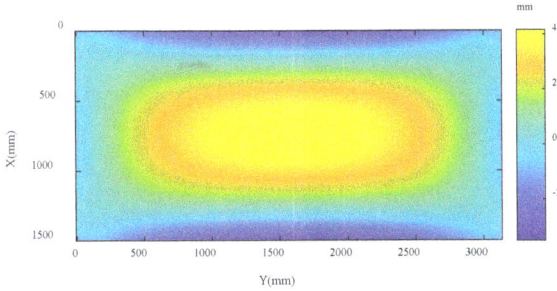

Fig. 2 FEM displacement along the Z-axis

3.2 Identification of spring-supported boundary constraint

A variable-size discretization [8] based on the displacement gradient of the honeycomb plate is shown in Fig. 3. The inverse elements at the four fixtures each have four constrained element nodes, with each node subjected to elastic constraints. The boundary parameters $k_{elastic}$ and $f_{elastic}$ of each constrained element node are optimized based on the genetic algorithm. The parameters of the genetic algorithm are set as follows: population size N is 100, evolution number G_{max} is 50, crossover probability is 0.6, mutation probability is 0.1. The changes in the average value and the best value of the objective function of each generation in the optimization process are shown in Fig. 4.

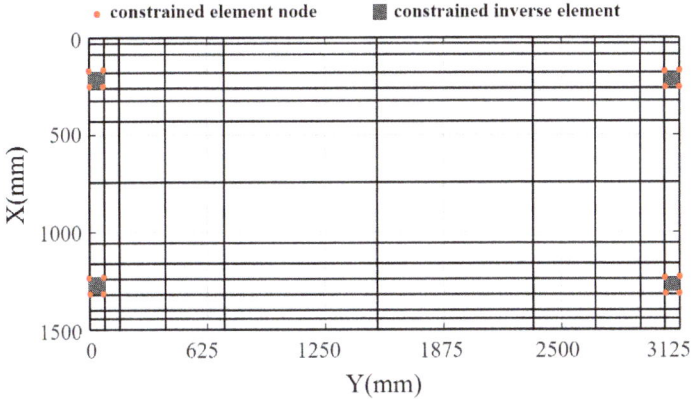

Fig. 3 variable-size discretization using iQS4

Structural Health Monitoring: 10APWSHM   Materials Research Forum LLC
Materials Research Proceedings 50 (2025) 15-23   https://doi.org/10.21741/9781644903513-3

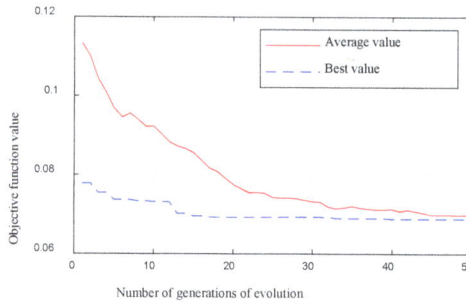

Fig. 4 convergence tendency of the genetic algorithm

When the population evolves to 50 generations, the function value converges to 0.07 and the optimization value k_{spring} compared with the setting values in the FEM model is shown in Table 2. It can be observed that the parameter optimization of based on the genetic algorithm effectively approximates and simulates the asymmetric characteristics of the elastic boundary constraints in the FEM model of the honeycomb plate structure.

Table 2 comparison of spring stiffness and the optimization result [N/mm]

	X-axis		Y-axis		Z-axis	
	Down	Up	Down	Up	Down	Up
Setting values	100000	100000	14555	11086	4709.1	4000
Optimization values	100000	100000	14543	11842	4636	4281

3.3 Validation of reconstruction accuracy

To verify the improvement of reconstruction accuracy, the reconstruction results of the boundary parameter-optimized iFEM are compared with those of iFEM based on ideal boundary conditions where the displacement boundary is defined and the boundary value is zero. The fixed boundary condition means that the rotations and displacements of the constrained element node are zero. The hinged boundary condition means that only the displacements of the constrained element node are zero. By extracting the strain in the direction of 0°, 90°, and 45° on the upper and lower surfaces at the centroid position of each inverse element as the iFEM input, the displacements along the Z-axis are reconstructed.

The reconstruction performance of iFEM with different boundary conditions is shown in Fig. 5 to Fig. 7. The maximum reconstructed error of iFEM with elastic boundary conditions is 0.3mm, while the maximum error of iFEM with hinged and fixed boundary conditions are 0.4mm and 3.0mm. The results show that the proposed boundary parameter-optimized iFEM has higher reconstruction accuracy.

Structural Health Monitoring: 10APWSHM
Materials Research Proceedings 50 (2025) 15-23

Materials Research Forum LLC
https://doi.org/10.21741/9781644903513-3

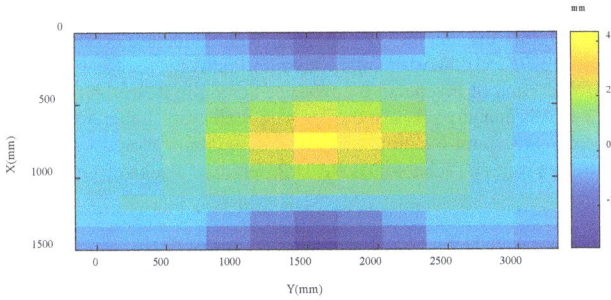

Fig. 5 reconstructed deformation of iFEM with elastic boundary conditions

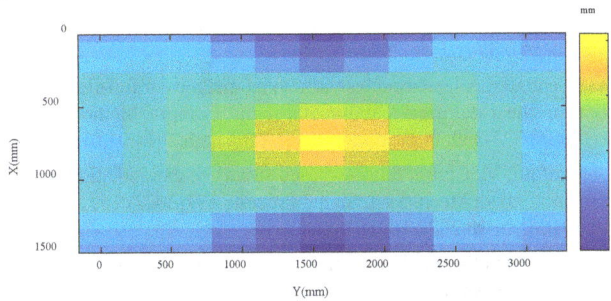

Fig. 6 reconstructed deformation of iFEM with hinged boundary conditions

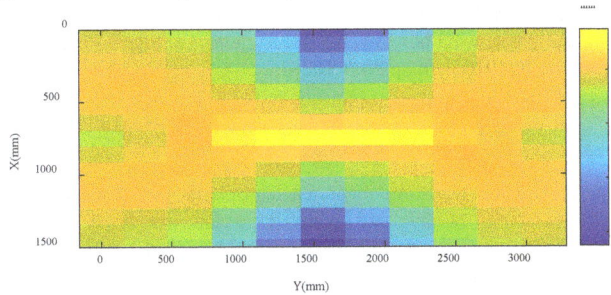

Fig. 7 reconstructed deformation of iFEM with fixed boundary conditions

4. Conclusions

A boundary parameter-optimized iFEM is proposed to improve the shape-sensing accuracy of structures with complex boundary constraints. Numerical verification of the antenna sub-panel with spring-supported boundary constraints is carried out. The parameter optimization results based on the genetic algorithm show that the proposed method can effectively approximate and simulate the non-ideal boundary constraints. By comparing with the reconstruction performance of iFEM using ideal boundary conditions, the proposed method can achieve higher accuracy in the deformation reconstruction of structures with complex boundary constraints.

Funding

The authors are grateful for the support from National Natural Science Foundation of China [Grant No. 51921003, No. 52275153]; Fundamental Research Funds for the Central Universities [Grant No. NI2023001]; The Fund of Prospective Layout of Scientific Research for Nanjing University of Aeronautics and Astronautics; Research Fund of State Key Laboratory of Mechanics and Control for Aerospace Structures (Nanjing University of Aeronautics and Astronautics) [Grant No. MCAS-I-0423G01]，Frontier Technologies R&D Program of Jiangsu，Grant No. BF2024068.

References

[1] Rai E, Nishimoto S, Katada T, et al. Historical overview of phased array antennas for defense application in Japan, Proceedings of International Symposium on Phased Array Systems and Technology, IEEE. 1996: 217-221. https://doi.org/10.1109/PAST.1996.566088

[2] LENG Guojun，BAO Hong，DU Jingli，et al. Topology optimization with multiple discrete variables in large phased array radar, Journal of Mechanical Engineering. 2013.49 (3) 174-179. https://doi.org/10.3901/JME.2013.03.174

[3] A. Tessler, J.L. Spangler, A least-squares variational method for full-field reconstruction of elastic deformations in shear-deformable plates and shells, Comput. Methods Appl. Mech. Engrg. 194 (2-5) (2005) 327-339. https://doi.org/10.1016/j.cma.2004.03.015

[4] D. Oboe, C. Sbarufatti, M. Giglio, Physics-based strain pre-extrapolation technique for inverse finite element method, Mech. Syst. Signal Process. 177(2022) 109167. https://doi.org/10.1016/j.ymssp.2022.109167

[5] Li T, Liu M, Li J, et al. Full-field deformation reconstruction for large-scale cryogenic composite tanks with limited strain monitoring data, Smart Materials and Structures. 2023, 32(11): 115021. https://doi.org/10.1088/1361-665X/acfde4

[6] Esposito M, Gherlone M. Material and strain sensing uncertainties quantification for the shape sensing of a composite wing box, Mechanical Systems and Signal Processing. 2021 160: 107875. https://doi.org/10.1016/j.ymssp.2021.107875

[7] Huang T, Dong T, Yuan S. A piecewise inverse finite element method for shape sensing of the morphing wing fishbone, Smart Materials and Structures. 2024. https://doi.org/10.1088/1361-665X/ad2c70

[8] Dong T, Yuan S, Huang T. Real-time shape sensing of large-scale honeycomb antennas with a displacement-gradient-based variable-size inverse finite element method, Composite Structures. 2024: 118320. https://doi.org/10.1016/j.compstruct.2024.118320

[9] Tessler A, Roy R, Esposito M, et al. Shape sensing of plate and shell structures undergoing large displacements using the inverse finite element method. Shock Vib. 2018:1-8. https://doi.org/10.1155/2018/8076085

[10] M.A. Abdollahzadeh, H.Q. Ali, M. Yildiz, A. Kefal, Experimental and numerical investigation on large deformation reconstruction of thin laminated composite structures using inverse finite element method, Thin-Walled Struct. 178 (2022)109485. https://doi.org/10.1016/j.tws.2022.109485

[11] Colombo, Luca, C. Sbarufatti, and M. Giglio. "Definition of a load adaptive baseline by inverse finite element method for structural damage identification." Mechanical Systems and Signal Processing 120 (2019): 584-607. https://doi.org/10.1016/j.ymssp.2018.10.041

[12] Li, Mingyang, et al. "Dent damage identification in stiffened cylindrical structures using inverse Finite Element Method." Ocean Engineering 198 (2020): 106944. https://doi.org/10.1016/j.oceaneng.2020.106944

[13] Kefal, Adnan, et al. Coupling of peridynamics and inverse finite element method for shape sensing and crack propagation monitoring of plate structures, Computer Methods in Applied Mechanics and Engineering. 391 (2022): 114520. https://doi.org/10.1016/j.cma.2021.114520

[14] Ganjdoust, Faraz, Adnan Kefal, and A. Tessler. A novel delamination damage detection strategy based on inverse finite element method for structural health monitoring of composite structures, Mechanical Systems and Signal Processing. 192 (2023): 110202. https://doi.org/10.1016/j.ymssp.2023.110202

[15] Roy R, Tessler A, Surace C, et al. Efficient shape sensing of plate structures using the inverse Finite Element Method aided by strain pre-extrapolation[J]. Thin-Walled Structures, 2022, 180: 109798. https://doi.org/10.1016/j.tws.2022.109798

[16] D. Poloni, D. Oboe, C. Sbarufatti, M. Giglio, Towards a stochastic inverse finite element method: A gaussian process strain extrapolation, Mechanical Systems and Signal Processing 189 (2023) 110056. https://doi.org/10.1016/j.ymssp.2022.110056

[17] Oboe D, Colombo L, Sbarufatti C, et al. Shape sensing of a complex aeronautical structure with inverse finite element method[J]. Sensors, 2021, 21(4): 1388. https://doi.org/10.3390/s21041388

[18] Colombo L, Oboe D, Sbarufatti C, et al. Shape sensing and damage identification with iFEM on a composite structure subjected to impact damage and non-trivial boundary conditions[J]. Mechanical Systems and Signal Processing, 2021, 148: 107163. https://doi.org/10.1016/j.ymssp.2020.107163

Structural Health Monitoring: 10APWSHM
Materials Research Proceedings 50 (2025) 24-43

Materials Research Forum LLC
https://doi.org/10.21741/9781644903513-4

Research on vibration monitoring and vibration control of buildings near subways

Pei LIU[1,a], Honglei WU[1,b*], Changjia CHEN[1,c], Zi WANG[2,d]

[1]Tongji Architectural Design (Group) Co., Ltd., Shanghai, China

[2]Wuxi Fuyo Tech Co., Ltd., Wuxi, China

[a]5lp@tjad.cn, [b]8whl@tjad.cn, [c]52ccj@tjad.cn, [d]wangzi2@tjad.cn

Keywords: Building Structure, Subway Vibration, Vibration Monitoring, Vibration Control, Vibration Isolation Design

Abstract. With the acceleration of urbanization, urban subway networks have become increasingly dense. Subway operations generate significant vibrations and noise, which are transmitted through tunnels, foundations and bases into interiors of buildings. It reduces the serviceability and comfort level of buildings. Field monitoring and data analysis were conducted on several sites of buildings during construction, revealing notable impacts from both vibrations and noise. Analysis and summary of subway vibration characteristics and propagation features within buildings indicate that subway vibrations possess wide frequency bands, small amplitudes, and are predominantly vertical in nature. Within buildings, the high-frequency components of subway vibrations attenuate rapidly along the floors, while the mid-to-low-frequency components, which are similar to the dominant frequencies of structure, attenuate slowly or even amplify. To mitigate the impact of subway vibrations, three approaches could be adopted: vibration control at the source, vibration control along the external transmission path of buildings and vibration isolation bearings within structures. Taking a kindergarten project in Shanghai as background, steel spring vibration isolation bearings were utilized in the subway vibration isolation design. The design resulted in a reduction of up to 14dB in the maximum vibration acceleration level of superstructure and a reduction of up to 19dB in the maximum secondary radiated noise of superstructure, demonstrating a significant effect of vibration control. The design could provide valuable references for related projects.

1. Introduction

With the acceleration of urbanization, the global subway mileage is growing rapidly, and urban subway networks are becoming increasingly dense, resulting in a gradual reduction in the distance between vibration sources and buildings. Furthermore, driven by the trends of new urbanization and carbon reduction in the future, the Transit-Oriented Development (TOD) model for rail transportation will further develop, leading to an increasing number of residential and commercial centers being built near subway stations. However, subway operations generate significant vibrations and noise, which are transmitted into buildings through tunnels, foundations, and bases, thereby reducing the usability and comfort of the buildings and causing disturbances to residents. Subway vibrations also have a considerable impact on buildings with strict requirements for vibrations and noise, such as concert halls, laboratories, and ancient architectures. Existing research has shown that subway vibrations can cause cumulative damage to ancient architectures and disrupt the normal operation of precision instruments [1,2].

To mitigate the adverse effects of subway vibrations on superstructures, scholars at home and abroad have initiated studies on the characteristics of subway vibration sources and their propagation patterns within buildings through field measurements, numerical simulations, and other methods. Based on these studies, various vibration reduction measures have been developed,

Structural Health Monitoring: 10APWSHM   Materials Research Forum LLC
Materials Research Proceedings 50 (2025) 24-43   https://doi.org/10.21741/9781644903513-4

such as vibration-absorbing fasteners, floating slab track beds, and vibration isolation bearings. With the deepening of research, subway vibration reduction technologies have achieved certain results and are gradually being applied internationally.

This paper summarizes and organizes current research on subway vibration characteristics, propagation features and vibration control measures based on field measurement data. It identifies shortcomings in current research and application, thereby providing a reference for the direction and approach of future research and further promoting the development of vibration monitoring and control technologies for building structures. Additionally, using a kindergarten project in Shanghai as an engineering background, this paper conducts field monitoring and data analysis of subway vibration responses at the proposed site and designs vibration control measures, aiming to offer insights and guidance for similar projects.

2. Research on Subway Vibration Characteristics and Propagation Features within Buildings Based on Measured Data

2.1 Characteristics of Subway Vibrations

Subway vibrations are generated by the friction between subway wheels and steel rails. These vibrations propagate through the tracks in the form of waves, further transmitting into the soil layers and eventually reaching the building foundations, thereby significantly affecting the comfort of buildings. Compared to seismic vibrations, subway vibrations differ significantly in terms of frequency, amplitude, and control direction.

The energy distribution of seismic vibrations is relatively concentrated, typically within the range of 1-20Hz, whereas vibrations induced by subways exhibit a broader frequency range with distinct broadband vibration characteristics. Sheng Tao et al. arranged KD1100LC vibration sensors at a free-field site to test the vibration acceleration time history. The vertical vibration time history curve and power spectrum indicate that when a subway passes, the vibration of fasteners is distributed within 0-100Hz, with amplitudes at various frequencies being relatively similar [3]. Sanayei et al. conducted vibration tests on the foundation slab of a building, and the test results showed that most of vibration energy is within the range of 10-200Hz, with peak vibrations typically distributed within the range of 20-40Hz [4]. According to ISO2631, vibrations within the frequency range of 1-80Hz are most sensitive to human body, and subway vibrations are distributed within this range, readily causing discomfort [5].

Furthermore, subway vibrations exhibit relatively small amplitudes, with peak acceleration generally ranging from 0.01 to 0.2 m/s² [6,7], whereas the peak horizontal acceleration during rare earthquakes can reach 2 to 3 m/s². The control direction of subway vibration also differs from that of earthquakes. Studies have shown that for subway vibrations, the maximum vertical velocity range on building foundation slabs caused by vertical vibrations is 63-70 VdB, and the maximum horizontal velocity range caused by horizontal vibrations is 51-57 VdB [4]. Shi et al. obtained the peak acceleration values of one-third octave band vibration levels for each floor within a building through measurement data and found that the peak vertical acceleration was significantly greater than the lateral and longitudinal accelerations [8]. Therefore, subway vibration energy is mainly concentrated in the vertical direction, and the vertical component of subway vibrations should be prioritized in vibration reduction designs.

2.2 Transmission Law of Subway Vibration Response inside Buildings

In order to conduct reasonable vibration reduction and isolation design for buildings, it is necessary to study and summarize the transmission law of subway vibration response inside buildings. Generally speaking, the subway vibration signals inside buildings first propagate vertically upwards along the columns to the upper floors, and then transmit to the internal components through beams and slabs. The transmission law of subway vibration response inside buildings can be studied separately in horizontal and vertical directions.

Structural Health Monitoring: 10APWSHM   Materials Research Forum LLC
Materials Research Proceedings 50 (2025) 24-43   https://doi.org/10.21741/9781644903513-4

In the horizontal direction, the vibration amplitudes in different areas of the same floor exhibit certain patterns. Generally, the constraints at the corners and edges of columns are greater, resulting in higher stiffness compared to the center of floor. Therefore, when vibration is transmitted from the edges of columns or ends of beams to the middle of floor, a vibration amplification effect occurs [9]. Lu Dechun et al. provided comparative data on the peak vertical vibration acceleration at the mid-span and corner points of floor in the same room. It can be observed that the peak acceleration at the corner points is much smaller than that at the mid-span points. Further comparison using one-third octave band frequency-divided acceleration levels reveals that the acceleration levels at the corner points of floor are lower than those at the mid-span points in most frequency bands, demonstrating weaker vibration responses [10].

In the vertical direction, the frequency spectrum diagram reveals that vibrations of different frequencies exhibit distinct propagation characteristics along the vertical direction. Zhou Ying et al. obtained the corresponding frequency spectrum curves by performing Fourier transforms on the vibration time-history curves of each floor. By comparing the frequency spectrum characteristics of each floor, they found that as the floor level increases, subway vibration components with frequencies higher than the natural vibration frequency of structure are gradually filtered out during propagation, while vibration components close to the dominant frequency of structure remain almost unchanged or even amplified [11]. Similar conclusions can be found in the research of Ling Yuhong et al., whose test subject was a six-story framed structure with a natural vibration frequency range of 30-50 Hz. Due to the filtering effect of structure, the frequency bands close to the natural vibration frequency of structure decay relatively slowly, while the frequency bands of 50-80 Hz decay faster [12].

Subway vibrations primarily propagate vertically through components such as columns and shear walls, and the truncation of vertical components can play a certain role in weakening subway vibrations. Feng et al. selected a building above a subway track with a structural transfer floor and conducted vibration measurements on different floors. The results showed that significant vibration attenuation effects can be achieved in the frequency range above the cutoff frequency of transfer floor [13]. Tao et al. compared a low-rise building with a transfer floor to a high-rise building without a transfer floor, both located near subway lines. For the low-rise building, the vibration level at ground level was similar to that at the platform level for frequencies below 31.5 Hz, but the vibration level decreased significantly for frequencies above 40 Hz. However, the vibration reduction effect was not obvious in the high-rise building [14]. The reason is that the discontinuity of columns in the low-rise building inhibits the transmission of vertical vibration waves, reducing their transmission efficiency. The above conclusions can provide references for vibration reduction design of superstructures. When buildings are located near subway lines, transfer floor structures can be incorporated to reduce the transmission efficiency of subway vibrations.

3. Advances in Research on Vibration Control

Vibration control methods can be classified into vibration reduction at the vibration source, vibration reduction along the external transmission path of buildings, and vibration isolation using building structural bearings, based on the generation and propagation of vibrations. Common vibration reduction methods at the vibration source include fastener vibration reduction, sleeper vibration reduction, and ballast vibration reduction. For vibration reduction along the external transmission path of buildings, methods such as vibration isolation trenches, pile rows, and filled barriers can be used. Commonly used building structural vibration isolation bearings include steel springs, disc springs, and others. The classification of vibration control methods is shown in Figure 1.

Structural Health Monitoring: 10APWSHM
Materials Research Proceedings 50 (2025) 24-43

Materials Research Forum LLC
https://doi.org/10.21741/9781644903513-4

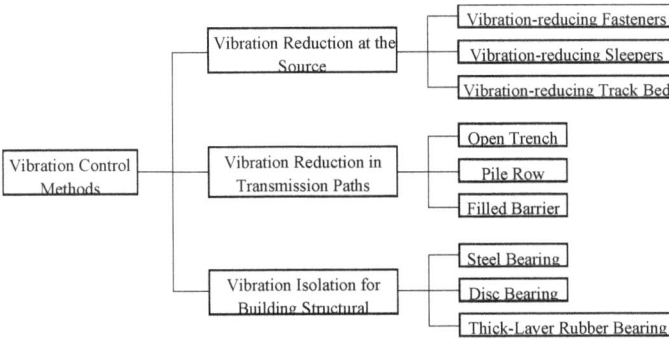

Fig.1 Classification of Vibration Control Methods

3.1 Vibration Reduction at the Source

During the operation of subway trains, friction occurs between the wheels and the tracks, generating significant interactions that lead to vibrations and noise. To reduce the interaction between the wheels and tracks and directly and effectively mitigate subway vibrations, vibration reduction measures can be implemented at three levels: the fasteners, the sleepers, and the track bed.

Vibration-reducing Fasteners

Vibration-reducing fasteners possess certain elasticity, which can dissipate the energy of subway vibrations, thereby achieving vibration reduction. Common types of vibration-reducing fasteners include the DT series fasteners, double-layer nonlinear vibration-reducing fasteners, shock absorber fasteners, spring fasteners, and others.

Currently, vibration-reducing fasteners have been widely applied in China. For instance, the Wenzhou suburban railway employs double-layer nonlinear vibration-reducing fasteners [15]. The installation of these fasteners is illustrated in Figure 2. Tests conducted using equipment such as MTS testing machines have shown that the vibration reduction amplitude of these fasteners is between 4.9dB and 5.2dB, with good safety and fatigue performance, meeting the design requirements of suburban railways. The design of Beijing Subway Line 5 also incorporates vibration-reducing fasteners, including DTVI2 fasteners and Type III track vibration isolators. Field tests indicate that after installing DTVI2 fasteners, the vibration transmission between the rail and track bed is reduced. Moreover, Type III track vibration isolators have a broader effective frequency range and provide a 10-15dB improvement in vibration reduction compared to DTVI2 fasteners [16].

Fig.2 Double Layer Non-linear Fasteners

Although vibration-reducing fasteners are economical and easy to install, they only function to reduce vibrations within specific frequency ranges and may even resonate near their natural frequencies, exacerbating vibrations in the rail and track bed. Liu Xuefeng established finite element models for various types of tracks equipped with vibration-reducing fasteners. Modal analysis revealed that anti-resonance occurs between the rail and the fastener's iron baseplate at specific frequencies, leading to ineffective vibration reduction [17].

Vibration-reducing Sleepers

Apart from vibration-reducing fasteners, scholars have also researched new types of sleepers to enhance their energy-absorbing capacity and cushioning effect, thereby mitigating subway vibrations. Deng Yushu et al. conducted field vibration tests based on trapezoidal sleepers installed between elevated tracks. The structure of the trapezoidal sleeper track is shown in Figure 3. The sleeper is designed in a trapezoidal shape, and the horizontal forces in both the longitudinal and lateral directions of track are balanced by vibration-reducing pads at the bottom and cushioning pads on the sides. This design allows the track to effectively distribute loads and exhibit excellent vibration reduction capabilities. The test results indicate that the average vibration reduction for the trapezoidal sleeper track across all frequency bands is 8.6dB [18]. Ma Meng et al. established numerical models for trapezoidal sleeper tracks and conventional tracks. The model analysis shows that trapezoidal sleepers can effectively distribute track impact loads and reduce the peak values of vertical acceleration time histories of vibrations [19].

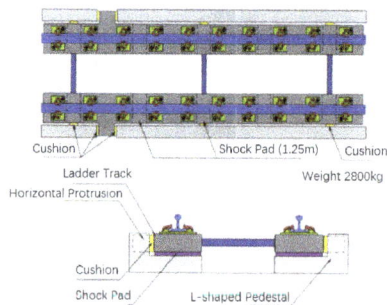

Fig.3 Components of a Trapezoidal Sleeper Track System

Structural Health Monitoring: 10APWSHM   Materials Research Forum LLC
Materials Research Proceedings 50 (2025) 24-43   https://doi.org/10.21741/9781644903513-4

Vibration-reducing Track Bed

To further reduce the vibration transmitted to the foundation, polyurethane or rubber cushion layers can be placed under the track bed to dissipate energy through the damping effect of track bed. Wang Yigan et al. selected a depot of Beijing Subway to conduct an in-situ replacement experiment. They tested the vibration reduction effects of six types of vibration-reducing track beds with different materials and laying methods. The test results revealed that the vibration reduction range of six track beds was between 6.4 and 14.6 dB [20].

The steel spring floating slab, as a monolithic track bed, offers good elasticity and can effectively isolate vibration transmission between the track and the foundation. Tu Qinming conducted vibration field tests on four subway lines in Guangzhou, each equipped with ordinary tracks, medium vibration-reducing fastener tracks, trapezoidal sleeper tracks, and steel spring floating slab tracks, respectively. The vibration reduction performance of three vibration-reducing tracks was evaluated using the insertion loss of tunnel wall. The comparison results, as shown in Figure 4, reveal that the steel spring floating slab track achieves the best average vibration reduction effect. Except for the frequency range near 160Hz, it outperforms the other two vibration-reducing tracks in all other frequency bands [21]. Lei et al. established a finite element model of steel spring floating slab track and conducted harmonic response analysis to study the impact of different parameters on the vibration reduction rate. The results indicate that the stiffness and damping of steel spring floating slab track are the controlling factors. When the stiffness is controlled within 5-10 MN/m and the damping is controlled within 50-100 kN·s/m, satisfactory vibration reduction effects can be achieved [22].

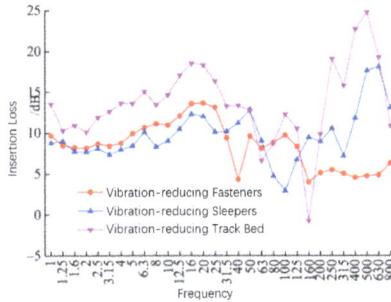

Fig.4 Damping Effect in Different Frequency Bands for Three Types of Damping Tracks

The three vibration reduction measures for vibration sources can all achieve varying degrees of vibration reduction in subway systems, but they also have certain usage limitations. Although vibration-reducing fasteners have economic advantages, they have the lowest vibration reduction capacity and may exacerbate the corrugation effect of track after use. While vibration-reducing sleepers outperform the former in terms of performance, their structure is more complex, and their vibration suppression effect is inferior to that of vibration-reducing track beds. Although vibration-reducing track beds have significant advantages in vibration isolation efficiency, their construction is complex, and their vibration isolation effect in the low-frequency range is not ideal, which may easily lead to resonance [23,24]. In order to achieve vibration reduction in the full frequency range of subway systems, these three methods should be appropriately combined, or the vibration reduction effect can be further improved through vibration reduction in the transmission path and vibration isolation of supports.

3.2 Vibration Reduction in Transmission Paths on the Exterior of Buildings

Soil serves as the medium of vibration signal transmission from subway tunnels to buildings. Before the vibration waves reach the building foundation, isolation facilities such as open trenches, pile rows, and filled barriers can be installed in the soil layer to achieve vibration reduction in the transmission paths on the exterior of buildings.

Open Trench

Vibration reduction using an open trench involves excavating a trench in the soil medium. Due to its ability to block the propagation of vibration waves in the soil, the vibration reduction effect is relatively significant. The effectiveness of vibration reduction using an open trench is influenced by parameters such as depth and width. Deng Yahong et al. established a finite element model for the foundation and open trench, conducted elastoplastic numerical analysis, and found a positive correlation between the depth and width of open trench and the vibration reduction efficiency. They also concluded that the depth has the greatest impact on the vibration reduction efficiency [25]. Yao Jinbao et al., through mutual verification of formula calculations and numerical analysis, also drew the conclusion that the vibration reduction effect becomes more significant as the depth of open trench increases [26]. Zou et al. employed a numerical model consisting of a train-track dynamics model and a track-soil-building finite element model, and determined the vibration reduction effect of open trench based on the insertion loss of each layer of superstructure. Numerical simulations showed that when the depth of open trench is less than 0.5m, the vibration reduction efficiency is poor, and while the width has a certain impact on the vibration reduction effect, it is not significant [27].

Pile Row

Inserting a pile row between a building and a railway track can alter the direction and intensity of vibration wave propagation, thereby dissipating the energy of subway vibrations. Similar to open trenches, studies have shown that increasing the length and width of pile rows can enhance the vibration reduction effect [28]. However, the vibration reduction efficiency of pile rows is inferior to that of open trenches. Kattis et al. conducted numerical solutions for vibration isolation problem of single-row piles using the frequency-domain boundary element method and compared it with open trenches. The numerical analysis indicated that the ratio of vertical displacement component of soil after vibration isolation by an open trench to that before vibration isolation was 0.172, while the corresponding ratio for a single-row of concrete piles was 0.521, significantly higher than that of the open trench [29]. Gao et al. also studied the vibration reduction efficiency of multi-row piles compared to single-row piles. With the increase in the number of pile rows, the vibration reduction effect can be further improved, with the most significant vibration reduction effect occurring near the middle of the multi-row piles, similar to that of open trenches. This suggests that multi-row piles can function as an integrated system to isolate subway vibrations. However, it should be noted that the distance between each row of piles should not be too large to prevent them from behaving as a set of independent piles [29].

Filled Barrier

Filled barriers generally refer to filled trenches, where various materials are used to fill the trench to achieve vibration reduction within a certain frequency range based on the different properties of the filling materials. Chen Gongqi and Gao Guangyun established a 2.5D finite element model of foundation-barrier system and studied the impact of material stiffness on vibration reduction effectiveness. Taking a train speed of 200 km/h as an example, the vibration attenuation coefficients for different filling materials are shown in Figure 5. It can be observed that the concrete-filled trench has the worst vibration reduction effect, while fine sand, rubber, and foam,

Structural Health Monitoring: 10APWSHM   Materials Research Forum LLC
Materials Research Proceedings 50 (2025) 24-43   https://doi.org/10.21741/9781644903513-4

as flexible materials with higher damping ratios, can achieve better vibration reduction effects on both sides of filled trench [30]. Zou et al. also studied the sensitivity of vibration to the properties of filling materials using a track-soil-building finite element model. The results showed that in the main frequency range of 16-40 Hz, the vibration reduction effect of geomembrane trenches (flexible material) was more significant compared to lightweight aggregate concrete trenches (rigid material). Thus, it can be seen that the stiffness characteristics of filling material have a significant impact on the vibration reduction effectiveness of filled trenches [27].

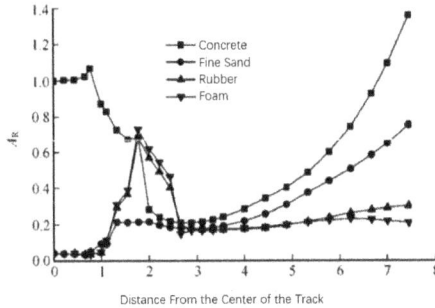

Fig.5 Vibration Attenuation Coefficients For Different Filling Materials

Regarding vibration reduction measures for external transmission paths of buildings, some projects in China have already applied such measures. For example, a double vibration reduction trench was set up next to a new campus of a university in Jinzhong [31], a vibration isolation project in the eastern suburbs of Beijing and a precision laboratory of an institute in Luoyang utilized multi-row piles for vibration reduction [32,33], and a subway depot in Guangzhou employed a combination of open trenches and rigid-filled trenches for vibration isolation [34]. All these measures have achieved good vibration reduction effects. However, there are certain limitations to the application of open trenches, filled trenches, and pile rows. For open trenches, their construction and maintenance costs are high, and it is difficult to achieve excavation to great depths. As for filled trenches and pile rows, their vibration isolation effects are limited, and the vibration isolation frequency range cannot cover the main frequency range of subway vibrations. In addition, the implementation of above three vibration reduction measures is easily affected by the building red line, making them unsuitable for jointly constructed subway tunnel projects and greatly limiting their application scope.

3.3 Vibration Isolation for Building Structural Bearings
Buildings with specific functions, such as museums, concert halls, and laboratories, have strict requirements for vibration and noise control. External vibration reduction solutions, such as floating slab track beds, may not meet the vibration reduction demands, necessitating further adoption of integral structural vibration reduction measures.

Steel Bearing
Currently, research has focused on vertical vibration isolation bearings, including those made of steel springs, disc springs, and others. Among them, steel spring bearings exhibit high vertical vibration isolation efficiency, good economy, and ease of installation, leading to their widespread application both domestically and internationally. Wang Tianyou et al. achieved overall vibration isolation for a building by installing spring vibration isolation bearings at the base of Shanghai Symphony Orchestra Hall, meeting the vibration reduction target of an average reduction of 8.5 dB in the Z-vibration level. A schematic diagram of bearings is shown in Figure 6 [35]. Besides the

Structural Health Monitoring: 10APWSHM   Materials Research Forum LLC
Materials Research Proceedings 50 (2025) 24-43   https://doi.org/10.21741/9781644903513-4

Shanghai Symphony Orchestra Hall, a structural vibration isolation project for a Beijing courtyard residence also utilized steel spring vibration isolation bearings. During the design phase, the SAP2000 software was used for modeling, where the target frequency was first set, and based on this, the vertical compression was calculated to determine the number of steel springs in the device, ultimately achieving a vertical natural frequency of 3 Hz [36]. Overseas projects applying steel spring vibration isolation include a cinema project at the British Film Institute [37] and a five-story office building project in Berlin, Germany [37], both of which have achieved effective vertical vibration isolation from subway vibrations.

Fig.6 Steel Spring Vibration Isolator

Disc Bearing

Compared to steel springs, disc springs also exhibit relatively low vertical stiffness while possessing high vertical load-bearing capacity. Jia Xiaodong utilized the SAP2000 finite element software to establish an integrated model encompassing the disc spring bearing and the superstructure. By simulating subway operations using the ADINA software to obtain the vibration acceleration near the building, and inputting this into the structural model, the analysis revealed that after vibration isolation, the natural frequency of superstructure decreased, the amplitude of vertical acceleration of floors decreased, and the vibration isolation effect could reach up to 80% [38]. Ling Yuhong et al. [39] used MIDAS software to establish a finite element model of the tunnel-soil-superstructure system and compared three vibration isolation methods: setting a sand cushion layer, filling with a vibration isolation layer, and using a disc spring bearing. The study found that the disc spring bearing provided the best vibration isolation effect, achieving an average vibration isolation rate of 11.23%. Chen et al. also proposed a nonlinear vertical vibration isolation system (NES), in which a disc spring and a viscous damper are connected in series, as shown in Figure 7. This system has almost zero tangent stiffness at the gravitational equilibrium position. Finite element analysis showed that the NES system can effectively reduce the displacement and force transmission rates of vibrations, thereby absorbing the kinetic energy from external excitations and inhibiting vertical vibrations from subways [40].

Structural Health Monitoring: 10APWSHM
Materials Research Proceedings 50 (2025) 24-43

Materials Research Forum LLC
https://doi.org/10.21741/9781644903513-4

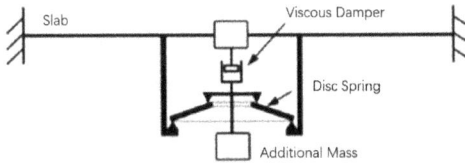

Fig.7 Components of an NES System Using Disc Springs

Thick-Layer Rubber Bearing

Traditional laminated rubber bearings have low horizontal stiffness and are generally used for horizontal seismic isolation. To reduce the vertical stiffness of traditional rubber bearings, the number of layers in the laminated rubber bearing can be decreased while increasing the thickness of each individual rubber layer. This new type of bearing is known as a thick-layer rubber bearing. Due to its low vertical stiffness, it can isolate vibrations from subways in the vertical direction and has therefore gradually become a research focus in recent years.

Pan et al. conducted tests on thick-layer rubber bearings and conventional rubber bearings under the same designed compressive stress. The results indicated that the isolation effect of thick-layer rubber bearings was significant, while conventional rubber bearings exhibited a multimodal phenomenon after vibration isolation, meaning they had multiple resonant frequency ranges and low vibration isolation efficiency [41]. Chen Haowen applied thick-layer rubber bearings in a subway vibration isolation project in Wuhan. Finite element analysis showed that after using these bearings for vibration isolation, the vibration acceleration at multiple measurement points of structure was reduced, with an average vibration reduction efficiency reaching approximately 90%. The vibration response of structure after vibration isolation met the requirements of the regulatory limits [42].

Currently, the application of vertical vibration isolation bearings in China is still in its developmental stage. Although the design and practice of steel spring bearings have become increasingly mature, their relatively low vertical load-bearing capacity limits their application to small and medium-sized buildings. Compared to steel spring bearings, vertical vibration isolation bearings such as disc springs exhibit superior mechanical properties and significantly enhanced vertical load-bearing capacity. However, research has mainly focused on numerical simulation analysis and lacks support from actual data, necessitating further advancements at the application level in the future. Although thick-layer rubber bearings can achieve the goal of vertical vibration isolation, certain defects limit their application. Firstly, studies have shown that due to the large compressive deformation of thick rubber layer, there is a significant deviation between the theoretical vertical stiffness and the experimental value of thick-layer rubber bearings [43,44]. The stiffness calculation formula for traditional rubber bearings is not applicable to thick-layer rubber bearings. Secondly, while thick-layer rubber bearings reduce vertical stiffness, they also decrease vertical load-bearing capacity [45]. In practice, it is difficult to balance the needs for load-bearing capacity and vertical stiffness. Further exploration is required in areas such as theoretical stiffness calculation, mechanical property coupling, design processes and schemes.

4. Vibration Monitoring and Design for a Kindergarten

4.1 Project Overview

Shanghai Xinzha Road Kindergarten is located on Xinzha Road in Huangpu District, Shanghai. It has a total gross floor area of 4,628m², a building height of 15.6m, two underground floors, and four floors aboveground. The architectural rendering is shown in Figure 8.

Fig.8 Architectural Rendering of Xinzha Road Kindergarten

The project site is located at Xinzha Road Subway Station on Shanghai Metro Line 1. Upon completion of development, it will comprise a prestigious community including schools, multi-story residences, and high-rise residences. In particular, the kindergarten is situated adjacent to the outer wall of subway tunnel, and it is anticipated that the vibration impact will be significant.

4.2 Vibration Control Standards

To limit the impact of environmental vibration on people's daily lives, studies, and rest, China formulated the "Standard for Environmental Vibration in Urban Areas" (GB10070-1988) and its supporting "Method for Measurement of Environmental Vibration in Urban Areas" (GB10071-1988) in 1988. The vibration evaluation index in this standard is the Z-vibration level, and the weighting curve adopts the recommended values from ISO2631-1:1985, with a frequency weighting range of 1 to 80Hz. The regulation stipulates that the maximum value during each train passage should be read during measurement and used as the evaluation quantity. The standard limits for Z-vibration levels in different areas are shown in Table 1.

Table 1 Standard Values of Vertical Z-Vibration Levels in Urban Areas

Category of Vibrational Environment Functional Zone	Daytime /dB	Nighttime /dB	Description of Applicable Zones
Special Residential Areas	65	65	Residential areas where tranquility is particularly required
Residential and Cultural-Educational Areas	70	67	Areas for residents, cultural-educational institutions, and government agencies
Mixed Zones and Commercial Center Areas	75	72	Mixed zones of general commerce and residents; mixed zones of industry, commerce, limited traffic, and residents; bustling commercial areas
Industrial Concentration Zones	75	72	Industrial zones clearly defined in the urban or regional planning
Both Sides of Major Traffic Routes	75	72	Both sides of roads with a traffic volume of over 100 vehicles per hour
Both Sides of Railway Main Lines	80	80	Residential areas located 30 meters outside the outer tracks of railways with a daily traffic volume of not less than 20 trains, on both sides

This project is controlled according to the vibration limit requirements for special residential areas specified in the "Standard for Environmental Vibration in Urban Areas" (GB 10070-88), which stipulates that the weighted Z-vibration level should not exceed 65dB.

Structural Health Monitoring: 10APWSHM   Materials Research Forum LLC
Materials Research Proceedings 50 (2025) 24-43   https://doi.org/10.21741/9781644903513-4

4.3 Vibration Monitoring of the Proposed Subway Site

The measurement points were selected at the test site where Line 1 crosses the plot. Systematic testing was conducted at various measurement points within the existing school on the plot and at the vacant land above the subway on the proposed site. The layout of the measurement points is shown in Figure 9.

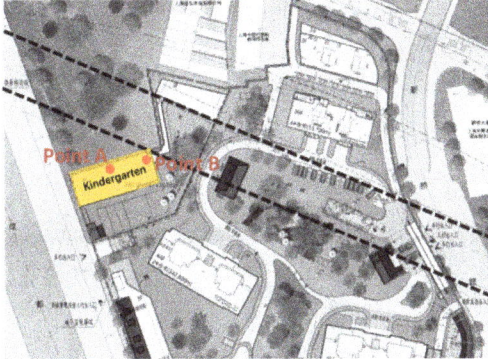

Fig.9 Measurement Points Layout

The acquisition chassis used in this experiment were the 4-channel chassis cDAQ9191 and the 16-channel chassis cDAQ9184, both from National Instruments (NI) of the United States, as shown in Figure 10. The sensors employed were piezoelectric triaxial acceleration sensors from DYTRON, as depicted in Figure 10. The time-history curves and spectrum analysis curves of the Z-direction vibration acceleration at Points A and B are provided in Figure 11.

Fig.10 Signal Acquisition Chassis and Triaxial Acceleration Sensor and Integrated Acquisition Instrument

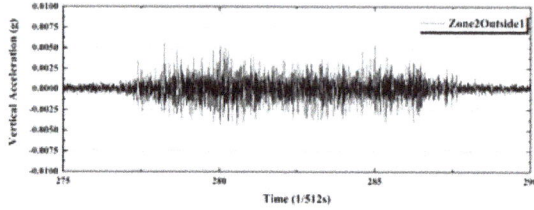

Time-history Curve of Z-direction Vibration Acceleration at Surface Measurement Point A

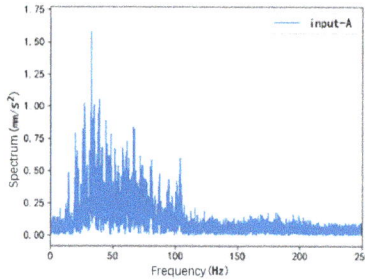

Vibration Frequency Spectrum Curve in Z-direction at Measurement Point A

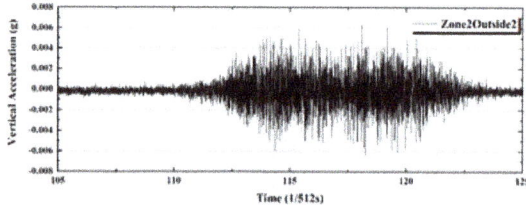

Time-history Curve of Z-direction Vibration Acceleration at the Surface of Measurement Point B

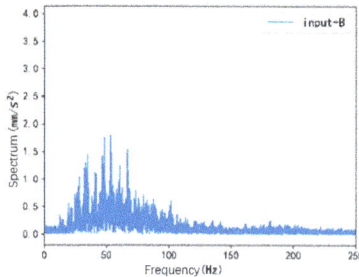

Vibration Frequency Spectrum Curve in Z-direction at Measurement Point B
Fig.11 Time-history Curves and Spectrum Analysis Curves of the Z-direction Vibration Acceleration

The test data and frequency spectrum analysis results indicate that the maximum Z-direction vibration acceleration at both measurement points A and B is approximately 0.06 m/s², with relatively small vibration amplitudes. The vibrations at measurement point A are mainly concentrated between 10~70 Hz, while those at measurement point B are mainly concentrated between 20~80 Hz. After calculation, the maximum weighted Z-vibration level at measurement point A is 68.2 dB, and at measurement point B is 68.8 dB, both exceeding the target control limit of 65 dB. There is a significant risk of exceeding the structural vibration limit. Therefore, it is recommended to adopt vibration control measures.

4.4 Vibration Control Scheme

Due to the high vibration control requirements of this project, a spring bearing vibration isolation system is proposed. The vibration isolation layer will be installed between the first floor of the kindergarten and the basement. The vertical arrangement of vibration isolation bearings is shown in Figure 12.

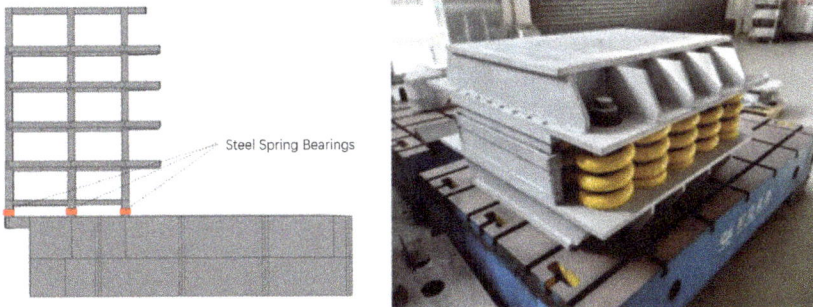

Fig.12 Vertical Setups of Spring Isolators for Vibration Isolation Floor

The design vertical vibration isolation frequency for this project is 3.0Hz. To enable rapid attenuation of structural vibrations, vertical viscous dampers are incorporated into the vibration isolation layer. The viscous dampers should be uniformly arranged within the vibration isolation layer, and it is optimal for them to be integrated with the spring vibration isolation bearings. Under corresponding subway vibration conditions, the vibration isolation system provides a vertical damping ratio of 3%. The parameters of vibration isolation system are provided in Table 2.

Table 2 Parameters of Vertical Vibration Isolation System

Vertical Weight on Top of Bearing (kN)	58071
Total Vertical Stiffness (kN/mm)	2077
Horizontal Stiffness for Vibration (kN/mm)	1283
Vertical Damping Coefficient (kN·s/m)	6658
Vertical Frequency (Hz)	2.98
Horizontal Frequency for Vibration (Hz)	2.37
Vertical Static Displacement (mm)	27.9

4.5 Damping Effect

The evaluation points on different floors are selected for comparative analysis, and the distribution of measurement points is shown in Figure 13. Using the time-domain analysis method, the measured vibration acceleration time histories are input, and the Z-vibration level results at different measurement points are compared between the two schemes: without vibration isolation and with vibration isolation.

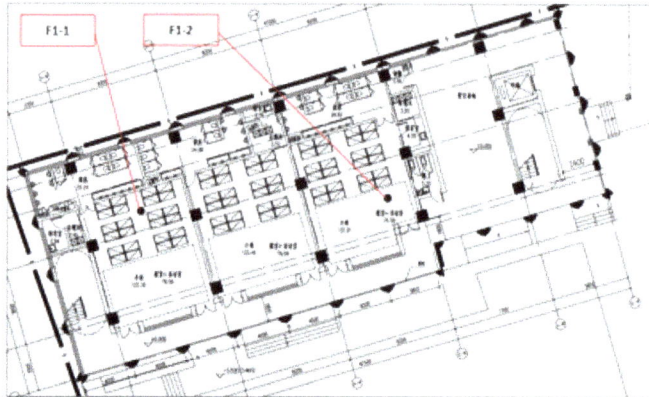

Fig.13 Distribution Map of Measurement Points on Different Floors

According to Table 3, under the input time histories of input-A and input-B, the weighted Z-vibration levels at the measurement points on each floor of non-isolated structure all exceed 65dB, with the maximum Z-vibration level reaching 87.6dB. After adopting the vibration isolation system, the weighted Z-vibration levels at the measurement points on each floor all meet the limit values, with the maximum Z-vibration level being 62.6dB, demonstrating a significant vibration isolation effect.

To more intuitively reflect the vibration conditions at various locations within the floor, Figure 14 presents the contour plot of Z-vibration level distribution on the F1 floor under different operating conditions. It can be seen that after adopting the vibration isolation system, the overall vertical vibration response of structure is effectively controlled.

Table 3 Calculation Results of Vertical Z-Vibration Level (dB)

Condition	Limit	Scheme	Input	F1-1	F1-2	F2-1	F2-2	F3-1	F3-2	F4-1	F4-2
input-A	65	Isolated	68.2	62.6	61.8	61.6	61.6	61.1	61.3	61.1	61.2
		Traditional	68.2	87.5	84.7	85.7	81.4	81.5	78.4	80.4	76.2
		Difference	0.0	24.9	23.0	24.0	19.8	20.4	17.1	19.3	15.0
input-B		Isolated	68.8	61.1	60.4	60.2	59.9	59.3	59.6	59.4	59.5
		Traditional	68.8	87.6	84.5	85.5	82.0	81.4	78.6	80.4	77.2
		Difference	0.0	26.4	24.1	25.3	22.1	22.0	19.0	21.0	17.8

Note: *1: "Standard for Environmental Vibration in Urban Areas" (GB 10070-88)

Non-isolated Structure under Input-A Condition

Isolated Structure under Input-A Condition

Non-isolated Structure under Input-B Condition

Isolated Structure under Input-B Condition

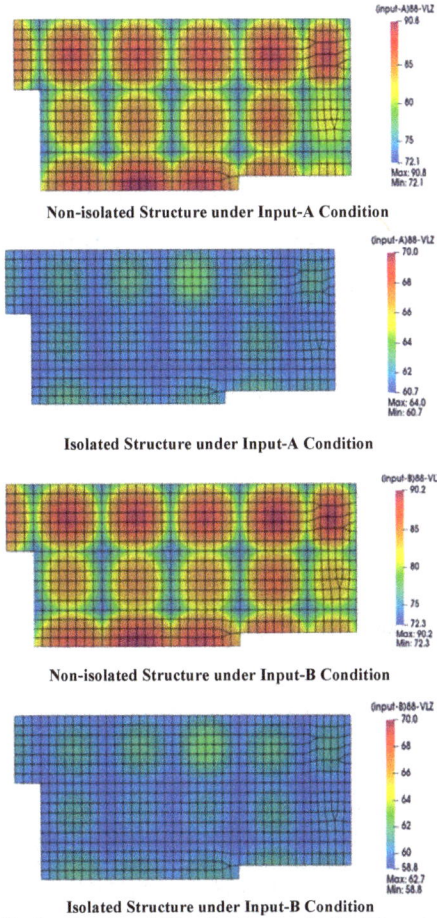

Fig.14 Contour Plot of Z-vibration Level Distribution on the F1 Floor under Different Operating Conditions

Conclusion

This paper summarizes and reviews the vibration characteristics and propagation features of subway vibrations, as well as vibration control measures. It discusses the issues in current research and applications, and takes an actual project as an example to conduct subway vibration monitoring analysis and vibration control design, providing a reference for further research. The main conclusions are as follows:

(1) Unlike horizontal earthquakes, subway vibrations exhibit characteristics of a wide distribution of frequencies, small amplitudes, and a primary vertical orientation. Within buildings, in the horizontal direction, due to differences in component constraints and stiffness, the vibration acceleration response at the edges of walls and columns is generally smaller than that at the center of floor. In the vertical direction, according to spectral analysis, the high-frequency components

of subway vibrations decay rapidly along the floors, while the mid-to-low-frequency components, which are similar to the natural frequencies of structure, decay slowly or may even be amplified.

(2) In response to the need for vibration isolation in buildings, many researchers have proposed various vibration reduction measures from the perspectives of vibration source and transmission path, and analyzed the vibration reduction effects and influencing parameters of different measures. Various vibration reduction devices generally have limitations such as limited vibration reduction frequency bands and low efficiency in reducing low-frequency subway vibrations. To address these issues, different vibration reduction measures can be combined, or improvements can be made based on existing vibration reduction methods to propose new vibration reduction methods for the vibration source and the external transmission path of buildings.

(3) Vertical vibration isolation bearings include steel spring bearings, disc spring bearings, and thick-layer rubber bearings, among others. Currently, vertical vibration isolation bearings still have certain application limitations, such as the relatively low vertical bearing capacity of steel spring and thick-layer rubber bearings, the limited practical application of disc spring bearings, and discrepancies between the theoretical and experimental vertical stiffness values of thick-layer rubber bearings.

(4) At the application level, vibration reduction measures at the vibration source, such as vibration-reducing fasteners and floating slab track beds, are widely used. However, their efficiency in reducing low-frequency subway vibrations is relatively low, and they are difficult to meet the stringent vibration control requirements of special-function buildings such as museums and concert halls. For vibration reduction measures targeting the external transmission path of buildings, implementation is prone to being affected by building setback lines, and they face difficulties in construction, low economic efficiency, and relatively poor practicability. Regarding vibration isolation bearings, steel spring bearings and steel spring-damper devices have certain engineering applications both domestically and internationally, but due to their low vertical bearing capacity, they are mostly used in low-rise and mid-rise buildings. There are fewer application cases for other types of vibration isolation bearings, and most research remains at the numerical simulation and experimental stages. To promote the application of vibration isolation bearings, future efforts should be made to improve the design methods of different devices, conduct studies on economic efficiency and practicability, and drive the construction of demonstration projects.

(5) Taking a kindergarten project adjacent to a subway in Shanghai as an example, field monitoring and analysis of subway vibrations were conducted to study their impact on neighboring buildings. By adopting a spring vibration isolation system for vibration isolation design, the maximum weighted Z-vibration level of the superstructure was reduced from 87.6dB to 62.6dB, demonstrating a significant vibration control effect. This provides a design reference for related projects.

References

[1] Han Guangsen. The micro vibration of urban rail transit impact on ancient building[D]. Xi'an University of Architecture and Technology, 2011: 69-72. (in Chinese)

[2] Geng Wanli, Liu Dunyu, Caiyongen, et al. Prediction of the influence of the proposed Beijing metro line 16 on a precise instrument of Peking University[J]. Earthquake Engineering and Engineering Dynamics, 2014, 34(6): 19-25. (in Chinese)

[3] Sheng Tao, Zhang Shanli, Shan Jiazheng, et al. In situ measurement and analysis of subway vibration's transmission and the influence to nearby buildings[J]. Journal of Tongji University(Natural Science), 2015, 43(1): 54-59. (in Chinese)

[4] Sanayei M, Maurya P, Moore J A. Measurement of building foundation and ground-borne vibrations due to surface trains and subways[J]. Engineering Structures, 2013, 53: 102-111. https://doi.org/10.1016/j.engstruct.2013.03.038

[5] Mechanical vibration and shock - Evaluation of human exposure to whole-body vibration - Part 1: General requirements: ISO 2631-1[S]. Geneva: International Organization for Standardization, 1997.

[6] Zhao Yanhui. Study on influences of vibration induced by metro depot trains on the buildings nearby and isolation measures[D]. Beijing Jiaotong University, 2019: 20-25. (in Chinese)

[7] Xu Defeng, Daijunwu, Yang Yongqiang, et al. Study vibration measurement and numerical simulation analysis of buildings along subway[J]. Building Structure, 2021, 51(Suppl.2): 697-701. (in Chinese)

[8] Shi W, Bai L, Han J. Subway-Induced Vibration Measurement and Evaluation of the Structure on a Construction Site at Curved Section of Metro Line[J]. Shock and Vibration, 2018, 2018: 1-18. https://doi.org/10.1155/2018/5763101

[9] Yuan Kui. Vibration characteristics analysis and vibration isolation research of metro depot over-track buildings[D]. Wuhan University of Technology, 2019: 19-24. (in Chinese)

[10] Lu Dechun, Gao Zejun, Kong Fanchao, et al. Numerical study on vibration of ground building adjacent to metro induced by operation of subway train[J]. Journal of Civil and Environmental Engineering, 2023, 44(09): 113-124. (in Chinese)

[11] Zhou Ying, Guo Qihang, Zhang Zengde, et al. Measurement and simulation analysis of vehicle-induced vibration of isolated over-track building[J]. Journal of Building Structures, 2023, 44(09): 83-92. (in Chinese)

[12] Ling Yuhong, Huangfu Chanyuan, Wu Jingzhuang, et al. Test and numerical analysis of subway-Induced vibrations of metro-line nearby buildings[J]. Journal of South China University of Technology(Natural Science Edition), 2015, 43(2): 33-40. (in Chinese)

[13] Shijin F, Fuhao L, Xiaolei Z, et al. In-situ experimental investigation of the influence of structure characteristics on subway-induced building vibrations[J]. Earthquake Engineering and Engineering Vibration, 2021, 20(3): 673-685. https://doi.org/10.1007/s11803-021-2046-3

[14] Tao Z, Wang Y, Sanayei M, et al. Experimental study of train-induced vibration in over-track buildings in a metro depot[J]. Engineering Structures, 2019, 198: 109473. https://doi.org/10.1016/j.engstruct.2019.109473

[15] Zeng Fei, Wang Yanfei, Wu Fei, et al. Design and application of damping fasteners for line S1 of Wenzhou regional railway[J]. Urban Rapid Rail Transit, 2020, 33(4): 104-108. (in Chinese)

[16] Li Kefei, Liu Weining, Sun xiaojing, et al. In-situ test of vibration attenuation of underground line of Beijing metro line5[J]. Journal of the China Railway Society, 2011, 33(4): 112-118. (in Chinese)

[17] Liu Xuefeng. Vibration transmission characteristic of rail fastenings in metro track[D]. Beijing Jiaotong University, 2015: 43-47. (in Chinese)

[18] Deng Yushu, Xia He, Zenda Yasuo et al. Experimental study of ladder track on a rail transit elevated bridge[J]. Engineering Mechanics, 2011, 28(3): 49-54. (in Chinese)

[19] Ma Meng, Liu Weining, Liu Weifeng, et al. Theoretical analysis and tests for vibration mitigation characteristics of ballasted ladder track[J]. Journal of vibration and shock, 2015, 34(6): 62-66. (in Chinese)

[20] Wang Yigan, Liu Penghui, Li Teng, et al. Tests for effect of track vibration reduction measures in a depot on vibration and noise reduction of a superstructure[J]. Journal of vibration and shock, 2020, 39(21): 284-291. (in Chinese)

[21] Tu Qinming. Comparative study on vibration characteristics and vibration-reducing effects of metro general vibration-reducing tracks[J]. Railway Engineering, 2020, 60(5): 135-138. (in Chinese)

[22] Lei X, Jiang C. Analysis of vibration reduction effect of steel spring floating slab track with finite elements[J]. Journal of Vibration and Control, 2016, 22(6): 1462-1471. https://doi.org/10.1177/1077546314539372

[23] Gupta S, Liu W F, Degrande G, et al. Prediction of vibrations induced by underground railway traffic in Beijing[J]. Journal of Sound and Vibration, 2008, 310(3): 608-630. https://doi.org/10.1016/j.jsv.2007.07.016

[24] Balendra T, Chua K H, Lo K W, et al. Steady-State Vibration of Subway-Soil-Building System[J]. Journal of Engineering Mechanics, 1989, 115(1): 145-162. https://doi.org/10.1061/(ASCE)0733-9399(1989)115:1(145)

[25] Deng Yahong, Xia Tangdai, Chen Jingyu. Analysis of efficiency of vibration isolating groove subjected to vehicle load[J]. Rock and Soil Mechanics, 2007(5): 883-887+894. (in Chinese)

[26] Yao Jinbao, Xia He, Hu Jingliang. Study on vibration isolation effect of open trench on environmental vibration induced by train operation[J] China Railway Science, 2018, 39(2): 44-51. (in Chinese)

[27] Zou C, Wang Y, Zhang X, et al. Vibration isolation of over-track buildings in a metro depot by using trackside wave barriers[J]. Journal of Building Engineering, 2020, 30: 101270. https://doi.org/10.1016/j.jobe.2020.101270

[28] Kattis S E, Polyzos D, Beskos D E. Vibration isolation by a row of piles using a 3-D frequency domain BEM[J]. International Journal for Numerical Methods in Engineering, 1999, 46(5): 713-728. https://doi.org/10.1002/(SICI)1097-0207(19991020)46:5<713::AID-NME693>3.0.CO;2-U

[29] Gao G Y, Li Z Y, Qiu Ch, et al. Three-dimensional analysis of rows of piles as passive barriers for ground vibration isolation[J]. Soil Dynamics and Earthquake Engineering, 2006, 26(11): 1015-1027. https://doi.org/10.1016/j.soildyn.2006.02.005

[30] Chen Gongqi, Gao Guangyun. Vibration screening effect of in-filled trenches on train dynamic loads of geometric irregular track in layered grounds[J]. Chinese Journal of Rock Mechanics and Engineering, 2014, 33(1): 144-153. (in Chinese)

[31] Guo Hong, An Ming. Application of double vibration reduction trench in reducing dynamic compaction vibration influence[J]. Construction Technology, 2013, 42(13): 59-60+63. (in Chinese)

[32] Li Zhiyi, Gao Guangyun, Qiu Chang, et al. Analysis of multi-row of piles as barriers for isolating vibration in far field[J]. Chinese Journal of Rock Mechanics and Engineering, 2005(21): 192-197. (in Chinese)

Structural Health Monitoring: 10APWSHM
Materials Research Forum LLC
Materials Research Proceedings 50 (2025) 24-43
https://doi.org/10.21741/9781644903513-4

[33] Sun Chenglong, Gao Liang. Field measurement and analysis on vibration isolation effects of multi-rows of piles for railway engineering[J]. Railway Engineering, 2016(3): 163-167. (in Chinese)

[34] Zheng Hui. Research and design application of vibration damping ditch technology in metro depot with upper property development[J]. Railway Standard Design, 2018, 62(11): 150-154. (in Chinese)

[35] Wang Tianyou, Zhang Zheng, Ding Jiemin. Study on vibration and noise control of Shanghai symphony orchestra music hall[J]. Environmental Engineering, 2012, 30(Suppl.1): 13-18. (in Chinese)

[36] Zhao Peng, Fu Weiqing, Han Yanyan, et al. The applicable design method of isolated structure withsteel spring isolator[J]. Low Temperature Architecture Technology, 2020, 42(5): 44-48. (in Chinese)

[37] Wagner H G. Vibration control systems for trackbeds and buildings using coil steel springs[J]. Proceedings of acoustics, 2004: 99-104.

[38] Jia Xiaodong, Zhang Yumin, Lan Zhuangzhi. Vertical vibration mitigation effect for subway surrounding buildings with disc springs[J]. Journal of North China University of Science and Technology (Natural Science Edition), 2016, 38(4): 106-112. (in Chinese)

[39] Ling Yuhong, Wu Shan, Gu Jingxin, et al. A comparative study on vibration isolation methods of metro superstructure[J]. Journal of North China University of Science and Technology (Natural Science Edition), 2021, 49(4): 1-8. (in Chinese)

[40] Chen P, Wang B, Zhou D, et al. Performance evaluation of a nonlinear energy sink with quasi-zero stiffness property for vertical vibration control[J]. Engineering Structures, 2023, 282: 115801. https://doi.org/10.1016/j.engstruct.2023.115801

[41] Pan P, Shen S, Shen Z, et al. Experimental investigation on the effectiveness of laminated rubber bearings to isolate metro generated vibration[J]. Measurement, 2018, 122: 554-562. https://doi.org/10.1016/j.measurement.2017.07.019

[42] Chen Haowen. Application of thick rubber bearing in vibration isolation for metro surrounding building structures[D]. Tsinghua University, 2014: 67-71. (in Chinese)

[43] He Wenfu, Liu Wenguang, Yang Yanfei, et al. Basic mechanical properties of thick rubber isolators[J]. Journal of PLA University of Science and Technology(Natural Science Edition), 2011, 12(3): 258-263. (in Chinese)

[44] Yabana S, Matsuda A. Mechanical properties of laminated rubber bearings for three-dimensional seismic isolation[C]// 12th World Conference on Earthquake Engineering. Auckland, New Zealand, International Association for Earthquake Engineering. 2000: 2452.

[45] Imbimbo M, De Luca A. F.E. stress analysis of rubber bearings under axial loads[J]. Computers & Structures, 1998, 68(1): 31-39. https://doi.org/10.1016/S0045-7949(98)00038-8

Structural Health Monitoring: 10APWSHM
Materials Research Proceedings 50 (2025) 44-51

Materials Research Forum LLC
https://doi.org/10.21741/9781644903513-5

Long-term prediction of tunnel primary lining deformation based on wireless monitoring

Yuxuan Xia[1,a], Dongming Zhang[1,b]*, Bo Zhang[1,c], Changze Li[1,d], Yue Tong[2,e]

[1]Key Laboratory of Geotechnical Engineering, Dept. of Geotechnical Engineering, Tongji Univ., Shanghai 200092, China

[2]Broadvision Engineering Consultants, Kunming 650041, China

[a]2232581@tongji.edu.cn, [b]09zhang@tongji.edu.cn, [c]2132587@tongji.edu.cn, [d]2232580@tongji.edu.cn, [e]tongyue2014@yeah.net

Keywords: Long-Term Prediction, Wireless Monitoring, Large Deformation, Tunnel Primary Lining

Abstract. Due to the harsh construction environment such as blasting excavation and dust, conventional monitoring methods are challenging to be continuously deployed at the primary lining section of the drilling and blasting tunnel, which makes it difficult to capture the large deformation evolution trend and warn the deformation risk. Therefore, this paper takes a biased construction section of Xujiacun tunnel as a site case to implement a wireless sensor network, including gateway and laser distance senor, which is embedded into the primary lining to protect sensors in harsh construction environment for high frequency of primary lining deformation monitoring. Subsequently, a *DLinear* model is applied, which can be used for the long-term prediction of deformation time-series by splitting deformation sequence into trend sequence and residual sequence. Its prediction performance under different prediction steps is tested, so as to provide accurate and long-term risk early-warning in real-time for construction sites.

1. Introduction

The rapid expansion of transportation networks has led to a growing number of drill-and-blast tunnels, essential for urban transit, railways, and highways[1]. However, with larger projects and more advanced techniques, the geological complexity of tunnel sites has increased, creating significant challenges. A key risk is large-scale deformation of primary lining structures during construction, threatening both tunnel integrity and worker safety[2]. Ensuring safety against these deformations is now a critical concern in drill-and-blast tunnel engineering.

During the construction of drill-and-blast tunnels, the most commonly used method for monitoring lining deformations is manual measurement with a total station. However, this approach has significant limitations, particularly in terms of monitoring frequency, as measurements are typically taken only once per day. This low frequency makes it challenging to continuously track the development of deformations in real-time[3]. Recently, various advanced automatic monitoring techniques, such as fiber optic sensors, have been increasingly applied to monitor structural deformations in tunnels[4]. While these technologies offer potential for improving monitoring precision and frequency, they face significant challenges in drill-and-blast tunnel environments. The dust, rock splinters, and debris generated during blasting operations hinder the long-term deployment of these sensors, limiting their effectiveness in continuous deformation monitoring. Thus, the need for robust, accurate, and reliable monitoring solutions in such environments remains an ongoing challenge.

In addition to improving the monitoring of deformation development during drill-and-blast tunnel construction, there is a critical need for accurate prediction of deformation trends. Predicting these changes in advance provides construction teams with sufficient time to manage and mitigate

Structural Health Monitoring: 10APWSHM
Materials Research Proceedings 50 (2025) 44-51

Materials Research Forum LLC
https://doi.org/10.21741/9781644903513-5

potential risks. Deep learning techniques have shown remarkable potential in this area, particularly in capturing hidden patterns within time-series data, enabling accurate predictions of future states[5]. As deep learning technologies continue to advance, predictive models have evolved from traditional Convolutional Neural Networks (CNN) to more sophisticated architectures such as Transformer[6]. These models are also increasingly being applied to underground engineering for structural safety control. Their ability to model complex, sequential data makes them powerful tools for forecasting deformations, offering promising solutions to enhance safety management in tunnel construction projects.

Therefore, this paper takes the Xujiacun Tunnel, an under-construction tunnel in Yunnan Province, China, as a case study. Laser sensors were installed in the primary lining construction section to collect high-frequency deformation data over a period of time. A new deep learning time-series prediction framework, *DLinear*, was then applied to forecast the deformation, with its performance tested across different prediction timesteps.

2. Site experiment and the Collection of Deformation Dataset

2.1 Project Background

The site study is based on the construction of the Xujiacun tunnel of the Xuanhui expressway construction project, Yunnan Province, China. The total length of the tunnel is 1271 m. It is designed with a tunnel span of 12.2 m and a tunnel height of 9.3m. The Xujiacun Tunnel is primarily located in fully to moderately weathered basalt, with fractured surrounding rock and shallow burial depth. The self-stability of the rock is poor, and tunnel excavation is prone to collapse accidents. As shown in Fig. 1, the section most deserving of attention during the construction of the Xujiacun Tunnel is the shallow-buried section at mileage ZK78+280 to ZK78+380. The green lines represent the tunnel excavation outlines as per the design, while the red region highlights the shallow-buried section that carries significant construction risks. The surrounding rock in this section is classified as Grade V_2, characterized by fully to strongly weathered rock with well-developed joints and fissures. The rock mass is highly fragmented, presenting a shattered structure with poor self-stabilizing capacity during tunnel excavation. This section faces a substantial risk of large-scale deformation during excavation.

This study implemented a WSN (wireless sensor network) monitoring system in this high-risk region to mitigate the large deformation construction risks. To ensure that the monitoring system can operate continuously in the blasting environment, we have also designed a box structure to protect sensors from interference caused by flying rocks and dust during blasting. All details will be presented in the subsequent monitoring scheme.

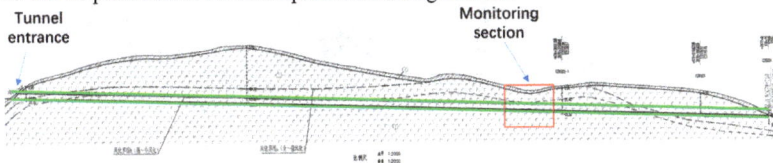

Fig. 1 Schematic diagram of construction design and monitoring section of the Xujiacun tunnel

2.2 Tunnel Monitoring Scheme

The longitudinal arrangement of the WSN monitoring system is shown in **Fig. 2**. The WSN monitoring system measuring the primary lining deformation includes two wireless distance laser sensor nodes and a wireless smart gateway (see in **Fig. 3**).

During the tunnel excavation, the deformation monitoring section is selected 10 meters from the tunnel face, where two wireless laser distance sensor points are installed in the arch waist of this transversal section. As the on-site condition shown in **Fig. 4**, one laser sensor node is installed

Structural Health Monitoring: 10APWSHM
Materials Research Proceedings 50 (2025) 44-51

Materials Research Forum LLC
https://doi.org/10.21741/9781644903513-5

at an upward diagonal angle, projecting towards the top of the monitoring section to measure the crown settlement of the primary lining. The other sensor is installed horizontally, projecting towards the other side of the monitoring section to measure the horizontal convergence. The distance laser sensor nodes are buried within the primary lining, providing protection against rock debris generated during tunnel blasting, thereby safeguarding the sensors from potential damage.

The deformation data measured by the laser sensors will be wirelessly transmitted via the WSN AD hoc network to the wireless smart gateway installed on the secondary lining section. The gateway will upload the data to the online platform via 4G signal, allowing on-site personnel to access it in real time. Additionally, personnel can remotely adjust the monitoring frequency (ranging from once every 1 to 60 minutes) based on site requirements, enabling more timely insight into the large deformation state of the primary lining.

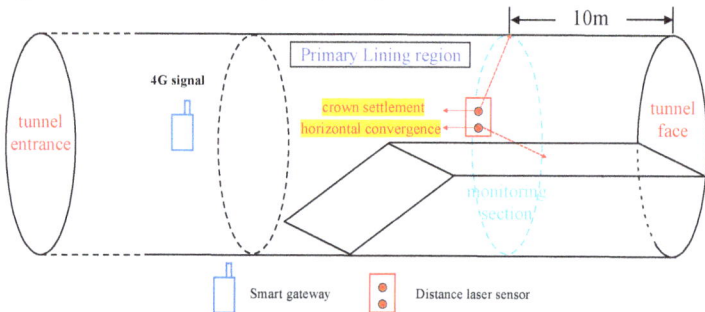

Fig. 2 Longitudinal schematic diagram of the WSN monitoring system layout

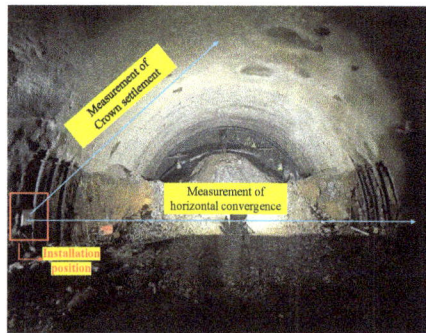

Fig. 3 WSN monitoring equipment Fig. 4 On-site layout in the transversal monitoring section

2.3 Collection of the primary lining deformation dataset
In the inner structure of laser distance sensor nodes, not only does a sensor include a laser ranging unit, but it also integrates an inclinometer unit. In the actual tunnel blasting excavation process, the harsh environment of the primary lining section is highly complex, and the monitoring posture of the sensors is significantly affected by blasting vibrations. Therefore, this adverse condition must be considered when calculating the deformation development of the primary lining. We assume a rigid connection between the sensors and the primary lining, ensuring that any movement or deformation of the lining is fully transmitted to the sensors for accurate measurement and analysis. This enables the sensor nodes to simultaneously detect both changes in the measured

Structural Health Monitoring: 10APWSHM   Materials Research Forum LLC
Materials Research Proceedings 50 (2025) 44-51   https://doi.org/10.21741/9781644903513-5

distance of the section and slight movements of the lining in contact with the sensor, based on the set monitoring frequency.

Let the tilt angle between the sensor and the horizontal plane be denoted as β, and the real-time measured distance between the sensor and the target point as d. For the measurement at time i, the laser distance value is d_i, and the tilt angle is β_i. Therefore, the crown settlement (S_i) and horizontal convergence (C_i) of the monitoring section at time i are given by equations (1) and (2), respectively.

$$S_i = d_i \sin \beta_i - d_0 \sin \beta_0. \qquad (1)$$
$$C_i = d_i \cos \beta_i - d_0 \cos \beta_0. \qquad (2)$$

where the subscript 0 represents the initial moment at time i. A negative value of S indicates that the tunnel crown is displacing in the direction of gravity, while a negative value of C signifies that the tunnel lining is converging horizontally.

The monitoring frequency of the entire WSN monitoring system was set to once every 15 minutes. As a result, a time-series dataset of horizontal convergence and crown settlement deformation of the monitoring section over a 10-day period was obtained. After this period, since the secondary lining was about to be constructed at this position, the monitoring system was dismantled. The deformation dataset obtained is shown in the Fig. 5.

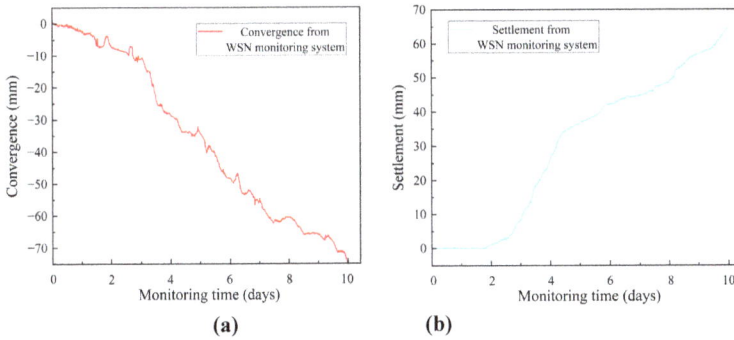

Fig. 5 Monitoring deformation dataset of the primary lining: (a) crown settlement; (b) horizontal convergence

3. Prediction of Primary Lining Deformation

3.1 *DLinear* model architecture

Accurate predictions of the primary lining deformation can provide on-site personnel with time to respond to impending risks. The longer the time step that can be accurately predicted, the more preparation time that the construction site will have, which is of great importance for managing construction safety risks. So, it is crucial to consider the appropriate approach to obtain accurate results.

In recent years, deep neural networks have emerged as powerful tools for time series prediction, and a newest deep learning framework, called *DLinear*, has been proposed[7]. The prediction framework of *DLinear* model is shown in Fig. 6(a). Unlike complex Transformer architectures, the *DLinear* model splits the historical time series data into seasonal and trend feature sequences, and maps them using only a few linear layers (see in Fig. 6(b)). The final prediction values are then generated. Even only by utilizing simple linear mapping matrices and a small number of training parameters, *DLinear* has achieved remarkable predictive performance on several public datasets, such as traffic and weather data. In this study, the *DLinear* model will be employed to

Structural Health Monitoring: 10APWSHM   Materials Research Forum LLC
Materials Research Proceedings 50 (2025) 44-51   https://doi.org/10.21741/9781644903513-5

test its prediction performance for the deformation of the primary lining in drill-and-blast tunnel projects.

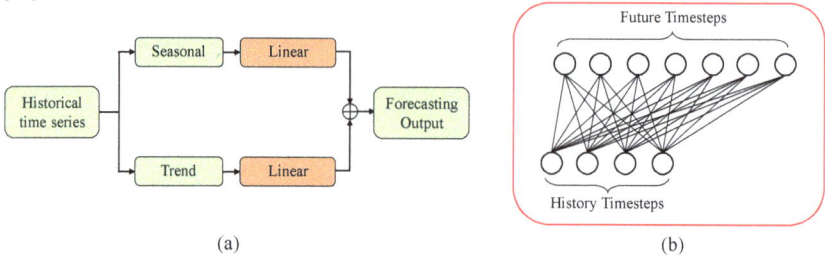

(a)   (b)

Fig. 6 Framework of DLinear: (a) Whole structure of DLinear; (b) Linear layer

3.2 Training and Testing Process

The *DLinear* model's input time step is set to 12, corresponding to the historical deformation data of the primary lining over the past three hours. The prediction time steps are set to 8, 16, 32, 64, 128, and 192, representing the deformation development for the next 2, 4, 8, 16, 32, and 48 hours, respectively. The *DLinear* model is trained on a desktop equipped with an Intel Core i7-14700KF CPU and an Nvidia RTX4060 GPU. The entire dataset is split into a 7:3 ratio for the training and testing sets. The initial learning rate is 1×10^{-4}. The Adam optimizer is used during training, along with an early stopping strategy (patience = 10). The loss function adopts the mean squared error (MSE) function (see in equation (3)). The performance of the *DLinear* model in the different prediction timesteps is evaluated quantitatively using the root mean square error (RMSE, see in equation (4)) between the predicted and actual values on the test set.

$$MSEloss = \frac{\sum_{i=1}^{n}(a_i^p - a_i)^2}{n}. \tag{3}$$

$$RMSE = \sqrt{\frac{\sum_{i=1}^{n}(a_i^p - a_i)^2}{n}}. \tag{4}$$

3.2 Prediction performance of *DLinear* model

The RMSE quantitative evaluation of the *DLinear* model's prediction performance at different time steps is shown in Table 1. Overall, both the crown settlement and horizontal convergence prediction results show smaller RMSE values at smaller timesteps, which indicate better prediction accuracy. As the prediction timesteps increase, the prediction performance gradually declines at first, followed by a rapid drop. The turning point for the rapid decline appears at the timestep of 32 for settlement prediction and at 64 for convergence prediction. By comparing the application of the *DLinear* model on the two deformation datasets—horizontal convergence and crown settlement—it is evident that, at the same time step, the RMSE between the predicted and actual deformation values is smaller for horizontal convergence. This suggests that the model performs better in predicting horizontal convergence on the construction site of Xujiacun tunnel.

Structural Health Monitoring: 10APWSHM
Materials Research Proceedings 50 (2025) 44-51

Materials Research Forum LLC
https://doi.org/10.21741/9781644903513-5

Table 1. Quantitative evaluation of the prediction performance at different prediction timesteps: (a) RMSE evaluation of settlement prediction; (b) RMSE evaluation of convergence prediction

Table (a)

Prediction timesteps	RMSE
8	1.60
16	1.96
32	2.6
64	4.54
128	16.54
192	19.11

Table (b)

Prediction timesteps	RMSE
8	0.63
16	0.97
32	1.89
64	2.48
128	7.28
192	11.33

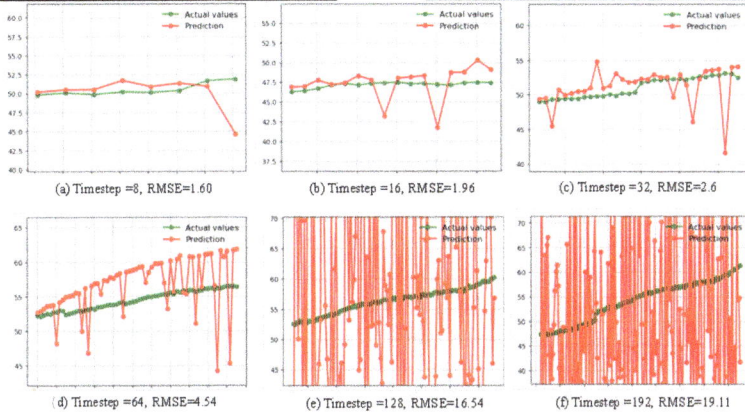

Fig. 7 Visualization of prediction result at different timesteps (for settlement)

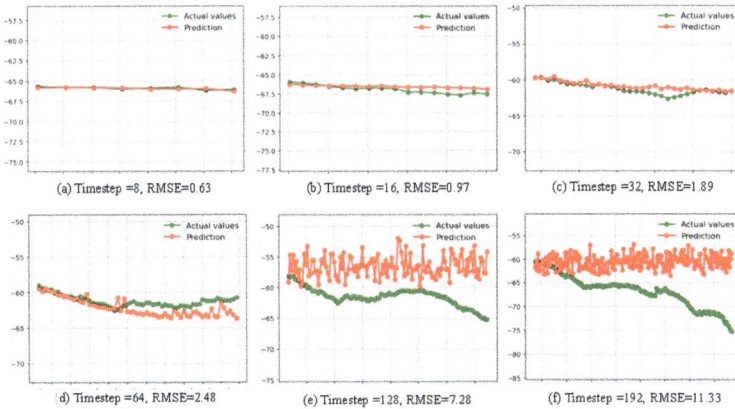

Fig. 8 Visualization of prediction result at different timesteps (for convergence)

Structural Health Monitoring: 10APWSHM   Materials Research Forum LLC
Materials Research Proceedings 50 (2025) 44-51   https://doi.org/10.21741/9781644903513-5

The predicted and actual values of horizontal convergence and crown settlement for the initial lining of the Xujiacun Tunnel, at different time steps, are visualized in Fig. 7 and Fig. 8. It can be observed that at fewer prediction timesteps, the predicted values closely match the actual deformation values. However, when the prediction timestep reaches 192, the prediction essentially loses accuracy. Additionally, it is evident that for horizontal convergence and crown settlement predictions, time steps of 64 and 32, respectively, offer the best balance between the longest prediction duration and reasonable accuracy. Therefore, 16 hours and 8 hours can be considered the effective prediction durations for these two key initial lining deformation control indicators.

4. Conclusions

This paper primarily designs a wireless sensor network (WSN) system to obtain real-time data on the deformation of the primary lining in a drill-and-blast tunnel, and applies deep learning methods to predict its future deformation state. The specific research works are concluded as follows:

(1) A WSN monitoring system, consisting of a wireless smart gateway and two wireless distance laser sensor nodes, was deployed at the drill-and-blast tunnel construction site. The two laser nodes were buried in the primary lining to resist the adverse effects of blasting and to measure changes in two key deformation indicators at the monitoring transection of the primary lining: horizontal convergence and crown settlement. The measurement data were transmitted in real-time to the smart gateway installed on the secondary lining section and uploaded to the online platform for viewing by on-site personnel.

(2) A novel deep learning time series prediction model, *DLinear*, was applied to predict the deformation of the primary lining in drill-and-blast tunnels. The *DLinear* model architecture achieves good prediction performance by decomposing time series features and using simple linear layers for processing. The deformation prediction performance of the *DLinear* model at different time steps is also discussed. The results showes that with increasing prediction steps, the model's performance first declined gradually and then dropped rapidly. Considering the balance between the length of prediction timestep and accuracy, 16 hours and 8 hours were identified as the optimal prediction times for horizontal convergence and crown settlement, respectively.

Finally, our future work will focus on improving the deployment methods of the sensors to better withstand the effects of blasting while ensuring more efficient installation. Additionally, we aim to incorporate more engineering information into the *DLinear* model structure to make it more applicable to drill-and-blast tunnel construction.

Reference

[1] B. He, D. J. Armaghani, S. H. Lai, X. He, P. G. Asteris, and D. Sheng, 'A deep dive into tunnel blasting studies between 2000 and 2023—A systematic review', *Tunnelling and Underground Space Technology*, vol. 147, p. 105727, May 2024. https://doi.org/10.1016/j.tust.2024.105727

[2] Y. Zhu, J. Zhou, B. Zhang, H. Wang, and M. Huang, 'Statistical analysis of major tunnel construction accidents in China from 2010 to 2020', *Tunnelling and Underground Space Technology*, vol. 124, p. 104460, Jun. 2022. https://doi.org/10.1016/j.tust.2022.104460

[3] C. Y. Gue, M. Wilcock, M. M. Alhaddad, M. Z. E. B. Elshafie, K. Soga, and R. J. Mair, 'The monitoring of an existing cast iron tunnel with distributed fibre optic sensing (DFOS)', *J Civil Struct Health Monit*, vol. 5, no. 5, pp. 573–586, Nov. 2015. https://doi.org/10.1007/s13349-015-0109-8

[4] D. Zhang, C. Nie, M. Chen, H. Huang, and Y. Wu, 'Wireless tilt sensor based monitoring for tunnel longitudinal Settlement: Development and application', *Measurement*, vol. 217, p. 113050, Aug. 2023. https://doi.org/10.1016/j.measurement.2023.113050

[5] N. Zhang, A. Zhou, Y. Pan, and S.-L. Shen, 'Measurement and prediction of tunnelling-induced ground settlement in karst region by using expanding deep learning method', *Measurement*, vol. 183, p. 109700, Oct. 2021. https://doi.org/10.1016/j.measurement.2021.109700

[6] A. Vaswani *et al.*, 'Attention Is All You Need', Aug. 01, 2023, *arXiv*: arXiv:1706.03762. Accessed: Jun. 22, 2024. [Online]. Available: http://arxiv.org/abs/1706.03762

[7] A. Zeng, M. Chen, L. Zhang, and Q. Xu, 'Are Transformers Effective for Time Series Forecasting?', Aug. 17, 2022, *arXiv*: arXiv:2205.13504. https://doi.org/10.48550/arXiv.2205.13504

Structural Health Monitoring: 10APWSHM
Materials Research Proceedings 50 (2025) 52-60

Materials Research Forum LLC
https://doi.org/10.21741/9781644903513-6

Post-earthquake rapid assessment of interconnected electrical equipment based on hybrid modelling

Beatriz Moya[1,a], Huangbin Liang[2,b *], Francisco Chinesta[1,3,c] and Eleni Chatzi[2,4,d]

[1]CNRS@CREATE LTD., CNRS, 1 CREATE Way, Singapore, 138602, Singapore

[2]Future Resilient Systems, Singapore-ETH Centre, 1 CREATE Way, Singapore, 138602, Singapore

[3]ESI Group Chair, PIMM Lab, ENSAM Institute of Technology, 151 Bvd. de l'Hôpital, Paris, 75013, France

[4]Department of Civil Environmental and Geomatic Engineering, ETH Zürich, Stefano-Franscini-Platz-5, Zurich, 8093, Switzerland

[a]beatriz.moya@cnrsatcreate.sg, [b]huangbin.liang@sec.ethz.ch, [c]francisco.chinesta@ensam.eu, [d]chatzi@ibk.baug.ethz.ch

Keywords: Hybrid Modelling, Rapid Assessment, Interconnected Equipment, Reduced Order Model, Graph Neural Network

Abstract. Electrical substations play an indispensable role in the power grid for adjusting power voltage and controlling power flow, ensuring normal power transmission and distribution. However, substation equipment is particularly vulnerable to seismic events due to the use of brittle porcelain materials and highly slender structures, increasing the risk of structural failure and cascading effects. While significant research has focused on seismic performance and innovative damping techniques for different standalone equipment, critical gaps remain, particularly in accounting for the interconnected nature of substation equipment and post-earthquake emergency response. This study addresses these gaps by proposing a hybrid post-earthquake rapid assessment model that utilizes monitored ground motion signals. Hybrid modeling combines physics-based models with data-driven approaches, leveraging the strengths of both to overcome their individual limitations. Specifically, the study integrates Graph Neural Networks (GNN) with mechanistic models of standalone equipment to capture the dynamic behavior and interaction forces of interconnected equipment under seismic loads. The spatial relationships of the interconnected equipment are represented through a graph structure, and temporal dependencies are learned through recurrent computations. Thus, this approach circumvents the need for costly and impractical sensor installations on every piece of equipment, offering a cost-effective and accurate assessment method. The efficacy of the proposed hybrid modelling technique is demonstrated with a case study on combinations of 800 kV post insulators interconnected by busbars. By providing a rapid assessment of each equipment's post-earthquake condition, the proposed model can be applied to inform emergency repair actions and enhance the resilience of power infrastructure.

Introduction

The electrical substation is integral to the power grid in that it regulates the voltages, controls the flow of power, and ensures the smooth transmission and distribution of electricity to the end users. But past seismic events have demonstrated that substations are highly susceptible to significant damage due to the high slenderness structural characteristics and brittle porcelain materials of substation equipment [1, 2]. The destruction of substation equipment during earthquakes can result in substantial consequences, such as direct financial losses from damaged devices or structural failures of equipment. Additionally, the indirect effects, including economic disruption, potential

Structural Health Monitoring: 10APWSHM
Materials Research Proceedings 50 (2025) 52-60

Materials Research Forum LLC
https://doi.org/10.21741/9781644903513-6

casualties, and the cascading loss of essential social services due to power outages, can be severe [3].

In view of the importance of substations in the functioning of power systems, extensive studies have been carried out to evaluate and improve the seismic performance of substation equipment. One common approach involves the installation of additional dampers [4] or isolators [5], which help to reduce or isolate vibrations, thereby minimizing structural damage and improving the resilience of substations. Despite significant progress in seismic research, most studies have primarily focused on the performance of individual pieces of equipment, often overlooking the interaction forces between them. However, in real-world applications, substation equipment are not isolated; they are interconnected by rigid pipes or flexible cables (seen in Fig. 1), which are essential for both power transmission and operational functionality. These connections can introduce complex dynamic interactions during seismic events, potentially amplifying the forces exerted on equipment, especially when the seismi input intensity is large. As a result, understanding how equipment behaves as part of an interconnected system, rather than in isolation, is critical for accurately assessing the overall seismic performance of substations and mitigating potential risks [6-7]. Furthermore, previous research mainly focus on pre-earthquake design, analysis, and retrofitting, while little attention has been paid to post-earthquake emergency response in the academic studies. In fact, timely and well-coordinated disaster relief can greatly mitigate the negative effects of an earthquake [8]. However, implementing an effective repair plan requires a thorough understanding of the condition of each piece of equipment after the event. Conventional manual inspections are not only time-consuming but also costly, especially when done for individual equipment. The rapid development of structural health monitoring (SHM) and data-based digital twinning (DT) techniques presents an opportunity to accurately and promptly assess the condition of equipment following an earthquake [9]. Nonetheless, many substations face restrictions on installing monitoring sensors on electrical equipment due to security, insulation and electromagnetic interference concerns. Moreover, the large number of devices involved in a substation make it impractical to install sensors on every piece of equipment.

Figure 1. Equipment connections within a substation

To address the above-mentioned issues in the post-earthquake evaluation of substation equipment, in this study, we propose a hybrid post-earthquake rapid assessment model that leverages just the monitored ground motion signals. The hybrid model combines physics-based approaches with data-driven techniques, allowing us to take advantage of the strengths of both while minimizing their respective limitations [10]. Specifically, we integrate Graph Neural Networks (GNN) with physics-based reduced order models (ROM) of standalone equipment to capture the dynamic response and interaction forces among interconnected equipment during seismic events. The spatial relationships between the equipment are represented through nodes and edges in a graph structure, while temporal dependencies of dynamic responses are learned in a data-driven manner. Then the trained hybrid model can accurately mimic the behavior of real systems under seismic loads through recurrent computations in a real-time manner, forming a feed forward hybrid digital twin (HDT). This method eliminates the need for costly and impractical sensor installations on each individual piece of equipment, providing a more efficient and accurate assessment solution with higher interpretability, and hence contributing to emergence response and resilience enhancement of the power infrastructures against earthquakes.

Physics-based ROM of equipment considering both ground motion and interaction force

ROMs are simplified versions of high-fidelity computational models designed to reduce computational cost while maintaining satisfying accuracy in simulations. They are widely used in fields such as structural dynamics and fluid mechanics to enable faster simulations and real-time predictions by capturing the dominant features of a system, where full-scale simulations can be time-consuming and computationally expensive [11]. According to the methodology used, ROMs can be classified into physics-based ROMs and data-driven ROMs. The former reduce the system's dimensionality by focusing on key modes or variables derived from equations like partial differential equations (PDEs), typical methods include propor orthogonal decomposition (POD) and Galerkin projection [11, 12]; the latter rely on machine learning or statistical methods to model system behavior directly from data, such as using Autoencoders [13] and gaussian process regression (GPR), when the physics of the system are complex or unknown.

Physics-based methods are particularly well-suited for developing ROM for substation equipment with cantilevered structures, such as transformers, circuit breakers, and insulators, since the dynamic behavior of these structures under seismic loads are well studied and best captured through known physical principles, e.g., Euler-Bernoulli beam partial differential equations. The detailed derivatives of such a physics-based ROM of equipment can be found in [12], this study further extends it to simultaneously consider the interaction force at the top of the equipment, providing a more accurate prediction of in-service equipment performance during earthquakes.

As shown in Fig. 2, the strucure of the equipment with a varing cross-section can be abstracted as multi-segment beams with distributed parameters connected by rigid nodes, whose vibration behavior is controlled by the PDE in Eq. 1 and its general solution can be represented as Eq. 2, where $u(h,t)$ denotes the lateral displacement that changes with the mode shape function $\phi(h)$ and the amplitude function $x(t)$. Note that $\phi(h)$ is a piecewise function in the global coordinate system,

$$m(h)\frac{\partial^2 u(h,t)}{\partial t^2}+\frac{\partial^2}{\partial h^2}\left[EI(h)\frac{\partial^2 u(h,t)}{\partial h^2}\right]=p(h,t) \tag{1}$$

$$u(h,t)=\phi(h)x(t) \tag{2}$$

$$\phi_i(h)=A_{i1}\cos a_i h+A_{i2}\sin a_i h+A_{i3}\cosh a_i h+A_{i4}\sinh a_i h \quad (a_i=\sqrt[4]{\frac{\omega^2 \overline{m}_i}{E_i I_i}} \text{ and } \omega=2\pi f, \ 0<h<L_i) \tag{3}$$

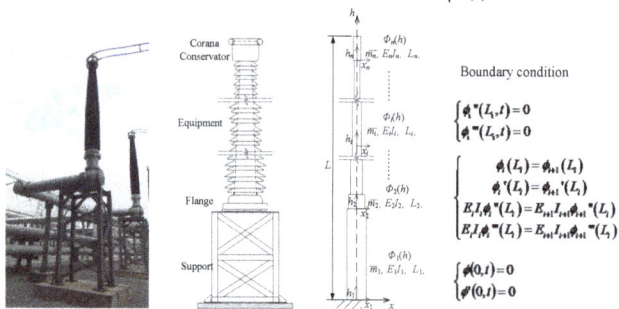

Figure 2. Analytical modelling of typical electrical equipment

and it can be expressed as Eq .3 for each segment in the local coordinate system, in f corresponds to the natural frequencies of the system and the other four constants (A_{i1}, A_{i2}, A_{i3}, A_{i4}) determine the mode shape of the i-th beam, which can be calculated by solving the eigenvalue problem for the system's mass and stiffness matrices based on the boundary conditions of each segmented

Structural Health Monitoring: 10APWSHM
Materials Research Proceedings 50 (2025) 52-60

Materials Research Forum LLC
https://doi.org/10.21741/9781644903513-6

beam. Then equations of motion can be transformed into a set of uncoupled equations, each corresponding to one of the natural modes of vibration, as described in Eq. 4 when considering both the seismic input $a_g(t)$ at the bottom and the interaction force $F(L,t)$ at the top of the equipment.

$$
\begin{aligned}
&M_n \ddot{x}_n(t) + C_n \dot{x}_n(t) + K_n x_n(t) = P_n(t), n = 1,2,3,\cdots\infty \\
&M_n = \int_0^L m(h)\left[\phi_n(h)\right]^2 dh \\
&C_n = 2\zeta_n \omega_n M_n \\
&K_n = \int_0^L \phi_n(h)\left[EI(h)\phi_n{''}(h)\right]'' dh \\
&P_n(t) = \int_0^L \phi_n(h)\, p(h,t)dh = -\int_0^L \phi_n(h) m(h) a_g(t)dh - \phi_n(L)F(L,t)
\end{aligned}
\tag{4}
$$

in which M, C, K, P is the modal mass, damping, stiffness and force. For each mode, the modal response $x_n(t)$ is computed independently using Newmark-β stepwise integration based on the dynamic force as follows with γ and β being 0.5 and 0.25 respectively for unconditional stability.

$$
\begin{aligned}
&x_{t+1} = x_t + \dot{x}_t dt + \frac{1}{2}[(1-2\beta)\ddot{x}_t + 2\beta\ddot{x}_{t+1}](dt)^2 \\
&\dot{x}_{t+1} = \dot{x}_t + [(1-\gamma)\ddot{x}_t + \gamma\ddot{x}_{t+1}]dt \\
&\ddot{x}_{t+1} = \frac{P_{t+1} - C\dot{x}_{t+1} - Kx_{t+1}}{M}
\end{aligned}
\tag{5}
$$

Subsequenly, the total dynamic response is obtained by superimposing the responses of all modes, and can be estimated by only considering the first few dominant modes, as presented in Eq. 6, thus significantly reducing computational efforts.

$$
u(h,t) = \sum_{j=1}^{\infty} u_j(h,t) \approx \sum_{j=1}^{k} \phi_j(h) x_j(t)
\tag{6}
$$

To verify the accuracy and efficiency of this physics-based ROM, a finite element model (FEM) of an interconnected equipment system composed of two pieces of equipment linked by a rigid pipe is established in the ABAQUS, as shown in Fig. 3(a) with some key parameters. Beam element are applied to model the equipment, Truss element is adopted to simulate the linking pipe, and all the joints are regarded as rigid connections. A sinusoidal signal with an amplitude of 4 m/s^2 and a frequency of 2 Hz is input to this FEM to generate the response dataset as ground truth, and then the same input and the derived $F(t)$ are input to the ROM to obtain the responses of interest for comparison. As provided in Fig. 3(c), the predicted results of the ROM show excellent agreement with the ground truth from FEM in terms of both the top displacement and bottom stress of Equipment 2. Moreover, the ROM computation time is less than 0.1 seconds, significantly faster than the time required for FEM analysis. In contrast, the ROM would, however, show substantial deviations from the actual results if it ignores the effect of the top interaction force, as presented in Fig. 3(d). The strong agreement between the ROM and FEM results highlights the effectiveness of the ROM in providing accurate and efficient predictions. On the other hand, the significant discrepancies that arise when interaction forces are neglected emphasize that such forces play a critical role in the dynamic response of the system, and their inclusion is essential for maintaining the reliability of the ROM.

Noteworthy, the interaction force is hard to measure in reality and assumed to be unknown in this study, and instead, it will be learned using GNN in a data-driven manner as introduced below.

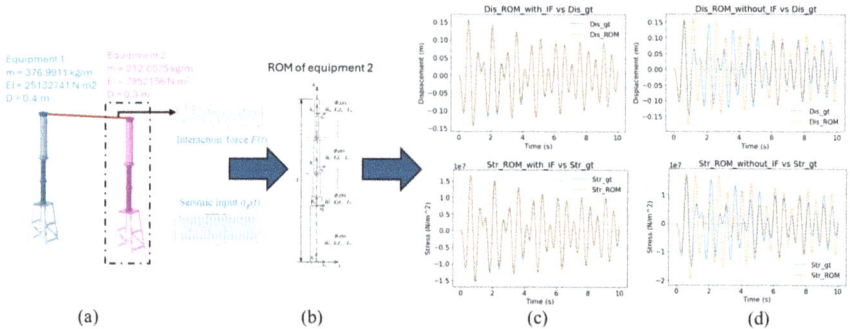

Figure 3. (a) FEM of the interconnected equipment system linked by a rigid pipe; (b) ROM of the equipment 2; (c) comparison between the ROM predictions considering both the seismic input and interaction force and the FEM ground truths in terms of top displacement and bottom stress; (d) comparison between the ROM predictions without considering the interaction force and the FEM ground truths in terms of top displacement and bottom stress

Dynamic behavior prediction of the interconnected system using ROM-embedded GNN

GNNs are an emerging class of machine learning models designed to process graph-structured data by capturing relationships and interactions between entities in a network, and they work by learning patterns from message passing, aggregating, and updating of the edge and node representations [14], as described in Eq. 7 and Eq. 8.

$$h_{ij}^{(k+1)} = \varphi\left(h_i^{(k)}, h_j^{(k)}, h_{ij}^{(k)}\right) \tag{7}$$

where $h_{ij}^{(k+1)}$ denotes the edge embeddings at the k-th layer between node i and node j, updated through a function φ(usually a neural network) using the embeddings of the nodes $h_i^{(k)}$, $h_j^{(k)}$ and the current edge embedding $h_{ij}^{(k)}$. Once the edges are updated, the nodes are updated as follows.

$$h_i^{(k+1)} = \sigma\left(h_i^{(k)}, \bigoplus_{j \in N(i)} f\left(h_j^{(k)}, h_{ij}^{(k)}\right)\right) \tag{8}$$

where f is the message passing function, $\oplus$ represents the aggregation function (e.g., weighted sum or average) applied to the neighboring nodes N(i), and σ is the activation function. This approach is effective for capturing topological relationships in domains like social networks and molecular graphs [15], but GNNs alone lack a grounding in physical laws, which is crucial for applications like structural dynamics or fluid mechanics.

Physics-Enhanced Machine Learning (PEML) paradigm [16] overcomes this by integrating domain-specific physics knowledge, such as conservation laws or governing equations like PDEs, directly into the GNN architecture. This hybrid approach improves the prediction of dynamic behaviors in complex systems while ensuring physical consistency. However, they also face challenges, such as the need for accurate physical models and increased computational complexity compared to purely data-driven GNNs. Within the PEML framework, in this work we propose a hybrid modelling tool to predict the dynamical behavior of interconnected equipment leveraging just the seismic input signals. Specifically, we compare two approaches to incorporate the physics into the learning process, as depited in Fig.4. Note that in neither of those cases we know the interaction force. Hence, both will learn from the indirect measurement, i.e., the seismic input signal (f_{GM}), and this mornitored information is cancatenated in the hidden node presentations. The first, termed Physics-informed GNN (PiGNN), imposes the physics in the loss function by adding the residual from evaluating the ROMs as the physics constrined loss term $\Lambda_{Physics}$ to the regular GNN loss Λ_{Data}, where the overall objective becomes:

$$L = L_{Data} + L_{Physics} \tag{9}$$

As shown in Fig. 4(a) in the encoder-processor-decoder PiGNN, the input node features include the acceleration (a), velocity (v), and type (u) of equipment, and the edge feature is the relative top displacement ($\|q_i\text{-}q_j\|$) of the equipment. After d passes of updating through Eq. 7 and Eq. 8, the output is the next time step responses of equipment $x_i^*(a_i^*,v_i^*,q_i^*)$, which will be compared with the ground truth $x_i^{(t+1)}$ to obtain the Λ_{Data} in Eq. 9; and the interaction force between equipment as the output edge feature (F_{ij}^*), will be further substituted into Eq.4 in the developed ROM to predict the new responses and also compared with $x_i^{(t+1)}$ to obtain the Λ_{Physics} in Eq. 9 for training.

The second approach, known as Physics-encoded GNN (PeGNN), introduces the physics as a hard constraint by creating a customed aggregation function based on the introduced ROMs of the equipment as presented in Eq. 10.

$$x_i^{(t+1)} = \text{ROM}\left(x_i^{(t)}, f_{GM}^{(t)}, \sum_{j \in N\,(i)} \text{MLP}\left(h_i^{(t)}, h_j^{(t)}, f_{GM}^{(t)}\right)\right) \qquad (10)$$

where MLP coresponds to the multilayer perceptron that incorporates the hidden representation of connected nodes to generate the interaction force along the corresponding edge direction, and then the sum of forces to the node i together with the ground motion as well as the node responses at current time step $x_i^{(t)}$ are input to the ROM to calculate the next time step responses $x_i^{(t+1)}$. In other words, we give a physical meaning to the generated hidden edge representation h_{ij}^* directly, and replace the decoder with the ROM model at the end for updating the node features $x_i^*(a_i^*,v_i^*,q_i^*)$.

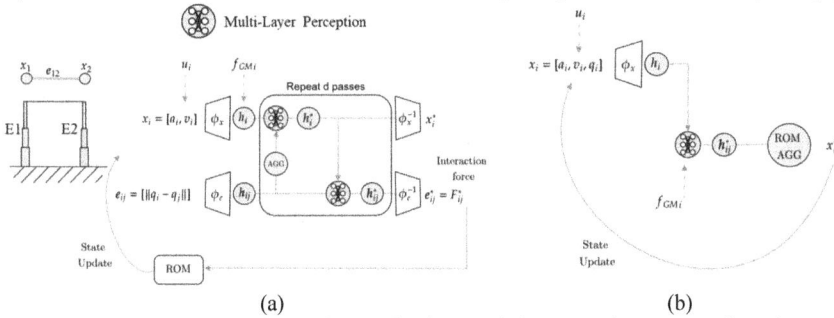

(a)   (b)

Figure 4. PEML structures for predicting the dynamic behavior and interaction force between equipment (a) PiGNN (b) PeGNN

Results

We use 30% of the dataset, generated by inputting a sinusoidal signal to the establishe FEM shown in Fig. 3(a), for training the two approaches, and their prediction performance is then evaluated through recurrent computations in rollout. The comparison results in terms of the critical responses of both pieces of equipment, including the top acceleration, top velocity, top displacement, bottom stress as well as the interaction force, are provided in Fig. 5, in which the black dash lines denote the ground truth, the solid lines in blue and in orange represent the GNN prediction responses for E1 and E2, respectively. As shown, the prediction results from the two trained models both agree well with the ground truth, demonstrating that the introduced PEGNN-based predictions can capture the dynamic response and the interaction force between the interconnected equipment structures.

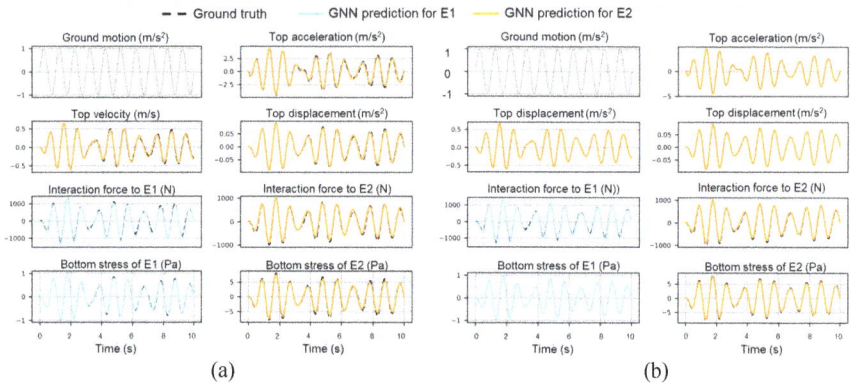

Figure 5. comparison results between the ground truth (dashed line) and the GNN predictions (solid lines) for the two pieces of equipment in terms of the critical responses under a sinusoidal signal input (a) PiGNN model (b) PeGNN model

For performance quantification, we further compare the maximum absolute errors and relative percentage errors between the ground truth from the FEM and the predictions from the repective two models with respect to the critical responses of E2 for instance, as listed in Table 1. Table 1 also presents the prediction results of a vanilla GNN model that does not incorporate any physical information as a baseline. As compared, we can conclude that integrating physics into the GNN learning process can improve prediction accuracy by about one order of the magnitude. Moreover, for this case study, the model that imposes physics in the aggregation function generally outperforms the one that incorporates physics in the loss function. This highlights the substantial advantages of embedding physical insights into GNN-based predictive models, resulting in superior accuracy compared to purely data-driven approaches.

Table 1. The maximum absolute errors and relative percentage errors between the ground truth and the predictions from the repective GNN models with respect to the critical responses of E2

Variables	Vanilla GNN	PiGNN	PeGNN
Top displacement (m)	9.68e-04(1.1%)	4.92e-05 (0.05%)	2.04e-05(0.02%)
Top velocity (m/s)	2.33e-04(0.4%)	7.09e-05(0.01%)	1.31e-04(0.02%)
Top acceleration (m/s2)	5.59e-03(1.1%)	3.86e-03(0.09%)	2.47e-03(0.05%)
Bottom stress (Pa)	9.59e+04(0.96%)	7.85e+03(0.07%)	2.03e+03(0.02%)
Interaction force (N)	118(9.1%)	13.7(1.1%)	11.8(0.91%)

Summary

Current studies often focus on pre-earthquake analysis of individual equipment but overlook the interconnected nature of substation components and the need for rapid post-earthquake assessment. To address this, this paper presents a hybrid post-earthquake rapid assessment model aimed at predicting the dynamic responses of interconnected substation equipment subjected to seismic loads in real time, leveraging just the ground motion signal. Specifically, we propose to integrate GNN with physics-based ROMs to capture the behavior and interaction forces between interconnected substation equipment. Two physics-integration approaches are introduced and compared: incorporating physics into the loss function (PiGNN) and incorporating physics into the aggregation function (PeGNN). A case study on 800 kV post insulators interconnected by busbars is conducted to validate the proposed approach. Our results show that integrating physics into the GNN learning process significantly improves prediction accuracy by about one order of magnitude compared to purely data-driven GNNs, with the PeGNN outperforming the PiGNN. Such hybrid models offer a cost-effective and accurate real-time assessment of equipment conditions without requiring extensive sensor deployment, contrbuting to informing emergency response efforts and enhancing the overall resilience of electrical substations in post-disaster recovery scenarios.

References

[1] Xie Q, Zhu R. Damage to electric power grid infrastructure caused by natural disasters in China. IEEE POWER ENERGY M, 2011, 9(2): 28-36.
https://doi.org/10.1109/MPE.2010.939947

[2] Kwasinski A, Eidinger J, et al. Performance of electric power systems in the 2010–2011 Christchurch, New Zealand, earthquake sequence. EARTHQ SPECTRA, 2014, 30(1): 205-230.
https://doi.org/10.1193/022813EQS056M

[3] Scawthorn C, Miyajima M, Ono Y, et al. Lifeline aspects of the 2004 Niigata ken Chuetsu, Japan, earthquake. EARTHQ SPECTRA, 2006, 22(1_suppl): 89-110.
https://doi.org/10.1193/1.2173932

[4] Xie Q, Liang H, Wang X. Seismic performance improvement of±800 kV UHV DC wall busing using friction ring spring dampers. EARTHQ SPECTRA, 2021, 37(2): 1056-1077.
https://doi.org/10.1177/8755293020970984

[5] Gökçe T, Yüksel E, Orakdöğen E. Seismic performance enhancement of high-voltage post insulators by a polyurethane spring isolation device. B EARTHQ ENG, 2019, 17: 1739-1762.
https://doi.org/10.1007/s10518-018-0494-6

[6] Liang H, Blagojevic N, Xie Q, et al. Seismic risk analysis of electrical substations based on the network analysis method. EARTHQ ENG STRUCT D, 2022, 51(11): 2690-2707.
https://doi.org/10.1002/eqe.3695

[7] Liang H, Blagojević N, Xie Q, et al. Seismic resilience assessment and improvement framework for electrical substations. EARTHQ ENG STRUCT D, 2023, 52(4): 1040-1058.
https://doi.org/10.1002/eqe.3800

[8] Liang H, Xie Q. Resilience-based sequential recovery planning for substations subjected to earthquakes. IEEE T POWER DELIVER, 2022, 38(1): 353-362.
https://doi.org/10.1109/TPWRD.2022.3187162

[9] Argyroudis S A, Mitoulis S A, Chatzi E, et al. Digital technologies can enhance climate resilience of critical infrastructure. CLIM RISK MANAG, 2022, 35: 100387.
https://doi.org/10.1016/j.crm.2021.100387

[10] Chinesta F, Cueto E, Abisset C E, et al. Virtual, digital and hybrid twins: a new paradigm in data-based engineering and engineered data. ARCH COMPUT METHOD E, 2020, 27: 105-134.
https://doi.org/10.1007/s11831-018-9301-4

[11] Vlachas K, Tatsis K, Agathos K, et al. A local basis approximation approach for nonlinear parametric model order reduction. J SOUND VIB, 2021, 502: 116055.
https://doi.org/10.1016/j.jsv.2021.116055

[12] Xie Q, Liang H. An analytical model for seismic response analysis of electrical equipment-steel support structure[J]. Journal of Constructional Steel Research, 2024, 213: 108379.
https://doi.org/10.1016/j.jcsr.2023.108379

[13] Pichi F, Moya B, Hesthaven J S. A graph convolutional autoencoder approach to model order reduction for parametrized PDEs. J COMPUT PHYS, 2024, 501: 112762. https://doi.org/10.1016/j.jcp.2024.112762

[14] Wu Z, Pan S, Chen F, et al. A comprehensive survey on graph neural networks. IEEE T NEUR NET LEAR, 2020, 32(1): 4-24. https://doi.org/10.1109/TNNLS.2020.2978386

[15] Park C W, Kornbluth M, Vandermause J, et al. Accurate and scalable graph neural network force field and molecular dynamics with direct force architecture. NPJ COMPUT MATER, 2021, 7(1): 73. https://doi.org/10.1038/s41524-021-00543-3

[16] Haywood-Alexander M, Liu W, Bacsa K, et al. Discussing the Spectrum of Physics-Enhanced Machine Learning; a Survey on Structural Mechanics Applications. DATA-CENTRIC ENG, 2024. https://doi.org/10.1017/dce.2024.33

Structural Health Monitoring: 10APWSHM
Materials Research Proceedings 50 (2025) 61-72

Materials Research Forum LLC
https://doi.org/10.21741/9781644903513-7

Predicting membrane strains with a deep learning encoder-decoder convolutional neural network architecture trained on synthetic finite element data

Benjamin Steven Vien[1,a]*, Thomas Kuen[2,b], Louis Raymond Francis Rose[1,c] and Wing Kong Chiu[1,d]

[1]Department of Mechanical and Aerospace Engineering, Monash University, Wellington Rd, Clayton, VIC 3800, Australia

[2]Melbourne Water Corporation, 990 La Trobe Street, Docklands, VIC 3008, Australia

[a]ben.vien@monash.edu, [b]thomas.kuen@melbournewater.com.au, [c]francis.rose@monash.edu, [d]wing.king.chiu@monash.edu

Keywords: Deep Learning, Membrane, Strain, Convolution Neural Network, Synthetic Data, Finite Element

Abstract. Melbourne Water Corporation's Western Treatment Plant, located in Werribee, Victoria, Australia, is pioneering the use of advanced technologies, including artificial intelligence, to enhance asset management practices. This is particularly important for the structural health monitoring (SHM) of anaerobic lagoon floating covers, which are crucial for biogas collection but can be compromised by scum accumulation underneath. This study introduces a novel encoder-decoder network within a convolutional neural network (CNN) framework, designed to predict strain distributions on deformed membranes. Due to the limited availability of real-world data, finite element analysis was utilised to generate synthetic data consisting of displacements and strain fields for model training. The study investigates the optimal quantity of synthetic samples needed for accurate predictions and discusses the proposed CNN architecture and data preparation techniques. The findings indicate that a dataset of at least 10,000 synthetic training samples is required to accurately predict strain distributions, which represents a significant improvement by orders of magnitude compared to using only 100 and 1000 samples. Furthermore, refinement learning methods were demonstrated, where a pretrained CNN model is further trained on a new dataset with lower strain variability. The results indicate that refining the pretrained model with frozen (fixed) weights in the encoder network yields better accuracy in predicting strain, at least 2.3 times better than those without frozen weights. However, the refined model without frozen weights retains more information from the original dataset and is more consistent in predicting strain profiles. The results suggest a high-quality, representative training dataset relating to the application of interest is essential for effective machine learning. These findings lay a fundamental basis for implementing practical deep learning approaches and further utilising unmanned aerial vehicle-based imagery for effective SHM of highly valuable assets.

Introduction

At the Western Treatment Plant (WTP) in Werribee, Victoria, Australia, operated by Melbourne Water, innovative floating covers have been installed over an area of approximately 450m by 170m to assist in the anaerobic digestion process of untreated sewage. These covers not only mitigate odours but also facilitate the capture of methane-rich biogas for renewable energy production. These floating covers are made from 2mm thick high-density polyethylene (HDPE) geomembranes, which are exceptionally durable and designed to withstand extreme conditions over decades [1]. However, challenges arise from the accumulation of solidified sewage, forming

Structural Health Monitoring: 10APWSHM   Materials Research Forum LLC
Materials Research Proceedings 50 (2025) 61-72   https://doi.org/10.21741/9781644903513-7

large mounds known as scumbergs, which may cause deformation and potentially compromise the covers' structural integrity and their capacity for renewable energy generation.

Over the last ten years, unmanned aerial vehicle (UAV)-based imagery has been deployed as a remote and efficient technique for the routine monitoring of WTP floating covers. This approach, particularly through the use of orthomosaics and digital elevation models (DEMs), facilitates the quantitative analysis of deformations (e.g., elevation changes) and the mapping of underlying scum accumulations [2]. Previous research [2-4] has highlighted the value of remote sensing imagery in the early identification of scum accumulation. More recent advancements include the application of machine learning (ML) to develop advanced diagnostic and prognostic models that maintain the integrity and optimise the functionality of these covers [2, 4-7]. An unsupervised ML algorithm was devised to facilitate the removal of artefacts (i.e. trapped rainwater, man-made structures and debris) from the floating cover for scum hardness analysis [7]. Moreover, Melbourne Water is pursuing advances in novel monitoring techniques to enhance both insight and quantifiable metrics, which are essential for effective asset management and decision-making processes. However, the extensive size of these structures, along with operational constraints, poses significant challenges in accurately evaluating engineering parameters, such as strain and stress. Furthermore, these challenges persist with existing commercial techniques, thereby highlighting the imperative for more robust methods for this domain-specific application.

Recently, there has been significant interest in adopting deep learning methods for full-field strain predictions over traditional and commercial methods. Nie et. al. [8] used cantilevered structure geometry, in-plane external loads and boundary conditions to predict the strain field with a mean relative error of 0.32% compared to ground truth. Yang et. al. [9] developed two tailored CNNs to determine localised and large strain and displacement fields of a pair of speckle images. However, a major challenge in practical applications of deep learning models is the limited availability of real-life datasets essential for effective model training. This challenge is particularly pronounced when dealing with unique and complex real-world assets, making it a primary obstacle to achieving machine learning readiness. A cost-effective and time-efficient solution, especially in the field of engineering, is generating training data from finite element (FE) simulations. This approach enables the simulation of a controlled environment under various conditions and outcomes, guided by theoretical physical laws. Although the data generated from FE simulations may not fully generalise to real-world scenarios, it is particularly useful as an initial step before further refining the machine learning models with real-life data.

As an extension of the research project, the present study introduces a deep-learning method for predicting membrane strain, coupled with a novel approach to overcome the limitations of real-world data for model training and refinement by employing synthetic data generated by FE simulation. The model's hyperparameters, including layers, activation functions, and batch size, are briefly discussed to assess their impact on the model's learning ability and prediction accuracy. Additionally, two refinement learning methods are presented and discussed. These preliminary results establish the foundation for a data-driven SHM strategy for floating covers, particularly by leveraging UAV-derived imagery for practical applications.

Method

Synthetic Data Generation
The MATLAB R2023b Partial Differential Equation Toolbox was used to generate the synthetic FE training data. A 500 mm square geometry with a thickness of 10 mm was considered, with 10 mm discretisation, as shown in Figure 1. The material properties were set to a Young's modulus of 30MPa and a Poisson's Ratio of 0.33. In order to generate training data, samples were randomly generated under the following conditions outlined in Table 1. Refer to Figure 1, the data generation process is as follows: the number of fixed boundary edges ranges from 1 to 4. If the number of

Structural Health Monitoring: 10APWSHM
Materials Research Proceedings 50 (2025) 61-72

Materials Research Forum LLC
https://doi.org/10.21741/9781644903513-7

fixed boundaries is less than 4, up to 3 boundary edge forces can be applied. Additionally, 1 to 6 point forces can be assigned to the vertices on the membrane. Two datasets were generated for model learning and refinement. For the initial model learning dataset, the vector forces range from -1000N to 1000N with a step of 10N in all three directions. For the model refinement learning dataset, the vector forces range from -100N to 100N with a step of 1N in all three directions.

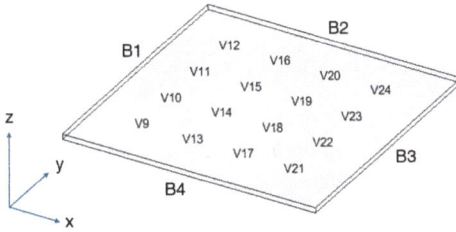

Figure 1: Locations of boundary edges and vertices on the membrane model

able 1: FE simulation conditions and range of values for generating random data.

Conditions	Range of values
Fixed boundary edge	[1,4]
Boundary Edge Force Vectors*	If fixed boundary edge < 4, [0,3]
Point Force Vectors*	[1,6] randomly assigned to the 16 vertices
*Force Vectors	**Original Model Learning Dataset:** Magnitude varying [-1000N,1000N] with step 10N in all three directions. **Refinement Learning Dataset:** Magnitude varying [-100N,100N] with step 1N in all three directions.

Considering the structure of interest is a membrane, it is assumed that out-of-plane strains are negligible, and therefore, only in-plane strains are considered. The primary goal is to obtain the displacement and strain fields on the membrane's top surface for use as sample data. The data of interest include displacements in all directions (x, y, and z) and in-plane strain (ε_{xx}, ε_{xy}, and ε_{yy}).

Data augmentation was conducted to rapidly increase the sample size. From a larger 50x50 grid, 19 samples can be derived from each 32x32 grid, requiring only 5,264 simulations to generate approximately 100,000 samples. This approach significantly reduces the computational time for simulations by a factor of at least an order of magnitude. It should be noted that the selection of the material properties, the conditions and the data augmentation process are to ensure that the absolute normal strains are approximately within the values of 2 and 0.2 for model learning and refining, respectively, and that each sample is unique, refer to Tables 2 and 3.

Materials Research Forum LLC

https://doi.org/10.21741/9781644903513-7

Table 2: Statistical measurements of strain values with different numbers of training samples for the original model learning dataset

SAMPLES	100			1,000			10,000			100,000		
	Range of Strain Extremities	25th-75th Percentiles	Median Strain Variability	Range of Strain Extremities	25th-75th Percentiles	Median Strain Variability	Range of Strain Extremities	25th-75th Percentiles	Median Strain Variability	Range of Strain Extremities	25th-75th Percentiles	Median Strain Variability
ε_{xx}	[-0.41,0.64]	[-0.06,0.07]	0.1578	[-1.38,1.5]	[-0.06,0.06]	0.1493	[-1.85,1.76]	[-0.06,0.06]	0.1505	[-1.85,1.84]	[-0.06,0.06]	0.1487
ε_{yy}	[-0.70,1.13]	[-0.04,0.06]	0.1359	[-1.44,1.34]	[-0.05,0.05]	0.1411	[-1.73,1.49]	[-0.06,0.06]	0.1481	[-1.86,1.86]	[-0.06,0.06]	0.1491
ε_{xy}	[-0.40,0.53]	[-0.05,0.05]	0.1128	[-0.75,0.72]	[-0.05,0.05]	0.1191	[-0.88,0.9]	[-0.05,0.05]	0.1227	[-0.94,0.94]	[-0.05,0.05]	0.1233

Table 3: Statistical measurements of strain values with different numbers of training samples for refinement learning dataset.

SAMPLES	100			1,000			10,000			100,000		
	Range of Strain Extremities	25th-75th Percentiles	Median Strain Variability	Range of Strain Extremities	25th-75th Percentiles	Median Strain Variability	Range of Strain Extremities	25th-75th Percentiles	Median Strain Variability	Range of Strain Extremities	25th-75th Percentiles	Median Strain Variability
ε_{xx}	[-0.118,0.108]	[-0.006,0.007]	0.02	[-0.16,0.15]	[-0.006,0.006]	0.016	[-0.190,0.182]	[-0.006,0.006]	0.0154	[-0.190,0.190]	[-0.006,0.006]	0.015

Refer to Tables 2 and 3, the random generation scheme statistics, which include the range of strain extremities (maximum and minimum strain values), the 25th and 75th percentiles of strain extremities and the median strain variability (difference between the maximum and minimum strains) of the deformed membrane, are shown for different sample sizes. It is shown that, regardless of the number of training samples, the 25th and 75th percentiles (-0.06 and 0.06) and median variability (approximately 0.14) remain very similar. As the sample size increased beyond 10,000, the range of strain extremities, with minimum and maximum values of approximately -0.190 and 0.190, respectively, remained unchanged. This consistency indicates that the range of strain values stabilised with larger samples. Similarly, for the refinement learning dataset, the median variability (approximately 0.015) and the 25th and 75th percentiles (-0.006 and 0.006) remain very similar, and the range of strain value stabilises once the sample size exceeds 10,000.

Python 3.10 and the TensorFlow library were utilised for the development of CNNs aimed at predicting in-plane strains. Figure 2 illustrates the architecture, which encompasses an encoder, a bottleneck layer, a decoder, and skip connections at various points between the encoder and decoder layers. This design draws inspiration from the U-Net architecture [10], renowned for its efficient image segmentation capabilities, particularly through the integration of skip connections that concatenate early-stage feature maps from the encoder network with feature maps in the decoder stage to mitigate the loss of spatial content details. Essentially, these connections act like shortcuts in the network, allowing it to leverage important information from earlier layers to help create more accurate and detailed results in later stages. In the preliminary investigation, it was observed that incorporating skip connections significantly enhances training speed and prediction accuracy.

The architecture consists of encoder and decoder networks comprising a series of 2D convolutional layers (Conv2D), 2D transposed convolution layers (Conv2DT), 2D max pooling layers (MaxPooling) and concatenative skip connection layers (Concat) that concatenate two sets of feature maps as the outputs. Furthermore, a *Conv2D block* comprises a sequence of a Conv2D layer with a kernel size of 3x3, followed by a batch normalisation layer to stabilise learning, and a ReLU activation function to introduce nonlinearity.

Structural Health Monitoring: 10APWSHM   Materials Research Forum LLC
Materials Research Proceedings 50 (2025) 61-72   https://doi.org/10.21741/9781644903513-7

Refer to Figure 2, the deep learning architecture begins with an input layer that takes a displacement array of dimensions 32x32x3. This is followed by an encoder network composed of two *Conv2D blocks*: the first with 32 kernel filters and the second with 64 kernel filters. Each block is followed by a MaxPooling operation of size 2x2, with no padding and a stride of 2, to reduce spatial dimensions. After the encoder network, the architecture features a bottleneck with a third *Conv2D block* containing 128 kernel filters, emphasising feature extraction at a reduced scale. Transitioning to the decoder network, it initiates with a Conv2DT layer with 64 kernel filters, followed by a Concat layer that merges the output of the second *Conv2D block* from the encoder and the output of the preceding Convo2DT layer to preserve spatial features and context. This is followed by a fourth *Conv2D block* with 64 kernel filters. Subsequently, another *Conv2DT layer* with 32 kernel filters is applied, followed by a Concat layer that integrates the outputs of the first *Conv2D block* and the previous Conv2DT layer, further enhancing feature recovery towards the input's spatial resolution. A fifth *Conv2D block* with 32 kernel filters is then used to refine the features. The architecture culminates in an output layer featuring a Conv2D layer with a linear activation function and a single kernel filter of size 3x3, designed to reconstruct the final output from the processed features. Overall, the total number of trainable model parameters (weights, bias, kernel, etc.) is 467,809 and for the refinement learning method with the frozen encoder network, the total of trainable model parameters is 179,905.

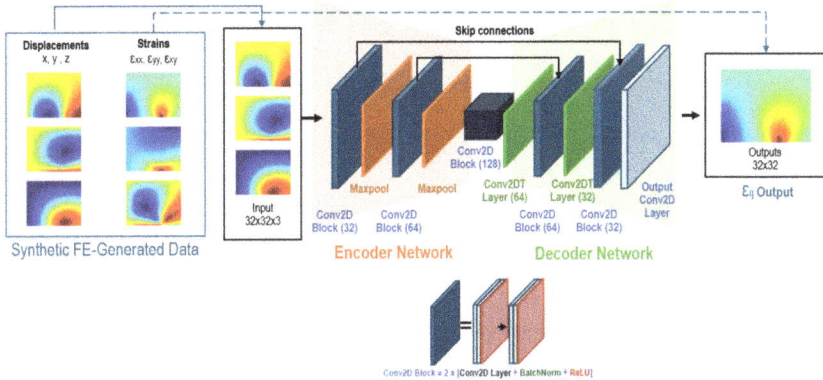

Figure 2: Schematic of the proposed encoder-decoder CNN architecture using synthetic FE-generated data for model learning.

The mean squared error (MSE) was employed as the objective loss function. The model is trained using the Adam optimiser with a learning rate of 0.001, using the exponential decay rate for the first- and second-moment estimates β_1 and β_2 with values of 0.9 and 0.999, respectively. Preliminary investigations into the optimal batch size indicated a range between 32 and 128 is ideal, with performance significantly degrading outside this range. The model was trained over 50 epochs with a batch size of 32. Additionally, an early stopping function was implemented to restore the best model weights if there was no improvement in the MSE of the validation loss after 5 iterations.

In this investigation, models are trained with varying sample sizes; 100, 1,000, 10,000, and 100,000, to assess the performance using the MSE of a separate validation dataset (MSE_{valid}) consisting of 100 samples. Additionally, the MSE of the training data (MSE_{train}) for each training sample size is reported. Three separate CNNs are constructed to predict each individual in-plane strain, ε_{xx}, ε_{yy} and ε_{xy}.

In refinement learning, only the CNN predicting ε_{xx} is fine-tuned using the refinement learning dataset. The refinement learning procedure begins by uploading all the weights from the pre-trained CNN model with the original 100,000 training samples, followed by re-training with the refinement learning dataset, which includes varying refined training sample sizes: 100, 1,000, 10,000, and 100,000. During the refinement phase, the Adam optimiser learning rate is adjusted to 0.0001. This method allows all weights to be updated, with this refined model denoted as CNN^R. Additionally, another refinement learning method commonly used in transfer learning is explored by freezing (fixing) the weights/parameters of the encoder network and the bottleneck layer, allowing only the decoder network weights to be updated. This method is referred to as the refined model with a frozen encoder network, denoted as $\text{CNN}^R_{\text{Frozen}}$. The refinement methods are optimised based on the MSE of a refined validation dataset ($\text{MSE}^R_{\text{valid}}$) consisting of 100 samples and the MSE of the refined training data ($\text{MSE}^R_{\text{train}}$) for each refined training sample. The refinement models are also compared with the original model learning dataset, such that $\text{MSE}_{\text{valid}}$ and $\text{MSE}_{\text{train}}$ of the refined models are reported, enabling a discussion of the comparison with the trained model and both refinement learning methods.

Results

Refer to Table 4, significant improvements in the MSE were achieved, with at least an order of magnitude reduction when using a training sample size of at least 10,000. For ε_{xx}, ε_{yy} and ε_{xy} validation datasets trained with 10,000 samples, the MSE improved by factors of 17.6, 29.3 and 32.5, respectively, compared to those trained with 100 samples.

Table 4: MSE of ε_{xx}, ε_{yy} and ε_{xy} validation and training data from model learning datasets for various numbers of training samples.

Training Samples		100	1,000	10,000	100,000
ε_{xx}	$\text{MSE}_{\text{valid}}$	3.53E-02	1.45E-02	1.09E-03	8.56E-04
	$\text{MSE}_{\text{train}}$	1.21E-02	1.13E-02	8.66E-04	7.56E-04
ε_{yy}	$\text{MSE}_{\text{valid}}$	3.13E-02	4.58E-03	1.77E-03	1.22E-03
	$\text{MSE}_{\text{train}}$	1.20E-02	3.24E-03	1.48E-03	1.18E-03
ε_{xy}	$\text{MSE}_{\text{valid}}$	5.22E-02	8.50E-03	1.78E-03	1.14E-03
	$\text{MSE}_{\text{train}}$	1.76E-02	6.12E-03	1.67E-03	7.27E-04

It is shown that the strain profile is accurately predicted using at least 10,000 training samples, refer to Figure 3. However, the models exhibit difficulties in accurately predicting highly localised strains, as indicated by the strain profiles of validation sample 2 in Figures 3-5. This outcome is anticipated due to the lack of representative samples, as this validation sample represents an extremity of the randomly generated data. For instances, ε_{yy} for samples 1 and 3 strain profiles, which exhibit lower strain variability and sample 4, which exhibits higher strain variabilities than the median variability of the dataset. Hence, these extremities are more accurately predicted with 100,000 training samples.

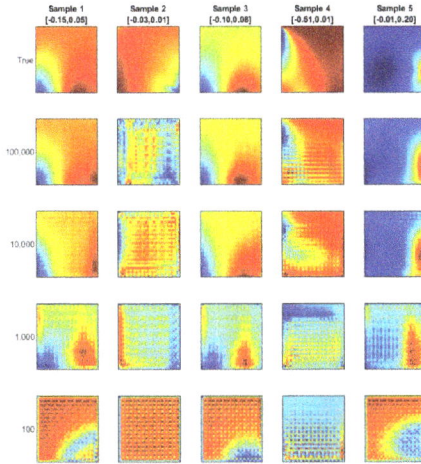

Figure 3: 5 samples of ε_{xx} of the validation data true and trained with 100,000, 10,000, 1,000 and 100 synthetic training samples, indicating minimum and maximum (i.e. [Min, Max]) strain values.

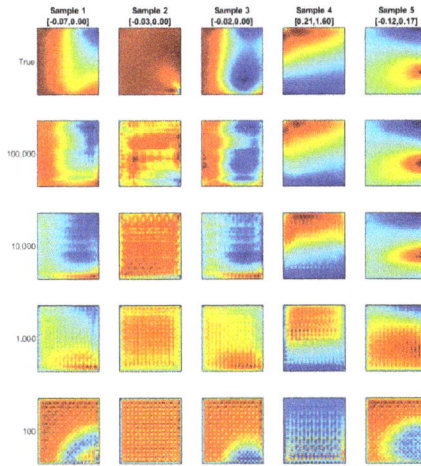

Figure 4: 5 samples of ε_{yy} of the validation data true and trained with 100,000, 10,000, 1,000 and 100 synthetic training samples, indicating minimum and maximum (i.e. [Min, Max]) strain values.

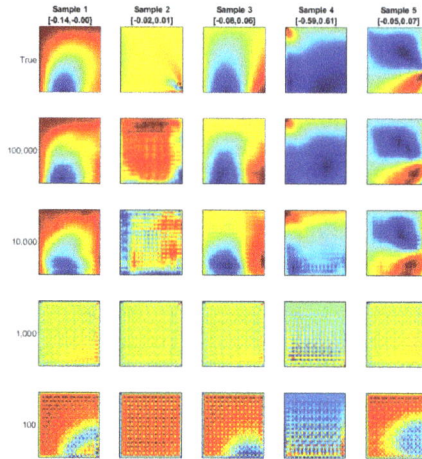

Figure 5: 5 samples of ε_{xy} of the validation data true and trained with 100,000, 10,000, 1,000 and 100 synthetic training samples, indicating minimum and maximum (i.e. [Min, Max]) strain values.

Refined Models

The MSEs of the validation and training data from the refinement learning and the original model learning datasets for the refinement learning models with frozen encoder network and without frozen weights, across varying training sample sizes, are shown in Table 5. It is shown that the CNN^R_{Frozen} yielded better accuracy in MSE^R_{valid} and MSE^R_{train} with approximately 2.3 to 5.7 times improvement than those of CNN^R for training sample sizes of 1000, 10,000 and 100,000. However, for a training sample size of 100, the MSE^R_{valid} and MSE^R_{train} remained similar. Conversely, the CNN^R yielded better accuracy in MSE_{valid} and MSE_{train} of at least an order magnitude improvement, approximately 27.3 and 17.7 times, respectively, than those of CNN^R_{Frozen}.

Table 5: Refined model with and without frozen encoder network refinement learning models: MSE of ε_{xx} validation and training datasets for various numbers of training samples.

Refined Training Samples			*100*	*1,000*	*10,000*	*100,000*
With Frozen Encoder Network	Refinement Learning Dataset	MSE^R_{valid}	3.61E-05	8.17E-05	2.32E-06	8.15E-07
		MSE^R_{train}	3.44E-05	8.81E-05	2.61E-06	8.44E-07
	Original Learning Dataset	MSE_{valid}	1.80E-03	1.65E-02	2.04E-02	7.21E-02
		MSE_{train}	7.40E-04	6.49E-03	7.90E-03	2.79E-02
No Frozen Weights	Refinement Learning Dataset	MSE^R_{valid}	3.87E-05	1.96E-05	6.51E-06	4.04E-06
		MSE^R_{train}	3.71E-06	2.07E-05	7.90E-06	4.77E-06
	Original Learning Dataset	MSE_{valid}	1.45E-03	4.22E-03	4.07E-03	2.64E-03
		MSE_{train}	7.66E-04	2.31E-03	2.10E-03	1.57E-03

As the number of refined training samples increases, the refined models' MSE_{valid} and MSE_{train} also increase and are larger than those observed for the original CNN models (refer to Tables 4 and 5). Specifically, for the CNN_{Frozen}^{R}, MSE_{valid} and MSE_{train} increased by 37.7 and 40.1 times, respectively, whereas for the CNN^{R}, MSE_{valid} and MSE_{train} increased by 2.1 and 1.8 times, respectively, relative to those refined using 100 refined training samples. Figures 6 and 7 illustrate that the refined models successfully predict smaller strains as intended while generally preserving the results of the original CNN models. However, it is observed that the refined models fail to accurately predict higher strain values, as demonstrated for sample 4, where CNN_{Frozen}^{R} performed the worst. Furthermore, while the predicted strain profiles of the refined models are similar, inconsistent strain predictions, such as impulse-like noises, were observed in CNN_{Frozen}^{R}.

As the number of refined training samples increases, the corresponding increase in MSE_{valid} and MSE_{train} suggests that the refined models are forgetting information learned from the previous dataset, especially in extreme strain cases, in order to improve performance on refinement learning dataset. Referring to Figures 6 and 7, it is shown that a minimum of 10,000 refined training samples is required to fine-tune the model to reliably predict the strain distribution of sample 2 as the larger dataset includes more data with smaller strain variability.

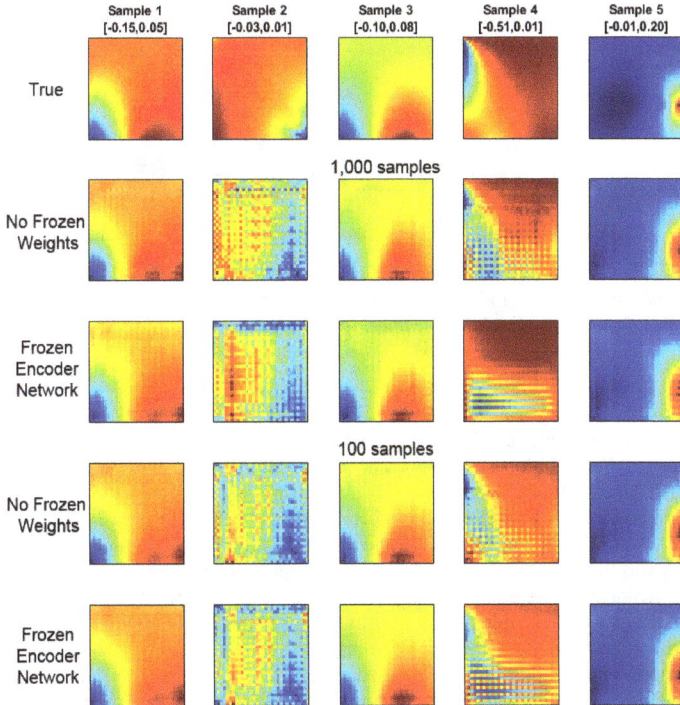

Figure 6: 5 samples of ε_{xx} of the validation data from the original model learning dataset showing true and refined model with and without frozen encoder network trained with 100 and 1000 synthetic refined training samples, indicating minimum and maximum (i.e. [Min, Max]) strain values.

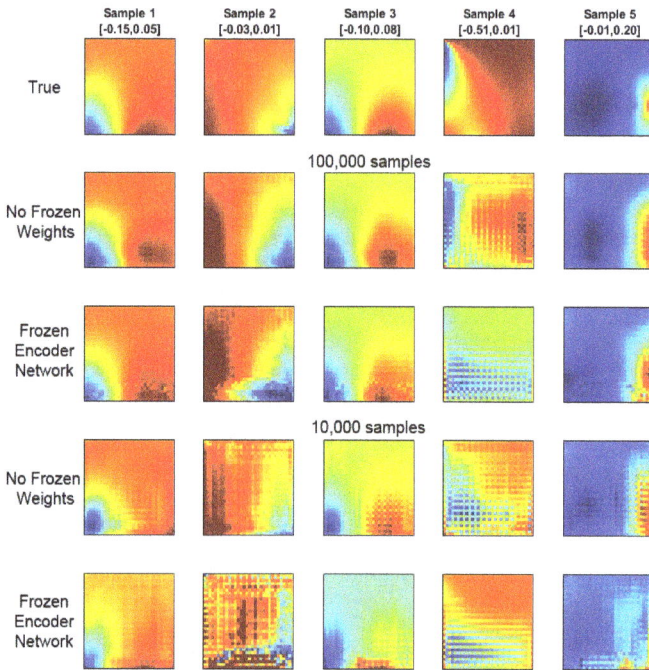

Figure 7: 5 samples of ε_{xx} of the validation data from the original model learning dataset showing true and refined model with and without frozen encoder network trained with 10,000 and 100,000 synthetic refined training samples, indicating minimum and maximum (i.e. [Min, Max]) strain values.

Discussion

The study demonstrates the use of deep learning models for predicting strain in deformed membrane structures. It has been shown that increasing the amount of training data improves model performance. For the proposed CNN architecture, the findings suggest that a minimum of 10,000 samples is needed to achieve acceptable strain distribution profiles, with at least an order of magnitude improvement in MSEs compared to using 100 or 1,000 samples. However, the prediction of low-value, highly concentrated, and/or highly variable strain values was inaccurate. This inaccuracy was likely due to a lack of data samples representing these features, i.e., those that differ from the majority of the training data (see Table 2). This is supported by refined models' results, which show improvement in predicting validation sample 2 when trained on a new dataset with similar strain features. In refinement learning, it is shown that the encoder network can effectively compress information in the latent space layer and perform better than those without frozen weight. However, refining the entire architecture (i.e., without fixing the encoder network weights) appears less prone to forgetting when predicting from the original dataset, as reflected by the lower MSE_{valid} and MSE_{train} values. Nevertheless, strategies such as training with a more diverse and representative dataset or selectively reducing learning rates on certain weights through an additional term in the loss function can help address these issues.

Structural Health Monitoring: 10APWSHM
Materials Research Proceedings 50 (2025) 61-72

Materials Research Forum LLC
https://doi.org/10.21741/9781644903513-7

It is important to note that changes to hyperparameters can significantly enhance the model. In the preliminary work on constructing the architecture, the choice of activation function for the Conv2D block was found to be crucial for accurate prediction. The tanh and sigmoid activation functions yielded highly erroneous results that did not represent the strain distributions at all. Overall, determining the required number of training samples, tuning hyperparameters, and designing the network architecture are essential procedures for each unique application.

Future research will advance into experimental investigations involving the deformation of membrane specimens, utilising these pretrained CNN models as a foundation. Moreover, subsequent studies will explore the practical extraction of node features and their robustness against noise, such as that encountered in feature extraction techniques. Additional efforts will focus on the practical deployment of deep learning models, including experimental validation and addressing challenges associated with the irregular discretisation of data points that represent real floating cover membranes.

Conclusion

This paper introduces encoder-decoder CNN models for predicting in-plane strains using the displacements of a deformed membrane. Given the scarcity of real-world data, synthetic data comprising displacements and strain fields was generated using finite element analysis. The results suggest that a minimum of 10,000 training samples is required for the models to reliably predict strain, representing a significant improvement by at least an order of magnitude compared to using fewer than 1,000 samples. Fine-tuning can be achieved by refining the original pre-trained model with a new refinement learning training dataset at a lower learning rate. The refinement learning method with frozen weights in the encoder network yields better results, with at least 2.3 times the accuracy, in predicting the refined training dataset, which consists of samples with lower strain variability, compared to the refined model without frozen weights. However, the refining without frozen weights retains more information from the original dataset and is more consistent in strain predictions compared to refining with frozen weights in the encoder network. This study validates the use of deep learning approaches for strain determination, advancing towards effective SHM of highly valuable assets. Furthermore, the findings suggest that a high-quality, representative training dataset tailored to the specific application is essential for effective machine learning. Ongoing work is currently underway to validate these findings with real-life samples and to address additional challenges associated with practical implementation.

Data Availability

Code and data can be found at: https://doi.org/10.5281/zenodo.14188874 [11]

References

[1] Rowe, R.K. and H.P. Sangam, *Durability of HDPE geomembranes.* Geotextiles and Geomembranes, 2002. **20**(2): p. 77-95. https://doi.org/10.1016/S0266-1144(02)00005-5

[2] Wong, L., B.S. Vien, Y. Ma, T. Kuen, F. Courtney, J. Kodikara, F. Rose, and W.K. Chiu. *Development of Scum Geometrical Monitoring Beneath Floating Covers Aided by UAV Photogrammetry.* in *Mater. Res. Proc.* 2021. https://doi.org/10.3390/rs12071118

[3] Wong, L., B.S. Vien, Y. Ma, T. Kuen, F. Courtney, J. Kodikara, and W.K. Chiu, *Remote Monitoring of Floating Covers Using UAV Photogrammetry.* Remote Sensing, 2020. **12**(7): p. 1118. https://doi.org/10.3390/rs12071118

[4] Vien, B.S., L. Wong, T. Kuen, and W.K. Chiu, *A Machine Learning Approach for Anaerobic Reactor Performance Prediction Using Long Short-Term Memory Recurrent Neural Network.* Structural Health Monitoring: 8APWSHM, 2021. **18**: p. 61. https://doi.org/10.21741/9781644901311-8

[5] Vien, B.S., L. Wong, T. Kuen, R.L.R. Francis, and W.K. Chiu, *Probabilistic prediction of anaerobic reactor performance using Bayesian long short-term memory artificial recurrent Neural Network Model*, in *International Workshop on Structural Health Monitoring (IWSHM) 2021: Enabling Next-Generation SHM for Cyber-Physical Systems*. 2021, DEStech Publications, Inc. p. 813-820. https://doi.org/10.12783/shm2021/36331

[6] Vien, B.S., T. Kuen, L.R.F. Rose, and W.K. Chiu, *Optimisation and Calibration of Bayesian Neural Network for Probabilistic Prediction of Biogas Performance in an Anaerobic Lagoon.* Sensors, 2024. **24**(8): p. 2537. https://doi.org/10.3390/s24082537

[7] Vien, B.S., T. Kuen, L.R.F. Rose, and W.K. Chiu, *Image Segmentation and Filtering of Anaerobic Lagoon Floating Cover in Digital Elevation Model and Orthomosaics Using Unsupervised k-Means Clustering for Scum Association Analysis.* Remote Sensing, 2023. **15**(22): p. 5357. https://doi.org/10.3390/rs15225357

[8] Nie, Z., H. Jiang, and L.B. Kara. *Stress field prediction in cantilevered structures using convolutional neural networks.* in *International Design Engineering Technical Conferences and Computers and Information in Engineering Conference.* 2019. American Society of Mechanical Engineers. https://doi.org/10.1115/DETC2019-98472

[9] Yang, R., Y. Li, D. Zeng, and P. Guo, *Deep DIC: Deep learning-based digital image correlation for end-to-end displacement and strain measurement.* Journal of Materials Processing Technology, 2022. **302**: p. 117474. https://doi.org/10.1016/j.jmatprotec.2021.117474

[10] Ronneberger, O., P. Fischer, and T. Brox. *U-net: Convolutional networks for biomedical image segmentation.* in *Medical Image Computing and Computer-Assisted Intervention–MICCAI 2015: 18th International Conference, Munich, Germany, October 5-9, 2015, Proceedings, Part III 18.* 2015. Springer.

[11] Vien, B.S., (2024). bensvien/ARC-Linkage-2024: ARC Linkage Project APWSHM 2024 Strain CNN Version 1.2 (Version APWSHM2024). Zenodo. https://doi.org/10.5281/zenodo.14188874

Structural Health Monitoring: 10APWSHM
Materials Research Proceedings 50 (2025) 73-81

Materials Research Forum LLC
https://doi.org/10.21741/9781644903513-8

Eddy current damper model identification using hybrid convolutional and recurrent neural network

Vitali Kakouka[1,a], Kohju Ikago[2,b] *

[1]Graduate School of Engineering, Tohoku University, Sendai 980-0845, Japan

[2]International Research Institute of Disaster Science (IRIDeS), Tohoku University, Sendai 980-8572, Japan

[a]kakouka.vitali.q5@dc.tohoku.ac.jp, [b]ikago@irides.tohoku.ac.jp

Keywords: Data-Driven Modeling, Damper, Eddy Current Effect, Machine Learning, Convolutional Neural Network, Recurrent Neural Network

Abstract. The eddy current damper is an energy dissipating device, one of the main advantages of which over conventional fluid dampers include the ability to produce resistive forces with no contact between the components wherein damping forces are generated, resulting in a less degradable device with less maintenance requirements. However, one of the challenges in its development process is identifying the human-interpretable model, the mathematical law, which describes its behavior. Therefore, in this paper several existing approaches to address this issue are discussed along with their advantages and disadvantages, and the new method involving the usage of a hybrid convolutional (CNN) and recurrent (RNN) neural network is presented. As a machine learning model deals with data, the approach of the test data preparation and mathematical equation representation, as well as the machine learning model training process, is discussed. Finally, the performance of the trained model is evaluated using the eddy current damper experiment data, showing its capability to identify the mathematical model even in presence of noise, and the conclusion on the effectiveness of the proposed approach is made.

Introduction

The problem of identification of laws that describe different physical phenomena have been one of the cornerstones in many fields of science, and the field of Earthquake Engineering is not an exception. One of the directions of study, that tries to address this issue, where active research is being done, is the model identification. In [1] the natural laws are identified from motion-tracking data captured from various physical systems, without any prior knowledge about its nature, with the use of a computational search. Another approach is conducting the polynomial regression, as done in [2] – given enough polynomial degrees it is possible to fit a curve into even the most complex data. For the identification of models in form of Partial Differential Equations (PDE), the SINDy algorithm was proposed in [3]. Despite the advantages of the approaches above, in each of them the identified model basically represents a polynomial function, that is not always human-interpretable and doesn't give insight into the nature of data. One of the attempts to produce a more complex model was done in [4], where the model was identified combinatorically, with the use of a genetic algorithm (GA) and a neural network, which required a huge number of computations. In the current study, an approach using a hybrid Convolutional (CNN) and Recurrent (RNN) Deep Neural Network is proposed with its further implementation and training, and its performance is evaluated for the model identification of the eddy current resistance component of the Inerter Eddy Current Damper (IECD), and the identified model is compared to the empirical model acquired in [5].

Structural Health Monitoring: 10APWSHM   Materials Research Forum LLC
Materials Research Proceedings 50 (2025) 73-81   https://doi.org/10.21741/9781644903513-8

Problem Statement

An IECD is a device, comprised of a screw shaft, a ball nut, a back iron, a conductor, and permanent magnets (Fig.1), in which the inertial resistive and damping forces are generated by the mass moment inertia of the rotary disk and the electromagnetic damping forces induced by the relative circumferential motion between the conductor and permanent magnets. ([5])

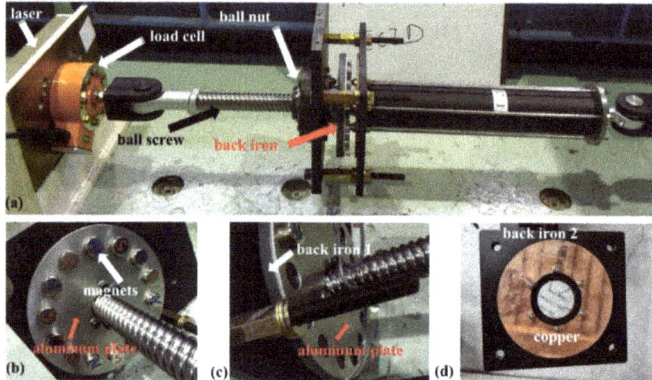

Fig. 1. Eddy Current Damper Test Setup: (a) Arrangement of the IECD test, (b) permanent magnets, (c) back iron and aluminum fixture of magnets, and (d) conductor.

Comprehensive study of this kind of device was conducted to identify the governing equation of motion of the IECD, and, particularly, to identify the empirical model for the eddy current resisting force.

The total axial resistive force F_d of the ball screw is expressed as the sum of the inertial force F_I, eddy current damping force F_e, and mechanical friction force F_f, as follows:

$$F_d = F_I + F_e + F_f \tag{1}$$

The inertial force F_I is expressed as

$$F_I = m_d \ddot{x}_d \tag{2}$$

where $\ddot{x}_d$ is the relative acceleration response of the two terminals in the IECD; $m_d = J_t a_d$ denotes the inertance, in which J_t is the mass moment of inertia of the disk, $a_d = (2\pi/L_d)^2$, and L_d is the lead length of the ball screw.

Mechanical friction occurs at the contact surface between screw threads and the ball bearings, so The Coulomb friction model was adopted to approximate the rolling contact friction F_f:

$$F_f = \tilde{F}_f sgn(\dot{x}_d) \tag{3}$$

where $\tilde{F}_f$ denotes the maximum friction force, identified using a very low-frequency test, $\dot{x}_d$ is the relative velocity response and $sgn(\bullet)$ denotes the signum function.

And finally, the equation for the eddy current damping force F_e was identified empirically and can be expressed as

$$F_e = c_0 \frac{v_0^2}{x_d^2 + v_0^2} \dot{x}_d \tag{4}$$

Structural Health Monitoring: 10APWSHM   Materials Research Forum LLC
Materials Research Proceedings 50 (2025) 73-81   https://doi.org/10.21741/9781644903513-8

where the coefficients describing the critical damping c_0 and critical velocity v_0 remain constant and are obtained from the computational results. This gives us a concise, human-interpretable model, that accurately describes the resisting force, generated in the device, as can be observed in Fig. 2.

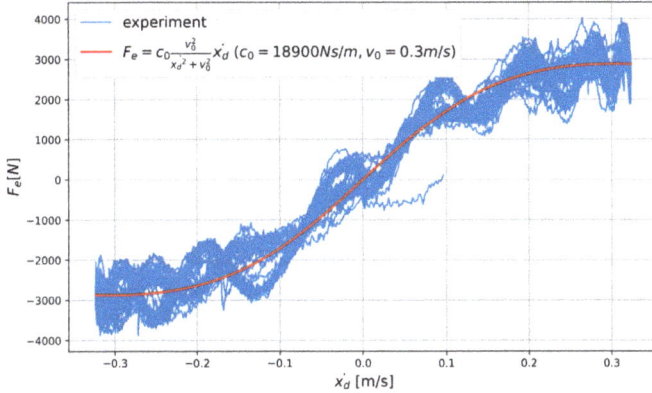

Fig. 2. Velocity $\dot{x}_d$ - Eddy Current Resistance Force F_e relation of the 5mm gap size IECD setup for the shake table experiment (sinusoidal wave excitation, $f = 50Hz$, $A = 30cm$),

However, since this model was acquired empirically, as a result of a trial-and-error process, a question arises on whether it is possible to automate such a process to determine such a model directly from the experimental data, in order to quickly give a researcher some insight into the physics behind it.

By the close examination of the Eq. 4, one can observe, that it represents a function of one variable and two parameters, in a form $f(\dot{x}_d, v_0, c_0)$, where $\dot{x}_d$ is the variable, and v_0 and c_1 are the model parameters. Thus, in order to develop an approach to acquire the human-interpretable model, the scope of current research would be the identification of an equation in a form $f(x, a_0, a_1)$, where x is the variable, and a_0 and a_1 are the arbitrary model parameters. This number of parameters is assumed to, on one hand, allow us to produce a balanced model, which does not overfit the data, and on the other hand have enough degrees of freedom to be fitted into the data. It is expected, that the approach can be extended for a larger number of parameters and find its applications in other fields of science as well.

Proposed Solution using Deep Neural Network (DNN)
In order to address the problem described above, a novel approach using Hybrid Deep Neural Network is proposed, presented in Fig. 3. The main idea of the approach is to, by recognizing the underlying patterns in the data, identify the mathematical expression in general form, without calculating the parameter values, and only after the target expression is identified, calculate the parameters using any of the optimization algorithms (e.g. Gauss-Newton algorithm). That helps to drastically decrease the computation requirements, as there's only one search in the parameter space required, as opposed to the "brute force" approaches like [4].

The neural network consists of the two major parts: the Convolutional Neural Network (CNN) and the Encoder-Decoder Recurrent Neural Network (RNN). The architecture is based on the sequence-to-sequence model, proposed in [6], which is used for the translation purposes, where the input sequence represents a sentence in one language and the output sequence a sentence in the target language. In the proposed network, however, the input sequence represents the sequence of measurements, acquired during the experiment, for which the mathematical law is to be identified, and the output sequence is the sequence of tokens, that represents a mathematical expression.

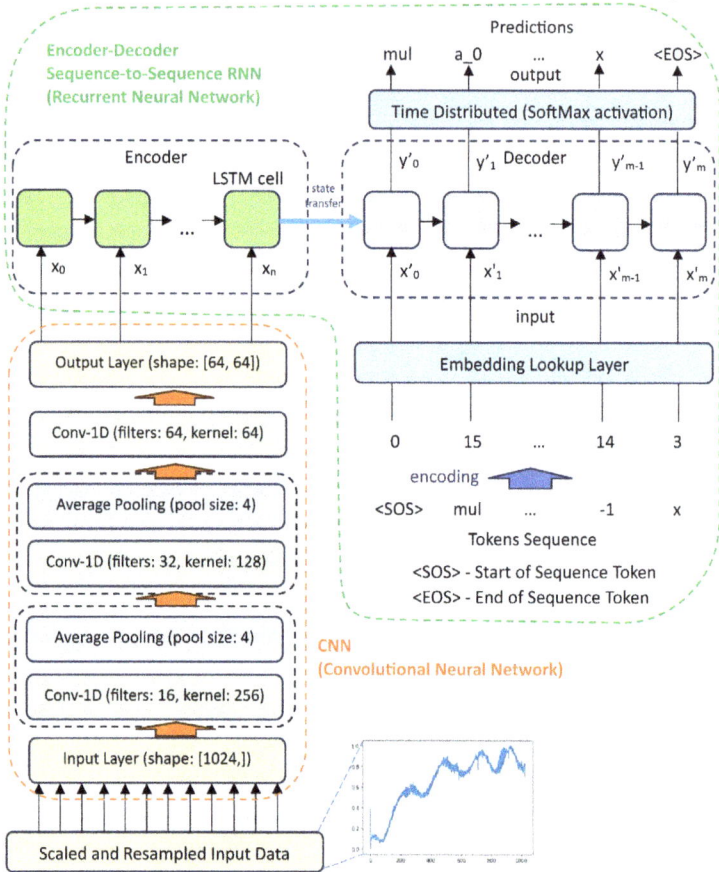

Fig. 3. Proposed Deep Neural Network architecture

The CNN is used to identify the patterns in the input sequence and enrich the RNN input with more features. The Convolutional Neural Network, or CNN, is a regularized type of feed-forward neural network that learns features via filter (or kernel) optimization, which makes it very powerful when it comes to pattern recognition. Even though 2-D CNNs found their application in modern image recognition neural networks and computer vision, 1-D CNNs found their niche in one-dimensional data processing [7].

Encoder-Decoder RNN in turn is used to "translate" the output of the CNN with the identified patterns, into an expression as the sequence of tokens - operators, parameters, constants, etc. RNN is a type of network where the output from the previous step is fed as input to the current step. In this way, unlike feedforward neural networks, which process data in a single pass, it processes data across multiple time steps, making it one of the best choices when it comes to processing time

Structural Health Monitoring: 10APWSHM
Materials Research Forum LLC
Materials Research Proceedings 50 (2025) 73-81
https://doi.org/10.21741/9781644903513-8

series and other sequences, such as text and speech [8]. Its configuration is summarized in Table. 1.

Table 1. Configuration of the Encoder-Decoder RNN

Encoder RNN	
Cell type	LSTM (Long-Short Term Memory)
Number of cells	256
Input size (CNN output size)	64 time steps x 64 features
Activation function	tanh
Recurrent activation	sigmoid
Kernel initializer	Glorot uniform ([9])
Regularization and constraints	None
Decoder RNN	
Cell type	LSTM
Number of Cells	256
Activation function	tanh
Recurrent activation	sigmoid
Kernel initializer	Glorot uniform
Regularization and constraints	None
Embedding layer size	10
Output dense layer size	Vocabulary size (number of unique tokens)

There are two major stages in the lifetime of a ML model can be defined: the training and the application. During the training, the training data is passed in small batches, multiple times (epochs), through the network to evaluate the model parameters (weights), and after that the model can be used to make prediction on the new data. In both cases, the input and the output data need to be preprocessed and represented in a certain way to be used by the network.

Data Preparation and Representation
Input Data (Features) Encoding. The input data represents the sampled function to be identified. As the size of the CNN input is constant and by trial and error was defined to have length 1024, resampling using Fast Fourier Transform ([10]) is applied, which not only allows to get the required input size, but also to reduce the undesired noise ([11]). After that, the values are rescaled to fit the range from 0 to 1 to make the model independent of the scale of the data – as we want the model to look at the shape of the function to identify distinctive patterns, the scale is not important.
Output Data (Label) Encoding. The output data represents the target mathematical expression. As the RNN outputs a sequence, that expression needs to be represented as a sequence of expression tokens, and thus the forward and backward conversion is used as shown in Fig. 4.

As part of the research, the method of expression string representation was developed, where a binary expression tree can be represented as a sequence of tokens in such a way, that during the backward conversion the same tree can be reconstructed, which is crucial during the inference.

Structural Health Monitoring: 10APWSHM   Materials Research Forum LLC
Materials Research Proceedings 50 (2025) 73-81   https://doi.org/10.21741/9781644903513-8

Only a binary tree can be represented in such a way, so ordinary-binary tree conversion step is also required.

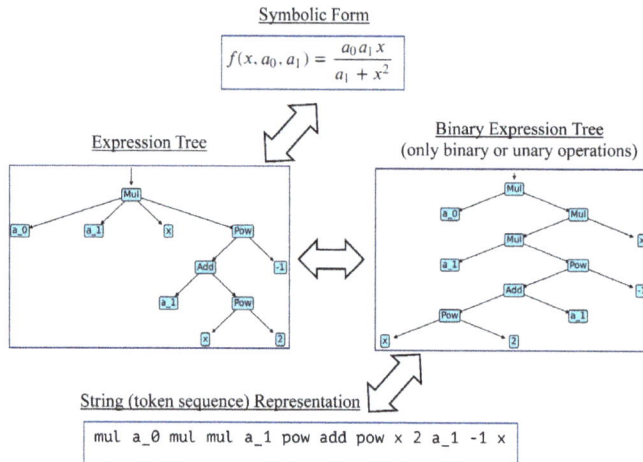

Fig. 4. Conversion between different representations of an expression

The forward conversion process can be described as follows: the expression in its symbolic form is first represented as an expression tree, which is then converted to a binary tree, and after that to the token sequence. Finally, it is encoded using One-Hot Encoding. In that way, during the training the training equations are converted using the forward conversion, and at the time of prediction (application), as the model outputs the tokens in the encoded form, the backward conversion is applied to convert the predicted sequence to the expression in the symbolic form.

Proposed Machine Learning Model Implementation, Training and Application
Model Implementation. The model described in Fig. 3 is implemented using Keras, an open-source, high-level deep learning framework that provides a Python interface for creating, training and application of the artificial neural networks. As the Keras backend, Tensorflow [12], an open-source software library for machine learning, was used.

Training Process. For the training, the set of training expressions was chosen and the data was sampled for each expression such that the parameter space is fully covered. It is important to note that no real experiment data was used for the training, and no noise was applied. As a result, dataset consisting of around 10000 rows with the total size of the dataset around 100 Mb was generated. The model was trained on the machine with 16 Gb of RAM and 8 core 3.2 GHz CPU (no GPU computing was used), thus it can be noted that a machine with moderate performance is enough for the training and inference. The early stopping was also implemented to stop the training as soon as the rate of validation loss decrease slows down significantly.

Machine Learning Model Application. Though the use of the proposed method and the model is not limited to the current field of study, in scope of this research it was applied to identify the mathematical law for the Eddy Current resisting force component (Eq. 4) of the IECD device described previously to evaluate its performance and make conclusions on its effectiveness.

The model application process is illustrated in Fig. 5 and can be described as follows: the experiment data in a 1-D vector format is first scaled and resampled, and then fed to the trained

Structural Health Monitoring: 10APWSHM
Materials Research Proceedings 50 (2025) 73-81

Materials Research Forum LLC
https://doi.org/10.21741/9781644903513-8

neural network. The CNN part identifies the patterns, that correspond to a specific expression, expands the data with new features thus creating the input for the Encoder.

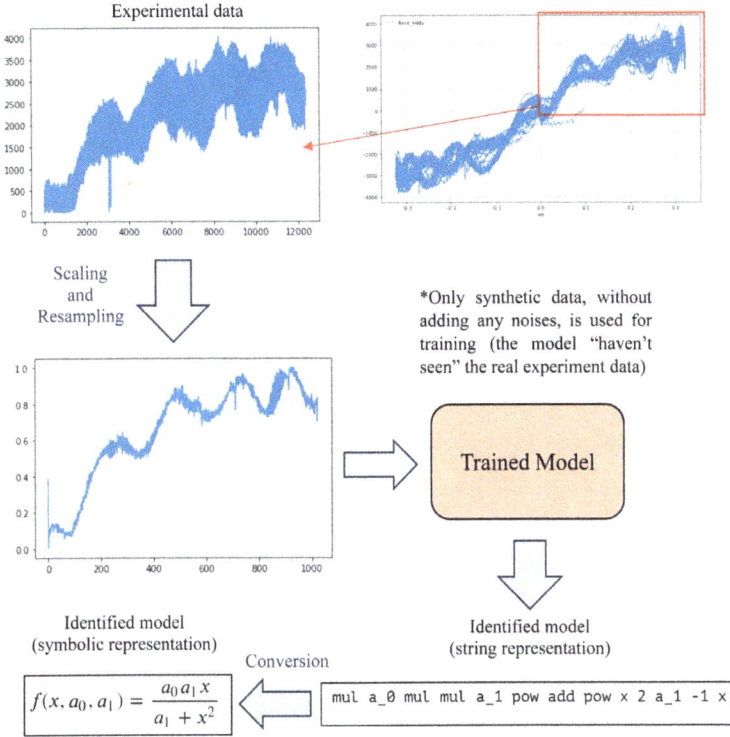

Fig. 5. Model prediction process

Encoder then reads the sequence and calculates the state, which encodes the way the patterns change throughout the experimental data (which is why this part is called the "encoder"). After that, this state is passed to the decoder RNN, which tries to interpret, or "decode", this state and represent it as a mathematical expression – at the beginning, the Start of Sequence (<SOS>) token is passed, and the decoder predicts expression tokens one after another, by concatenating the output tokens to the input, until the model outputs the End of Sequence (<EOS>) token, which means the prediction is over (Fig. 3). After that, the acquired sequence of tokens, that represents an expression, by using the backward conversion (Fig. 4), is converted to the symbolic form. The final result of the prediction is expressed as follows:

$$f(x, a_0, a_1) = \frac{a_0 a_1 x}{a_1 + x^2} \tag{5}$$

By comparing the predicted expression Eq. 5 with Eq. 4, it is clear that the ML model was able to successfully identify the equation in general form, where x corresponds to $\dot{x}_d$, a_0 corresponds to c_0, and a_1 corresponds to v_0^2.

Finally, Eq. 5 is fit into the original data by calculating the parameters a_0 and a_1 with the use of Gauss-Newton algorithm, and the final result is showed in Fig. 6. After comparing the result to the empirical model in Fig. 2, one can observe that the identified equation accurately represents the test data, which proves the validity of the proposed approach. Moreover, the whole prediction process, including the backward conversion, took 1.08 s in total on the above-mentioned machine, thus also demonstrating the high performance of the developed model.

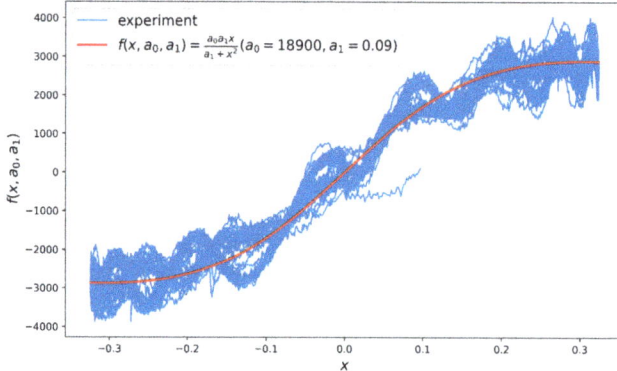

Fig. 6. The result of the predicted model fitting into the original test data

Conclusions

In scope of this research, the new approach using DNN to identify the equations in general form from the experiment data was proposed, successfully implemented and applied to identify the mathematical law of the eddy current resisting force component of the IECD. The method proved to be valid and have high computational performance. However, its performance in other applications and the influence of noise is still to be investigated in further research.

References

[1] S. Michael, L. Hod, Distilling Free-Form Natural Laws from Experimental Data. Science 324,81-85 (2009). https://doi.org/10.1126/science.1165893

[2] E. Ostertagová, Modelling using Polynomial Regression, Procedia Engineering, Volume 48, pp 500-506 (2012). https://doi.org/10.1016/j.proeng.2012.09.545

[3] S.L. Brunton, J.L. Proctor, J.N. Kutz, Discovering governing equations from data by sparse identification of nonlinear dynamical systems, Proc. Natl. Acad. Sci. U.S.A. 113, pp. 3932–3937 (2016). https://doi.org/10.1073/pnas.1517384113

[4] H.S. Suh, J.Y. Song, Y. Kim et al., Data-driven discovery of interpretable water retention models for deformable porous media, Acta Geotech. 19, 3821–3835 (2024). https://doi.org/10.1007/s11440-024-02322-y

[5] D. Li, K. Ikago, A. Yin, Structural dynamic vibration absorber using a tuned inerter eddy current damper, Mech Syst Signal Process 186:109915 (2023). https://doi.org/10.1016/j.ymssp.2022.109915

[6] S. Ilya, V. Oriol, V. Le Quoc, Sequence to Sequence Learning with Neural Networks, arXiv:1409.3215 [cs.CL] (2014)

[7] Y. LeCun et al., Backpropagation Applied to Handwritten Zip Code Recognition, Neural Computation, vol. 1, no. 4, pp. 541-551 (1989). https://doi.org/10.1162/neco.1989.1.4.541

[8] H. Sepp, S. Jürgen, Long Short-Term Memory. Neural Comput; 9 (8): 1735–1780. (1997). https://doi.org/10.1162/neco.1997.9.8.1735

[9] X. Glorot, Y. Bengio, Understanding the difficulty of training deep feedforward neural networks, Proceedings of the Thirteenth International Conference on Artificial Intelligence and Statistics, 9:249-256 (2010)

[10] E. O. Brigham, R. E. Morrow, The fast Fourier transform, IEEE Spectrum, vol. 4, no. 12, pp. 63-70 (1967). https://doi.org/10.1109/MSPEC.1967.5217220

[11] M. Farooq Wahab et al., Discrete Fourier transform techniques for noise reduction and digital enhancement of analytical signals, Trends in Analytical Chemistry, Vol 143 (2021). https://doi.org/10.1016/j.trac.2021.116354

[12] M. Abadi, P. Barham et al., Tensorflow: A system for large-scale machine learning. 12th Symp. on Op. Sys. Design and Impl., pp. 265–283 (2016)

Structural Health Monitoring: 10APWSHM
Materials Research Proceedings 50 (2025) 82-89

Materials Research Forum LLC
https://doi.org/10.21741/9781644903513-9

Manifold learning-based unsupervised feature selection for structural health monitoring

Tingna Wang[1,2,a] *, Limin Sun[1,2,3,b]

[1]Department of Bridge Engineering, Tongji University, Shanghai, China

[2]Shanghai Qi Zhi Institute, Shanghai, China

[3]State Key Laboratory of Disaster Reduction in Civil Engineering, Tongji University, Shanghai, China

[a]tina_wang@tongji.edu.cn, [b]lmsun@tongji.edu.cn

Keywords: Multivariate Filter Method, Unsupervised Feature Selection, Manifold Learning, Structural Health Monitoring, Anomaly Detection

Abstract. Feature selection plays an important role in enhancing the performance of a data-driven structural health monitoring (SHM) system by selecting more informative and lower-dimensional features from the original ones. In this paper, an unsupervised feature selection method based on manifold learning (MLUFS) is proposed to identify feature sets that are beneficial to preserve the structure of the data. This method leverages nonlinear dimensionality reduction methods and canonical correlation analysis to address the challenges associated with acquiring costly labels in the SHM field. It is applicable to high-dimensional data that lie on a low-dimensional manifold within the ambient space. A case study on the handwritten digit datasets is presented to compare the classification performances corresponding to the features selected by the MLUFS method and a supervised feature selection method. The results show that the proposed method can provide feature sets with higher generalization ability compared to the selection method using labels. Another simulated acceleration dataset is used to demonstrate the effectiveness of the proposed method in anomaly detection for an SHM system.

Introduction

For structural health monitoring based on statistical pattern recognition, feature selection serves as an essential step in addressing the challenges associated with high-dimensional data by enhancing model interpretability, improving generalization, and mitigating issues such as confounding effect caused by environmental or operational variations [1–3]. Given the difficulty and high cost of obtaining labels in the structural health monitoring (SHM) field, unsupervised feature selection (UFS) methods have attracted increasing interest in recent years due to their ability to identify features without label information [1,4,5].

According to the strategy used for selecting features, UFS methods can be categorised into three main types, including filter methods, wrapper methods and hybrid methods [6]. Since filter methods evaluate the utility of features based on the intrinsic properties of data without involving machine learning algorithms, they are more computationally efficient and scalable than the other two methods [7]. For unsupervised filter methods, multivariate techniques that assess feature sets as a whole tend to yield more effective feature subsets than univariate techniques that evaluate features individually [6]. However, the low utilization of multivariate methods compared to univariate methods may stem from the lack of user-friendly interpretability and applicability of current multivariate techniques.

Commonly used methods to extract intrinsic properties of data for UFS include information-based methods and manifold learning-based methods. Information-based methods are prevalent in multivariate filter methods and consider the relevance and redundancy of features by constructing new ranking criteria, such as in [8–10]. While manifold learning-based methods are typically combined with regression models for feature selection and regularised regression coefficients are used to rank features, such as in [11–14]. Note that manifold learning can be simply treated as a collection of nonlinear dimensionality reduction methods based on the assumption that the observed data lie on a low-dimensional manifold embedded in a higher-dimensional space [15]. Compared to information-based UFS, manifold learning-based methods are more suitable for data with complex, nonlinear structures. However, evaluating feature sets using a regression model with regularisation in the manifold learning-based UFS is complicated and time-consuming.

In this paper, a novel multivariate UFS is proposed by combining manifold learning techniques and canonical correlation analysis, making it an intuitive and fast approach. Manifold learning is used to derive pseudo labels for input data, providing a quantitative representation of the intrinsic geometric of the dataset. The canonical-correlation-based fast feature selection algorithm developed in [16] is combined with multivariate pseudo labels to select features, which are most beneficial to preserve the intrinsic structure of the data, in an unsupervised manner. Leveraging pseudo labels derived from manifold learning avoids the need for labelling data, and meanwhile, facilitates incorporating nonlinear data relationships in the feature selection process.

The next section will introduce the manifold learning-based unsupervised feature selection (MLUFS) method in detail. It will then be followed by experiments to demonstrate its advantages and one application for SHM. Conclusions are given at the end.

Proposed Method

Given the input feature matrix $\mathbf{X} \in \mathbb{R}^{N \times n}$, the pseudo labels $\mathbf{Y} \in \mathbb{R}^{N \times m}$ are generated by a manifold learning method. Then a set of features is selected based on the canonical correlation with the multivariate pseudo labels $\mathbf{Y}$, which can simultaneously filter out the redundant features. The main steps of the proposed method are formally described as follows.

Step 1 Generating the pseudo labels. Considering the computational feasibility and intuitiveness of the proposed method, six classical manifold learning methods, including MultiDimensional Scaling (MDS) [17], Isometric feature mapping (Isomap) [18], Locally Linear Embedding (LLE) [19], Spectral Embedding [20], t-distributed Stochastic Neighbor Embedding (t-SNE) [21] and Uniform Manifold Approximation and Projection (UMAP) [22], are adopted to provide the pseudo labels, which are list in Table 1.

Because the pseudo labels obtained by the methods in Table 1 are in a global coordinate system, all these manifold learning methods can also be called embedding algorithms [23]. It is necessary to point out that the goal of embedding algorithms is to create a smooth mapping of the input that minimally distorts the neighbourhood information from a weighted graph. These algorithms vary in the information they prioritize to preserve and the constraints they apply to ensure smoothness. Therefore, theoretically, it is impossible to say which embedding algorithm is generally better than another in providing pseudo labels. The next section will compare them empirically.

Moreover, some of the terms used to distinguish the six methods in Table 1 are explained to show their characteristics explicitly. Manifold learning that focusses on preserving global structure aims to maintain the overall shape and large-scale distances between points in the lower-dimensional embedding. In contrast, manifold learning that emphasises the preservation of local structure ensures that small-scale neighbourhoods are well-maintained after dimensionality reduction. Non-linear mapping means that the method aims to capture intrinsic geometry within high-dimensional data without assuming data lies along straight-line axes.

Structural Health Monitoring: 10APWSHM   Materials Research Forum LLC
Materials Research Proceedings 50 (2025) 82-89   https://doi.org/10.21741/9781644903513-9

Table 1 A summary of six classical manifold learning methods

Name	Information to preserve	Preserved Structure	Mapping
MDS (1952) [17]	pair-wise distance (Euclidean distances)	Global	non-linear
Isomap (2000) [18]	nearest-neighbour graph (Geodesic distances)	Both local and global	non-linear
LLE (2000) [19]	local neighbourhoods (Linear neighborhoods)	Local	local-linear
Spectral Embedding (2003) [20]	nearest-neighbour graph (Heat kernel weights)	Both local and global	non-linear
t-SNE (2008) [21]	probability-based pair-wise similarity	Local	non-linear
UMAP (2018) [22]	probability-weighted nearest-neighbour graph	Both local and global, but more local	non-linear

Step 2 Selecting the feature set. Considering the input features are multivariate and the pseudo labels are also multivariate, canonical correlation analysis can be used to evaluate the impacts of multiple features to manifest the relatively simple shape of the data without involving a regression model. In this case, the fast feature selection algorithm based on the sum of squared canonical correlation coefficients (SSC) [16] is used here to select features in a greedy manner, because this algorithm has been empirically validated for its effectiveness and efficiency in selecting useful features for both classification and regression tasks.

At Iteration $i \in \{0, 1, \ldots, t-1\}$, the $(i+1)$th useful feature $x_d \in \mathbb{R}^{N \times 1}$ from the matrix $X \in \mathbb{R}^{N \times n}$ is selected by greedy search, while retaining the features $X_s \in \mathbb{R}^{N \times i}$ selected in the previous iterations. By adopting the SSC as the ranking criterion and $Y \in \mathbb{R}^{N \times m}$ as the pseudo labels, the algorithm finds x_d, where,

$$d = \underset{j}{argmax} \sum_{k=1}^{(i+1) \wedge m} R_k^2\left(\left(X_s \middle| x_j\right), Y\right) \tag{1}$$

It should be clarified that although manifold learning helps to reduce the dimensionality of data and visualise the intrinsic data structure, the transformation obtained on the training dataset cannot usually be applied to future data directly unless some approximation techniques are used. The proposed MLUFS method is a good alternative that can reduce the input dimension while accounting for nonlinearity to some extent. The new observations of selected features can be directly used for future analysis.

Experiments and Results
In this section, two real datasets are used to compare the proposed MLUFS method with the SSC-based supervised feature selection method. In addition, since different embedding algorithms are used in MLUFS to generate pseudo labels, the corresponding effects will also be compared and

Structural Health Monitoring: 10APWSHM   Materials Research Forum LLC
Materials Research Proceedings 50 (2025) 82-89   https://doi.org/10.21741/9781644903513-9

analysed. Furthermore, the proposed method is applied to a simulated time series dataset with anomalies to demonstrate its ability to perform anomaly detection.

A case study for digits. The first dataset is the pen-based handwritten digit dataset with 64 features and 1797 samples collected from 44 writers [24]. The second dataset is the MNIST dataset with 784 features and 2500 samples from the testing dataset [25]. The key parameters for different embedding algorithms to provide the pseudo labels are the default values given in scikit-learn [26]. The dimension of the embedding space is two, that is $Y \in \mathbb{R}^{N \times 2}$. Given the selected features, a support vector machine with a radial basis function kernel is adopted for these two classification tasks. A Ten-fold cross-validation is implemented and the performance of feature selection is assessed by the averaged classification accuracy on the ten validation datasets. The results are shown in Fig. 1a and Fig. 1b.

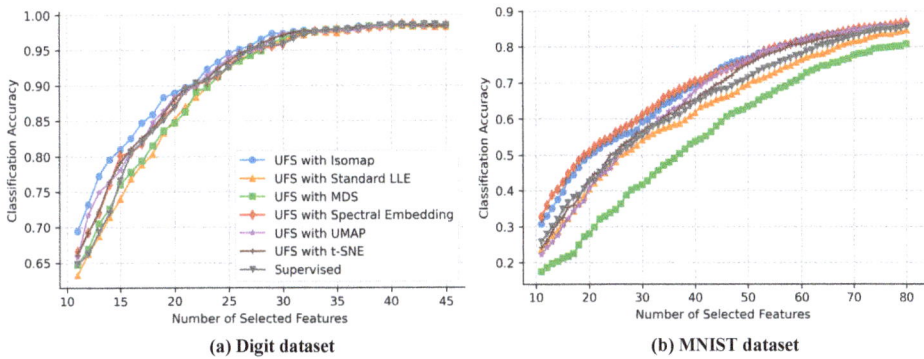

(a) Digit dataset   (b) MNIST dataset

Fig. 1 Comparison of the averaged ACC results between the supervised feature selection and the unsupervised feature selection with different pseudo labels.

It can be found that compared to the supervised feature selection using collected labels, the UFS methods using pseudo labels generated by Isomap or spectral embedding always give competitive results, whether the number of selected features is small or large. Table 1 shows that pseudo labels provided by these two manifold learning techniques can balance the preservation of global and local structures of the data. This phenomenon also empirically proves the rationality of the popularity of the spectral embedding techniques in the UFS field [6]. Moreover, especially from figure 1b, it can be seen that as the number of selected features increases, more UFS methods give results close to or better than the supervised feature selection. This phenomenon means that UFS using pseudo labels tends to provide feature sets with a better generalization ability.

A case study for SHM. To demonstrate the applicability of the proposed method in anomalous data detection, a simulated dataset is generated, providing three hundred minutes of time series data from ten sensors. Two different patterns of anomalies, namely the drift pattern and trend pattern, appear in different time windows (ten minutes as a window) across different sensor channels, as shown in Fig. 2. The normal time series is generated by a Gaussian distribution with zero mean and one standard deviation. The drift pattern and trend pattern are non-stationary vibration responses with random drift and monotonous trend respectively. Refer to [27] and [28] for further information on the anomalies shown in SHM data.

Structural Health Monitoring: 10APWSHM   Materials Research Forum LLC
Materials Research Proceedings 50 (2025) 82-89   https://doi.org/10.21741/9781644903513-9

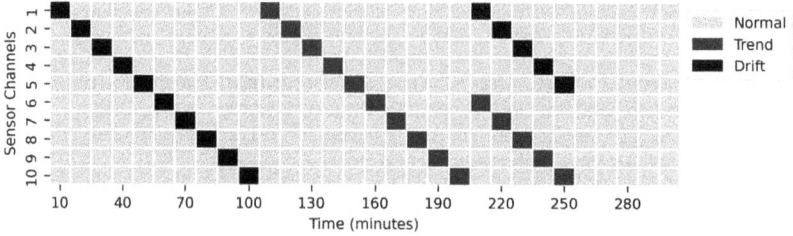

Fig. 2 Distribution of actual anomalies in time and space

It is determined every ten minutes whether there is an anomaly in the data. The specific method is as follows: The collected 10-minute data is concatenated with the 10-minute health data, and then the data from all sensors is reduced to two dimensions. Based on the findings of the digit case study, spectral embedding is used here to achieve nonlinear data dimensionality reduction. The commonly used linear dimensionality reduction method Principal Component Analysis (PCA) is used for comparison. Next, entropy is used to determine whether an anomalous pattern has occurred, because the presence of anomalies will change the data from a uniform distribution of a single cluster to a distribution of multiple clusters. The significance of the change depends on the extent to which the dimensionality reduction method can reveal the change in the local structure of the data.

It is found that the entropy corresponding to normal data is greater than five, so when the entropy value is lower than the threshold of five, the proposed method will be used to locate sensors with abnormal patterns. The results are shown in Fig. 3 and Fig. 4. The anomaly indicator is the correlation between the selected sensor data and the pseudo labels obtained by dimensionality reduction techniques, which represents the contribution of the data to the formation of the intrinsic structure of the current data. When the anomaly indicator is greater than 0.2, the corresponding data segment is treated as anomalous data. It can be seen that even if there is only one abnormal pattern occurs, the dimensionality reduction based on spectral embedding can reveal the occurrence of anomaly, which is missed by the PCA-based method. Moreover, the proposed MLUFS can also accurately locate the sensor channel that introduces the anomalous data, which can be directly removed before further analysis steps.

Conclusions

This paper proposes a MLUFS method, in which an embedding algorithm can be used to extract pseudo labels from the input data. Embedding algorithms that are able to balance the preservation of local and global data structures are recommended to generate pseudo labels. The SSC-based fast feature selection strategy is used to select useful features efficiently to consider the most general case, the multi-input multi-output case. Compared to the SSC-based supervised feature selection method, the features found by the proposed unsupervised method have a better generalisation ability for the upcoming data. The application of MLUFS on the anomaly detection task shows that the proposed method can successfully locate sensor channels when abnormal data occur, which is useful for pre-processing the data to distinguish anomalies caused by damage.

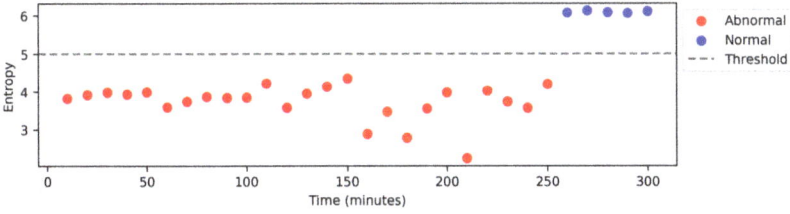

(a) Distribution of anomalies in time

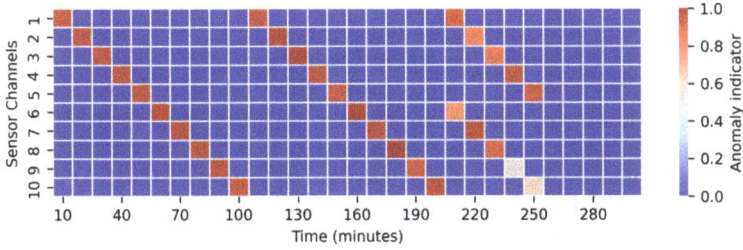

(b) Distribution of detected anomalies in time and space

Fig. 3 Anomaly detection results given by the spectral embedding-based UFS method

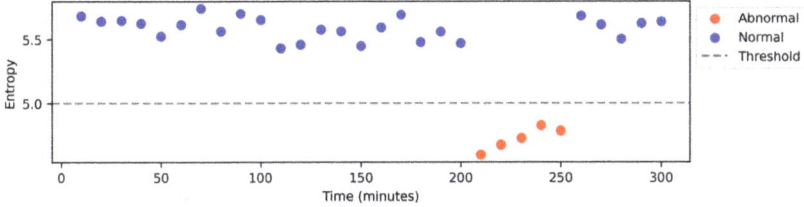

(a) Distribution of anomalies in time

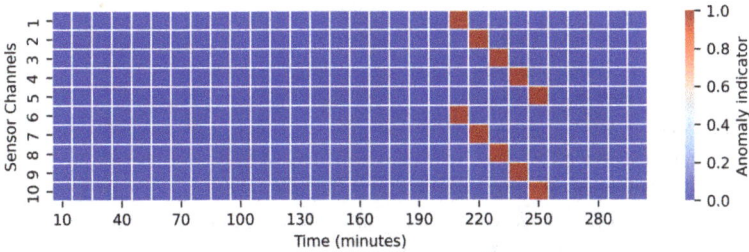

(b) Distribution of detected anomalies in time and space

Fig. 4 Anomaly detection results given by the PCA-based UFS method

References

[1] Alves V, Cury A. An automated vibration-based structural damage localization strategy using filter-type feature selection. Mechanical Systems and Signal Processing. 2023;190:110145. https://doi.org/10.1016/j.ymssp.2023.110145

[2] Bee S, Poole J, Worden K, Dervilis N, Bull L. Multitask feature selection within structural datasets. Data-Centric Engineering. 2024;5:e4. https://doi.org/10.1017/dce.2024.1

[3] Buckley T, Ghosh B, Pakrashi V. A Feature Extraction & Selection Benchmark for Structural Health Monitoring. Structural Health Monitoring. 2023;22(3):2082-127. https://doi.org/10.1177/14759217221111141

[4] Bull LA, Rogers TJ, Wickramarachchi C, Cross EJ, Worden K, Dervilis N. Probabilistic active learning: an online framework for structural health monitoring. Mechanical Systems and Signal Processing. 2019;134:106294. https://doi.org/10.1016/j.ymssp.2019.106294

[5] Gharehbaghi VR, Farsangi EN, Yang TY, Hajirasouliha I. Deterioration and damage identification in building structures using a novel feature selection method. Structures. 2021;29:458-70. https://doi.org/10.1016/j.istruc.2020.11.040

[6] Solorio-Fernández S, Carrasco-Ochoa JA, Martínez-Trinidad JF. A review of unsupervised feature selection methods. Artificial Intelligence Review. 2020;53(2):907-48. https://doi.org/10.1007/s10462-019-09682-y

[7] Guyon, I.; Elisseeff, A.: An Introduction to Variable and Feature Selection. , 26 (2003).

[8] Mitra P, Murthy CA, Pal SK. Unsupervised feature selection using feature similarity. IEEE Transactions on Pattern Analysis and Machine Intelligence. 2002;24(3):301-12. https://doi.org/10.1109/34.990133

[9] Ferreira AJ, Figueiredo MAT. An unsupervised approach to feature discretization and selection. Pattern Recognition. 2012;45(9):3048-60. https://doi.org/10.1016/j.patcog.2011.12.008

[10] Wang S, Pedrycz W, Zhu Q, Zhu W. Unsupervised feature selection via maximum projection and minimum redundancy. Knowledge-Based Systems. 2015;75:19-29. https://doi.org/10.1016/j.knosys.2014.11.008

[11] Cai D, Zhang C, He X. Unsupervised feature selection for multi-cluster data. In Proceedings of the 16th ACM SIGKDD international conference on Knowledge discovery and data mining. 2010;333-42. https://doi.org/10.1145/1835804.1835848

[12] Zhao Z, Wang L, Liu H, Ye J. On Similarity Preserving Feature Selection. IEEE Transactions on Knowledge and Data Engineering. 2013;25(3):619-32. https://doi.org/10.1109/TKDE.2011.222

[13] Hou C, Nie F, Li X, Yi D, Wu Y. Joint Embedding Learning and Sparse Regression: A Framework for Unsupervised Feature Selection. IEEE Transactions on Cybernetics. 2014;44(6):793-804. https://doi.org/10.1109/TCYB.2013.2272642

[14] Nie F, Zhu W, Li X. Unsupervised Feature Selection with Structured Graph Optimization. In Proceedings of the AAAI Conference on Artificial Intelligence. 2016;30(1). https://doi.org/10.1609/aaai.v30i1.10168

[15] Lin T, Zha H. Riemannian Manifold Learning. IEEE transactions on pattern analysis and machine intelligence. 2008;30(5):796-809. https://doi.org/10.1109/TPAMI.2007.70735

Structural Health Monitoring: 10APWSHM   Materials Research Forum LLC
Materials Research Proceedings 50 (2025) 82-89   https://doi.org/10.21741/9781644903513-9

[16] Zhang S, Wang T, Worden K, Sun L, Cross EJ. Canonical-correlation-based fast feature selection for structural health monitoring. Mechanical Systems and Signal Processing. 2025;223:111895. https://doi.org/10.1016/j.ymssp.2024.111895

[17] Torgerson WS. Multidimensional scaling: I. Theory and method. Psychometrika. 1952;17(4):401-19. https://doi.org/10.1007/BF02288916

[18] Tenenbaum JB. A Global Geometric Framework for Nonlinear Dimensionality Reduction. Science. 2000;290(5500):2319-23. https://doi.org/10.1126/science.290.5500.2319

[19] Roweis ST. Nonlinear Dimensionality Reduction by Locally Linear Embedding. Science. 2000;290(5500):2323-6. https://doi.org/10.1126/science.290.5500.2323

[20] Belkin M, Niyogi P. Laplacian Eigenmaps for Dimensionality Reduction and Data Representation. Neural Computation. 2003;15(6):1373-96. https://doi.org/10.1162/089976603321780317

[21] Van der Maaten L, Hinton G. Visualizing Data using t-SNE. Journal of Machine Learning Research. 2008;9(86):2579-605.

[22] McInnes L, Healy J, Melville J. UMAP: Uniform Manifold Approximation and Projection for Dimension Reduction. arXiv; 2020. Available from: http://arxiv.org/abs/1802.03426

[23] Meilă M, Zhang H. Manifold learning: what, how, and why. arXiv; 2023. Available from: http://arxiv.org/abs/2311.03757

[24] Alimoğlu F, Alpaydin E. Combining multiple representations for pen-based handwritten digit recognition. Turkish Journal of Electrical Engineering and Computer Sciences. 2001;9(1):1-2.

[25] LeCun Y, Bottou L, Bengio Y, Haffner P. Gradient-based learning applied to document recognition. Proceedings of the IEEE. 1998;86(11):2278-324. https://doi.org/10.1109/5.726791

[26] Pedregosa F, Varoquaux G, Gramfort A, Michel V, Thirion B, Grisel O, Blondel M, Prettenhofer P, Weiss R, Dubourg V, Vanderplas J. Scikit-learn: Machine learning in Python. Journal of Machine Learning Research. 2011;12:2825-30.

[27] Bao Y, Tang Z, Li H, Zhang Y. Computer vision and deep learning-based data anomaly detection method for structural health monitoring. Structural Health Monitoring. 2019;18(2):401-21. https://doi.org/10.1177/1475921718757405

[28] Ni F, Zhang J, Noori MN. Deep learning for data anomaly detection and data compression of a long-span suspension bridge. Computer-Aided Civil and Infrastructure Engineering. 2020;35(7):685-700. https://doi.org/10.1111/mice.12528

Structural Health Monitoring: 10APWSHM
Materials Research Proceedings 50 (2025) 90-97

Materials Research Forum LLC
https://doi.org/10.21741/9781644903513-10

Feature extraction method of stone curtain wall panels based on multi-source data

ZHUO Xiuqi[1,a], LU Wensheng[1,b], LI Zhiyu[1,c], WANG Shiteng[1,d], HUANG Jie[1,e] *

[1]Tongji University, Shanghai, China

[a]2211314@tongji.edu.cn, [b]wally@tongji.edu.cn, [c]2310098@tongji.edu.cn, [d]2052995@tongji.edu.cn, [e]huangjie@tongji.edu.cn

Keywords: Stone Curtain Wall Panel, Multi-Source Data, Feature Parameter, Feature Extraction, Performance Evaluation

Abstract. Various and diverse stone curtain walls serve as typical building facades in civil buildings, which were mainly composed of stone panels and supporting frames, have been widely used in various civil buildings around the world. During long-term service life, stone curtain wall might be subjected to earthquakes, strong winds, environmental changes and other effects, resulting in performance degradation. At present, the monitoring and sensing methods of stone curtain wall panel's safety and function states are always restricted at application level, and some safety hazards and functional degradation are hardly predicted with efficiency and accuracy, probably resulting in damage accidents, economic losses and negative social impacts during service time. In this paper, multi-source data were achieved through multiple monitoring means, characteristic parameters of mechanical and modeling dimensions were extracted, and the safety and functional performance evaluation of stone panels were investigated. Firstly, vibration signal data were collected by installing acceleration sensors on the curtain wall panel and supporting frame, and image data were analyzed by taking photos with high-definition cameras and drones. Then, the natural vibration frequency and crack length of the mechanical dimension, as well as the crack area and stain area of the molding dimension were extracted and analyzed. Finally, based on the monitoring data, the feature parameters in three dimensions were extracted to evaluate the safety and functional performance of stone panels. The safety and functional performance evaluation method based on multi-source data of stone curtain wall panel is more comprehensive and reliable than traditional methods, which can provide an innovation approach to evaluate and ensure the functional and safety performances of stone curtain wall.

Introduction

As a kind of building facades system, building curtain wall is composed of panel, supporting structure and anchoring system, and has certain bearing capacity, deformation and displacement ability relative to the main structure, and does not provide stiffness and strength contribution to the main structure[1]. Building curtain wall panels generally contain a variety of material types such as glass, stone, metal, and artificial panels[2]. Among them, Stone curtain wall panel (hereinafter called "stone panel") refers to a panel with stone as raw material and used for decoration. Stone panels have the advantages of beauty, durability and safety, and are increasingly used in all kinds of buildings[3,4]. However, effective monitoring and performance evaluation methods for the state of stone panels in actual projects are quite scarce, and the phenomenon of damage and fall often occurs, resulting in serious safety accidents and economic property losses, as well as adverse social impacts[5]. Typical damages of some stone curtain walls are shown in Figure 1.

Structural Health Monitoring: 10APWSHM   Materials Research Forum LLC
Materials Research Proceedings 50 (2025) 90-97   https://doi.org/10.21741/9781644903513-10

| (a) Stone panel off | (b) Stone panel stain | (c) Corrosion of anchor parts |

Fig.1 Typical damage of some stone curtain walls

Studies have shown that the state of stone panels is affected by a variety of characteristic parameters, including the flatness of the panel, the width of the crack, the width of the panel joint. Shi Luning et al.[6] analyzed the sensitivity of natural vibration frequency to parameters such as crack size and location through dynamic tests, and found that the natural vibration frequency decreased with the increase of cracks. Chiara et al.[7] evaluated the security state of the curtain wall by applying the possible effects of various curtain walls. Many scholars have proposed a variety of fast and accurate sensing methods to obtain multi-source data of curtain wall system and monitor or detect the performance changes of curtain wall system. Li et al.[8] installed accelerometers and anemometers in a curtain wall project to identify and monitor the dynamic indicators of the curtain wall, modify the finite element model and judge the safety state of the structure. Lee and Kouzehgar et al.[9] integrated the vision system on mobile platforms such as wall-climbing robots and unmanned vehicles, combined with deep learning algorithms such as YOLO and VGG-19, and improved the detection efficiency of curtain wall damage.

In view of the lack of effective monitoring methods and performance evaluation methods of stone panels, this paper proposes the means of information mining, visual image technology perception and sensor field perception to obtain multi-source data, and carries out feature extraction for the performance parameters of stone panels in a practical engineering case and proposes a comprehensive performance evaluation method.

Multi-source data acquisition method

Multi-source perception[10] refers to the acquisition of data related to the performance state of stone panels through a variety of different ways, and the acquired data is called multi-source data. Stone panel is a complex system formed by multi-stages and multi-modes such as design, construction, operation and maintenance. To evaluate it, a single monitoring method is not enough to extract sufficient curtain wall features, and multi-source sensing method is adopted to obtain multi-source data related to the performance state of stone panel. And through data fusion analysis, continuous accumulation of data in time and space dimension and iterative updating can form more accurate and comprehensive evaluation results.

The multi-source monitoring methods for stone panels can be divided into overall process information collection, geometric measurement perception, sensor field perception and visual image technology perception. Multi-source data monitoring method and corresponding acquisition modes are as shown in Table 1.

Table 1 Multi-source data acquisition methods

Multi-source data monitoring method	Multi-source data acquisition mode
Overall process information collection	Information mining
Geometric measurement perception	Physical measurement
Sensor field perception	Vibration test
Visual image technology perception	Image recognition

Structural Health Monitoring: 10APWSHM   Materials Research Forum LLC
Materials Research Proceedings 50 (2025) 90-97   https://doi.org/10.21741/9781644903513-10

Feature extraction method

Feature extraction of stone panels refers to obtaining key parameters to characterize their properties through parameter analysis. The main feature parameters of curtain wall system can be divided into three dimensions: mechanics D_1, construction D_2 and modeling D_3. The feature parameter selection principles are as followed: good perceptibility, high correlation with performance; good space-time continuity; relative independence; can be analyzed, evaluated and iteratively updated; conducive to digital expression.

According to the above principles, combined with relevant standard provisions [1,11,12] and engineering experience, the extracted feature parameters are as shown in Table 2.

Table 2 Feature parameters of stone panels

Dimension D_i	Feature parameter
Mechanics D_1	Elastic modulus, Poisson ratio, strength, frequency, deformation, stress
Construction D_2	Material type, weight per unit volume, type of construction
Modeling D_3	Length, width, thickness, flatness, flatness, stain

According to the multi-source data acquisition modes, the feature parameters related to the performance of stone panels, such as natural vibration frequency, crack length and stain area, can be extracted in Table 3.

Table 3 Feature parameter extraction

Multi-source data acquisition mode	Feature parameter
Information mining	Elastic modulus, Poisson ratio, weight per unit volume
Physical measurement	Length × width × thickness
Vibration test	Natural frequency
Image recognition	Crack length, crack area, stain area

Feature extraction based on traditional measurement

A large stone panel building is shown in Figure 3. As a brittle material, the destruction of stone panels has high complexity and uncertainty. The multi-source data related to safety and function feature of stone panels can be analyzed first, and the feature extraction can be carried out, so as to conduct the comprehensive performance evaluation.

Fig.3 A large stone panel building

For the actual curtain wall project, the state of the stone panels was observed on the spot, and the images of several stone panels with significant local differences were obtained, and the three complete stone panels were numbered 1, 2 and 3 from top to bottom, as shown in Figure 4. According to the collected stone panel design drawings and other information and traditional physical measurement methods, the basic feature parameter values of Table 2 are obtained.

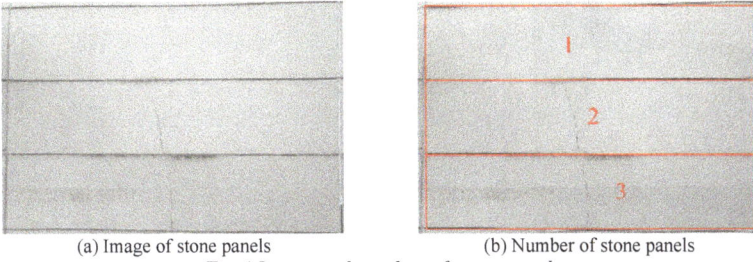

| (a) Image of stone panels | (b) Number of stone panels |

Fig.4 Image and number of stone panels

Table 4 Basic feature parameters of information

Feature parameter	Parameter value
Modulus of elasticity	55GPa
Poisson ratio	0.3
Unit volume weight	2600kg/m^3
Length × width × thickness	4.5m×0.9m×0.08m

Feature extraction based on vibration test

Ambient vibration will bring an incentive to the stone panels, and the parameter identification of natural vibration frequency can be carried out through the transfer function of the structure[13].

Based on the design data and field investigation, it can be seen that the actual supporting conditions of the short side of the stone panel are between the short side simply supported and the fixed support, while the long side has certain constraints with the adjacent plate. As shown in Figure 5, combined with the data in Table 2, the lowest natural vibration frequency of the stone panel is 8.85Hz calculated by ANSYS software according to the short side simply supported condition.

As shown in Figure 6, the sensors are arranged on the back of the three panels. Under ambient vibration, the vibration signal of the stone panels is collected by the sensors and analyzed by the transfer function method with the analysis software. It is obtained that the first-order natural frequency of panel 1, 2 and 3 are 10.49Hz, 9.78Hz and 9.88Hz respectively.The natural vibration frequency of the stone panel is normalized [14], and the vibration test feature value d_{11} is obtained, as shown in Table 3.

Fig.5 Minimum natural frequency calculated by ANSYS

Structural Health Monitoring: 10APWSHM   Materials Research Forum LLC
Materials Research Proceedings 50 (2025) 90-97   https://doi.org/10.21741/9781644903513-10

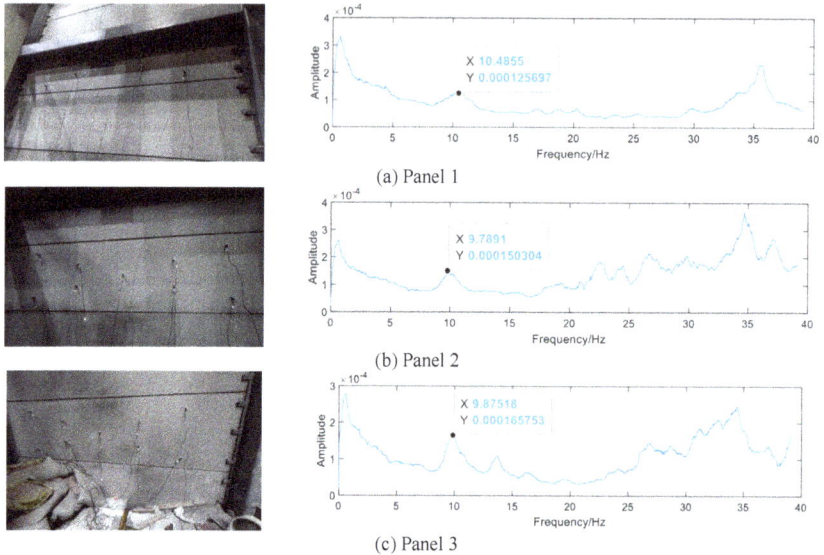

(a) Panel 1

(b) Panel 2

(c) Panel 3

Fig.6 Field sensor layout and spectrum diagram

Table 5 Feature values of vibration test

Panel number	1	2	3
Natural frequency d_{11}	1.000	0.683	0.728

Feature extraction based on image recognition

The stone panel images shown in Fig.1 and Fig.2 were captured, and SAM (Segment Anything Model)[15] and U-Net neural network[16] were used for image recognition. Among them, YOLO[17] object detection can divide the image into fixed size grids, and identify the position of the curtain wall panel in each grid; The SAM model is used to segment the curtain wall panel to obtain a single panel image. After training the UNET neural network, the model has the function of eliminating interference and identifying cracks specifically. The maximum tangent circle method based on KdTrees[18] was used to measure the data of curtain wall panels.

The recognition results of cracks on three stone panels and stains on the panel 3 can be obtained, as shown in Fig.7 and Fig.8. The data of crack length, area and stain area obtained from image recognition are shown in Table 4. The data obtained from image recognition is normalized to obtain the feature values of each parameter, which are listed in brackets in Table 6. Among them, the theoretical minimum value of each parameter data is 0, while maximum value of crack length is 2000, and theoretical maximum value of crack and stain area is 20000.

Table 6 Feature parameter values and feature values of image recognition

Panel number	1	2	3
Fracture length d_{12}	299(0.851)	738(0.631)	948(0.526)
Fracture area d_{13}	1887(0.906)	4957(0.752)	4348(0.783)
Stain area d_{31}	0(1.000)	0(1.000)	9922(0.454)

(a) Panel 1

(b) Panel 2

(c) Panel 3

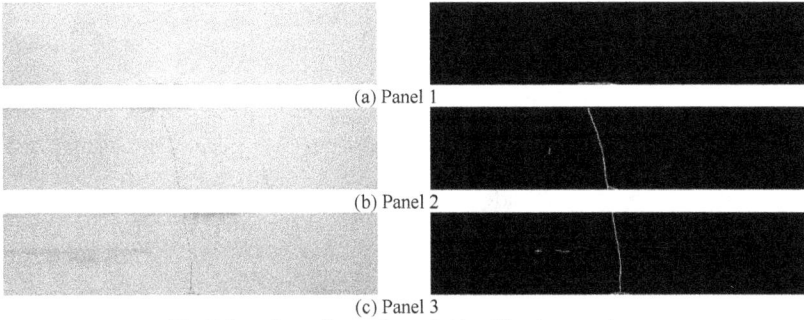

Fig.7 Curtain wall panel crack identification results

Fig.8 Stain identification results of panel 3

Performance evaluation of stone panels

Based on the extracted feature parameters, the safety and function performance status of stone panels can be evaluated from multi-dimensions. The feature parameters of the stone curtain wall panel have a clear logical correlation with its safety and function, so considering the two performance status of safety and function, the comprehensive performance evaluation theory of stone panels can be established. The evaluation method below is used to achieve the quantitative evaluation of the performance of stone panels:

(1) The feature parameter values c_{ij} are obtained by feature extraction;

(2) The score values d_{ij} are obtained by normalization;

(3) Calculate the evaluation parameter e_{mi} of each dimension;

$$e_{mi} = \sqrt[n]{d_{i1} \cdot d_{i2} \cdots d_{in}}. \tag{1}$$

(4) Calculate the comprehensive evaluation parameter p_m;

$$p_m = \alpha_1 e_{m1} + \alpha_2 e_{m2} + \alpha_3 e_{m3}. \tag{2}$$

Where, $\alpha_1, \alpha_2, \alpha_3$ are the influence weights of mechanics, construction and shape dimensions on panel safety and function, respectively. The influence weights of all dimensions on safety and function are shown in Table 5.

Table 7 Weight value of each dimension

	α_1	α_2	α_3
safety	0.5	0.3	0.2
function	0.2	0.3	0.5

(5) Calculate the comprehensive performance P;

When there are different target needs for the function and safety of stone panels, they can be evaluated according to formula (3).

$$P = \alpha \times p_1 + (1 - \alpha) \times p_2. \tag{3}$$

Where, α is the safety performance weight, $0 \leq \alpha \leq 1$, determined according to different comprehensive performance objectives.

(6) Evaluate the performance of stone panels combined with the given comprehensive performance target PT.

Due to the limitation of feature extraction method, all feature parameters cannot be obtained. The values of missing parameters are set to 1.000. In the further perception of the curtain wall

Structural Health Monitoring: 10APWSHM
Materials Research Proceedings 50 (2025) 90-97

Materials Research Forum LLC
https://doi.org/10.21741/9781644903513-10

panel, more data can be obtained, and the existing characteristic parameter values can be updated, so as to update the score, achieve real-time evaluation of the comprehensive performance of the stone panel, and timely discover and deal with the panels with risks. In actual engineering, the comprehensive performance objective is expressed as the curtain wall structure is safe and stable, corresponding to the safety performance weight $\alpha = 0.7$.

The comprehensive performance score of stone panels is shown in Table 6, and the comprehensive performance score threshold is set at 0.8. The score of panel 1 is close to 1.000, indicating that its comprehensive performance is good. The score of panel 2 is slightly higher than the threshold, so it is necessary to strengthen the monitoring of this panel to prevent further deterioration of its state. The score of panel 3 is lower than the threshold, and it has major safety and functional defects. Therefore, measures such as repair or replacement need to be taken to improve the performance. According to the field observation, panels 2 and 3 have penetrating cracks from top to bottom, panel 1 is relatively intact, and panel 3 has large area stains compared with panel 2. The results of comprehensive performance evaluation are consistent with the actual situation of stone panels.

Table 7 Comprehensive performance scores of stone panels

Panel number	1	2	3
$e_{11} = \sqrt[3]{d_{11} \cdot d_{12} \cdot d_{13}}$	0.917	0.687	0.669
$e_{21} = d_{13}$	0.906	0.752	0.783
$e_{23} = d_{31}$	1.000	1.000	0.454
$p_1 = \alpha_1 e_{11} + \alpha_2 e_{12} + \alpha_3 e_{13}$	0.958	0.843	0.835
$p_2 = \alpha_1 e_{21} + \alpha_2 e_{22} + \alpha_3 e_{23}$	0.981	0.950	0.684
$P = \alpha \times p_1 + (1 - \alpha) \times p_2$	0.965	0.876	0.789

Conclusion

In this paper, the feature extraction and comprehensive performance evaluation methods of stone panels are proposed, and the analysis and verification are combined with actual engineering cases, and the following conclusions are obtained:

(1) Feature parameters related to the performance of stone panels can be divided into three dimensions and obtained through feature extraction. Selecting suitable feature parameters can simply and efficiently express the performance state of stone panels;

(2) The comprehensive performance evaluation method of stone panels can synthesize multi-stage and multi-modal state information such as design, construction, operation and maintenance, and continuously accumulate data in time and space and iteratively update to form more accurate and comprehensive evaluation results;

(3) The proposed comprehensive performance evaluation method is a beneficial extension of the traditional safety performance evaluation method, and the safety and function status of stone panels can be accurately analyzed and evaluated through feature extraction and iterative updating, which is in good agreement with actual engineering cases.

(4) The feature extraction and performance evaluation methods of curtain wall panels were briefly investigated, and these methods can be further studied to evaluate the curtain wall supporting frame system accordingly.

References

[1] Standardization Administration of China. Curtain wall for building: GB/T21086-2007 [S]. Beijing: Standards Press of China, 2008. (in Chinese)

[2] HUANG Baofeng, LU Wensheng, CAO Wenqing. Discussion on safety assessment of existing architectural curtain walls[J]. Structural Engineers,2006,22(03):76-79. (in Chinese)

Structural Health Monitoring: 10APWSHM
Materials Research Proceedings 50 (2025) 90-97

Materials Research Forum LLC
https://doi.org/10.21741/9781644903513-10

[3] LIAO Lei, YIN Peng, FAN Yuan, et al. Mechanical Model and testing verification of the pull-out mechanism of the backbolt joints of stone panels[J]. Structural Engineers, 2022, 38(03): 110-116. (in Chinese)

[4] ZHU Qiankun, LIU Kaifang, RUI Jia, et al. Influence of curtain wall on vibration mode of cantilevered floor and equivalent simulation method[J]. Journal of Vibration, Measurement & Diagnosis, 2021,41(06): 1190-1198+1242. (in Chinese)

[5] BAO Yiwang, ZHENG Dezhi, LIU Xiaogen, et al. Rapid safety performance of stone curtain wall with laser vibration frequency method[J]. Journal of Shenyang University of Technology, 2021, 43(05): 595-600. (in Chinese)

[6] SHI Luning, YAN Weiming, HE Haoxiang, et al. Sensitivity of Natural Frequency and Dynamic Characteristics of the Beam with Open Cracks to Crack Parameters. Journal of Vibration, Measurement & Diagnosis, 2016, 36(05): 881-889+1022. (in Chinese)

[7] BEDON C, AMADIO C. Numerical assessment of vibration control systems for multi-hazard design and mitigation of glass curtain walls[J]. Journal of Building Engineering, 2018, 15: 1-13. https://doi.org/10.1016/j.jobe.2017.11.004

[8] Li H, Yang B, Li Z, et al. Wind-induced response monitoring and study of monolayer cable net [C]//Proceedings of IASS Annual Symposia. International Association for Shell and Spatial Structures (IASS), 2014, 2014(20): 1-6.

[9] Lee J, Hong J, Park G, et al. Contaminated Facade Identification Using Convolutional Neural Network and Image Processing [J]. IEEE Access, 2020, 8: 180010-180021. https://doi.org/10.1109/ACCESS.2020.3027839

[10] DONG D, MA C, WANG M, et al. A low-cost framework for the recognition of human motion gait phases and patterns based on multi-source perception fusion[J]. Engineering Applications of Artificial Intelligence, 2023, 120: 105886. https://doi.org/10.1016/j.engappai.2023.105886

[11] Shanghai, Engineering Construction Specification. Technical standard for curtain wall engineering: DG/TJ 08-56-2019 [S]. Shanghai, 2009. (in Chinese)

[12] Ministry of Construction of the People's Republic of China. Technical code for metal and stone curtain walls engineering: JGJ 133-2001 [S]. Beijing: China Architecture and Building Press, 2001. (in Chinese)

[13] HUANG Zhide. Study on safety-state evaluation model of building curtain wall and remote detection method[D]. University of Science and Technology Beijing,2019. (in Chinese)

[14] WANG Qin, LI Zhili. Application of dynamic method in safety detection of existing stone curtain wall [J]. Sichuan Building Science, 2023, 49(01): 56-61. (in Chinese)

[15] KIRILLOV A, MINTUN E, RAVI N, et al. Segment anything[C]//2023 Proceedings of the IEEE/CVF International Conference on Computer Vision (ICCV), 2023: 4015-4026. https://doi.org/10.1109/ICCV51070.2023.00371

[16] WENG W, ZHU X. INet: convolutional networks for biomedical image segmentation[J]. Ieee Access, 2021, 9: 16591-16603. https://doi.org/10.1109/ACCESS.2021.3053408

[17] JIANG P, ERGU D, LIU F, et al. A Review of Yolo algorithm developments[J]. Procedia Computer Science, 2022, 199: 1066-1073. https://doi.org/10.1016/j.procs.2022.01.135

[18] NARAYAN B, MURTHY C, PAL S. Maxdiff kd-trees for data condensation[J]. Pattern recognition letters, 2006, 27(3): 187-200. https://doi.org/10.1016/j.patrec.2005.08.015

Structural Health Monitoring: 10APWSHM
Materials Research Proceedings 50 (2025) 98-104

Materials Research Forum LLC
https://doi.org/10.21741/9781644903513-11

Damage localization in metallic plates through lamb wave frequency variation: A numerical study

Deepak Kumar[1,a] * and Sahil Kalra[1,b]

[1]Department of Mechanical Engineering, IIT Jammu, J&K, India

[a]2019rme0031@iitjammu.ac.in, [b]sahil.kalra@iitjammu.ac.in

Keywords: Lamb Wave, FEM, SHM, Damage Imaging, Dispersion Curve

Abstract. Lamb wave is extensively used in the structural health monitoring (SHM) of various structures for defect detection. However, the Lamb wave is constrained by two interlinked complexities: inherent dispersion and multimode characteristics. The presence of many dispersive modes poses challenges in their application for structural damage localization. The aim of this work is to examine Lamb wave behavior and determine the location of the damage when its central frequency varies in the range specified by its dispersion curve for a particular structural thickness. The analysis methodology is illustrated by twenty numerically simulated cases of aluminium plates using a three-dimensional finite element method (FEM) in Abaqus/CAE. A hole as a defect is included in the plate that extends through its entire thickness. The fundamental symmetric Lamb mode (S0) with increasing central frequency in the chosen range is actuated in the first ten cases as it exhibits minimal dispersion. History output for the out-of-plane components of the signal is obtained, which is used for damage location imaging using Matlab. A similar approach is repeated for the next ten cases using the fundamental anti-symmetric mode (A0). The images obtained for the damage locations show a clear distinction among the accuracy of Lamb wave frequencies that are used. As the central frequency is increased for S0 mode, the size of the located damage area increases, resulting in a decrease in the optimization of damage localization. Whereas a reverse output is observed with the A0 mode, where the localized area decreases with increases in frequency. In all cases, the location is accurately localized. However, the estimated area of damage is shifted from its central location. Another significant outcome from this investigation showed that at a higher frequency range, the localized area of the damage is approximately the same. The investigation concludes that low-frequency Lamb modes are better suited for damage localization using imaging techniques. It is also confirmed by the investigation that the low-frequency Lamb wave gives better localization results when applied to identifying comparatively large defects (in mm).

Introduction

Guided waves are rapidly gaining prominence as a method for structural health monitoring (SHM) and non-destructive evaluation (NDE) due to their broad application and minimal structural contact requirements. Lamb waves (LWs) are the guided waves that propagate within the surface boundary of thin-walled structures [1, 2]. Lamb waves-based SHM techniques are a promising NDE technique for detecting and localizing defects in metallic plates [3,4,5,6,7]. These guided elastic waves are capable of propagating over long distances and are sensitive to both surface and subsurface defects [8,9,10,11]. However, their interaction with defects is influenced by factors like wave frequency, defect size, and material geometry [12, 13]. Lamb waves exhibit multimodal behavior, indicating the simultaneous coexistence of at least two modes: the symmetric mode (S0) and the antisymmetric mode (A0) at any frequency of Lamb waves. With the increase in frequency, additional Lamb wave modes, including S1 and A1, become evident. These wave modes propagate at distinct velocities, rendering the interpretation of Lamb wave signals particularly difficult. Typically, LW-based SHM applications employ narrowband excitation. The frequency is

Structural Health Monitoring: 10APWSHM   Materials Research Forum LLC
Materials Research Proceedings 50 (2025) 98-104   https://doi.org/10.21741/9781644903513-11

meticulously chosen to permit the presence of only two essential modes within the designated frequency band.

Determining the wavelength, resolution, attenuation, and mode conversion characteristics of the Lamb wave as it passes through the material depends critically on its central frequency. Although they suffer from more attenuation, higher frequencies usually provide shorter wavelengths, which increases resolution for minor defect detection. On the other hand, lower frequencies provide less resolution but longer-range transmission free from attenuation. Lamb wave-based NDE procedures can thus be optimized by knowing how different frequencies affect damage detection and localization since the interaction of Lamb waves with defects like through-holes is quite sensitive to frequency.

This work focused on examining the effect of Lamb wave frequency on the identification and localization of defects in 2 mm aluminum plates. Lamb waves ranging from 100 kHz to 500 kHz are applied after a through-hole defect with a 10 mm radius is manually created on the plate. This work aimed to shed light on how frequency selection affects defect detection sensitivity, resolution, and propagation characteristics by means of analysis of the interaction between Lamb waves and the defect at several frequencies. The results of this work should guide the optimization of Lamb wave-based damage detection methods for uses in civil, mechanical, aeronautical, and automotive engineering sectors, where precise and dependable structural defect detection is of great relevance.

Methodology

The current study of the Lamb wave frequency variation is based on the dispersion curve for a 2 mm thick aluminum plate-like structure (Fig. 1). It is observed from Fig. 1 that the phase velocity of the fundamental symmetric mode decreases gradually as the frequency is increased, while for the fundamental antisymmetric mode, the phase velocity rapidly increases with increment in the frequency. A total of twenty finite element models are simulated in Abaqus/CAE [14] for the study of the effect of Lamb wave frequency variation on damage localization. The dimensions of the plate are 500 mm × 500 mm × 2 mm. A set of twenty-five points is arranged in a 5-by-5 configuration on the surface of the plate, depicting the actuating and sensing points for the data acquisition (Fig. 2). For the defective case, a through-hole of 10 mm radius is created at the location (25 mm, 25 mm), taking the centre of the plate as the origin (0,0) (Fig. 3). Aluminum has a modulus of elasticity (E) of 69 GPa, a density (ρ) of 2710 kg/m^3, and Poisson's ratio (ν) of 0.33. For the first ten models (five pristine conditions and five defective conditions), the symmetric fundamental mode of the Lamb wave with varying frequency is actuated, while for the remaining ten models, the antisymmetric mode of the Lamb wave with varying frequency is actuated. The technique for actuating the two distinct dominant modes in the simulation is explained in Fig. 4. Fig. 4(a) represents a symmetric mode which, in this study, is actuated by applying the LW from both the surfaces of the plate such that both signal actuation direction is into the plate. Similarly, Fig. 4(b) represents an antisymmetric mode, which is actuated by applying the LW on one of the plate surfaces going into the plate while on the opposite side, another load coming out of the plate. The direction of the application determined the mode of propagation. The same is verified from the dispersion curve by comparing the calculated phase velocity of the simulated wave with that obtained from the dispersion curve and discussed in the later section. The actuation signal is excited by a one-cycle Hanning windowed tone-burst. After applying these actuating loads, the models are meshed with a size of 1 mm C3D8 elements with linear type. The output is rendered from the history output, taking the U3 component of the load at all the specified sensing points. The data acquisition starts with taking transducer point T1 as the first actuation point and taking all other points T2 to T25 as the sensing points. The same step is repeated by changing the actuating points so that all the twenty-five points are used for actuating the LW signal. This results in a

Structural Health Monitoring: 10APWSHM   Materials Research Forum LLC
Materials Research Proceedings 50 (2025) 98-104   https://doi.org/10.21741/9781644903513-11

dataset of a total of 300 signals. These signals are then used for data localization algorithm in Matlab [15], which is explained below.

The imaging algorithm starts with determining the residual signal, which is the difference between the pristine state or baseline signals and the defective or current state signals of the same actuator-sensor pair. It is mathematically given as:

$$Signal\ _{Residual} = Signal\ _{Defective} - Signal\ _{Pristine.} \tag{1}$$

The non-zero residual signal signifies the presence of defect(s) in the structure. After identifying the presence of any defects, the next step is to localize it. The localization algorithm is based on the time-of-flight data. The plate is divided into pixels such that each pixel point 'p' denotes the probable location of the defect. The intensity of the signal is obtained at each pixel point for each actuator-sensor pair using the corresponding time-of-flight (t_{tof}) value. The point at which the highest intensity value is obtained denotes the defect location. The time-of-flight (t_{tof}) for each of the actuator-sensor pair is calculated as:

$$t_{tof} = \frac{\sqrt{(x_p-x_a)^2+(y_p-y_a)^2} + \sqrt{(x_s-x_p)^2+(y_s-y_p)^2}}{v}. \tag{2}$$

Where (x_a, y_a) and (x_s, y_s) are co-ordinates of actuator 'a' and sensor 's', whereas (x_p, y_p) is co-ordinates of any random point 'p' on the structure with a probability of being the defect point location, and v is the phase velocity of the actuated and sensed wave signal. The intensity of the signal is calculated at each of the pixel points 'p'. The point(s) at which the intensity value is high shows the location of the defect(s). The formulated algorithm is repeated for each of the simulated models for both the cases of symmetric and antisymmetric actuation signals.

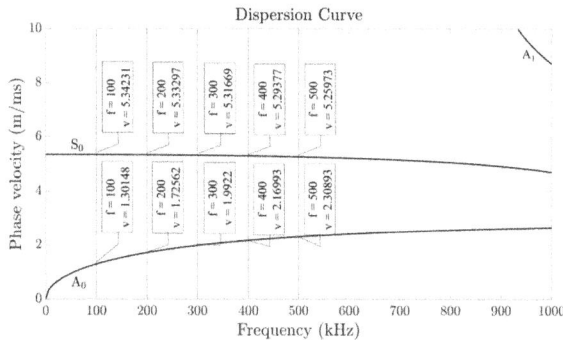

Fig. 1: Dispersion curve for 2 mm Aluminum plate. The wave velocity (v) for fundamental modes (A0 and S0) at varying central frequencies (f) of LW from 100 kHz to 500 kHz is notified.

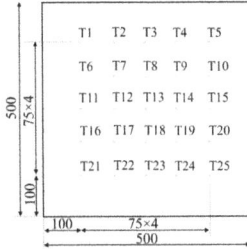

Fig. 2: Thin plate structure with 25 transducers placed in a square array in the form of a 5-by-5 arrangement (blue dots).

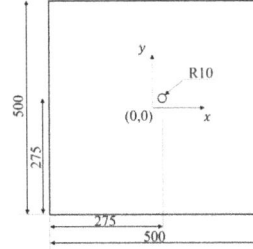

Fig. 3: Thin plate structure in the defective condition, where a circular through-hole is created as a defect at location (25,25).

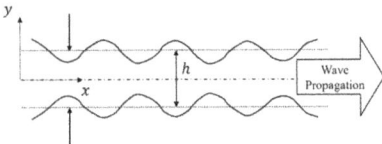

(a) Symmetric mode: LW is actuated from both sides of the plate in a direction going into the plate.

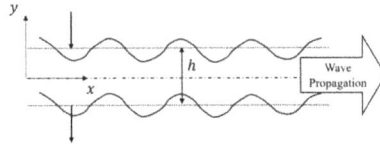

(b) Antisymmetric mode: LW is actuated in a direction going into the plate while the other one goes out of the plate.

Fig. 4: Actuation of the symmetric and antisymmetric dominant modes of Lamb wave.

Results and discussion

The defect imaging is done using the above-explained algorithm for localizing the defect. The algorithm's effectiveness for a particular case is determined by the location of the highest intensity in the plate and the area covered on the plate surface, which, in combination, forms the localization result. The algorithm is based on the time-of-flight value, which in turn is dependent on the phase velocity of the wave in the structure.

In all cases, the phase velocity of the wave is calculated after acquiring signal data from the FE simulation. The same is then compared with that obtained from the dispersion curve. The comparison of the phase velocity for symmetric and antisymmetric mode waves obtained through both procedures is compiled in Tables 1 and 2, respectively. It is evident from these two tables that the velocity calculated from the simulation data in each case is close to that shown in the dispersion curve. It proves that the simulated data can be used authentically for the study related to the Lamb waves on the aluminum plate. Also, as the frequency increases, the difference between the simulated and dispersion velocities increases with a maximum error of up to 2.5%, indicating that the localization results get variation from the actual location with frequency increment. All these phenomena are observed for both the symmetric and antisymmetric wave modes and hence indicate the consistency of the FE simulation used for the investigation. Using these velocity values in the algorithm produces the localized damage images.

The imaging results for the first five defective models i.e. for the symmetric wave mode case, are presented in Fig. 5, whereas the remaining five models of the antisymmetric Lamb mode case are presented in Fig. 6. From Fig. 5, it is observed that as the central frequency of the Lamb wave increases, the localized defect area increases up to a specific value (Fig. 5(a), 5(b), 5(c)) and then remains constant (Fig. 5(d), 5(e)). It is also observed that in all the cases, the algorithm showed the accurate location of the defect, varying only in the localized area. Contrastingly, from Fig. 6,

Structural Health Monitoring: 10APWSHM   Materials Research Forum LLC
Materials Research Proceedings 50 (2025) 98-104   https://doi.org/10.21741/9781644903513-11

it is observed that as the central frequency of the wave increases, the localized defect area decreases, although the centre of the localized defect area shows the accurate defect location.

Table 1: Comparison of the S0 velocity [mm/µs] from the simulated data and the dispersion curve- velocity calculated from the simulated data is slightly higher than that from the dispersion curve with error ≤ 2.5%

	100 [kHz]	200 [kHz]	300 [kHz]	400 [kHz]	500 [kHz]
Dispersion curve	5.3423	5.333	5.3167	5.2938	5.2597
Simulated data	5.4465	5.4381	5.4291	5.4104	5.3932
%Error	1.95	1.97	2.1	2.2	2.5

Table 2: Comparison of the A0 velocity [mm/µs] from the simulated data and the dispersion curve- velocity calculated from the simulated data is slightly higher than that from the dispersion curve with error < 2.5%

	100 [kHz]	200 [kHz]	300 [kHz]	400 [kHz]	500 [kHz]
Dispersion curve	1.3015	1.7256	1.9922	2.1699	2.3089
Simulated data	1.3249	1.7572	2.0302	2.2129	2.362
%Error	1.8	1.83	1.91	1.98	2.3

Out of all the cases for defect localization in 2 mm aluminum plate, the case with 100 kHz central frequency gives the best results for damage localization where the highest intensity values are well within the defect area, as represented by a white circle of radius 10 mm in Fig. 5(a) and 6(a). This is due to the fact that in spite of changing the mode of the LW from S0 to A0, the area obtained through the localized algorithm showed almost the same area covered with high damage intensity.

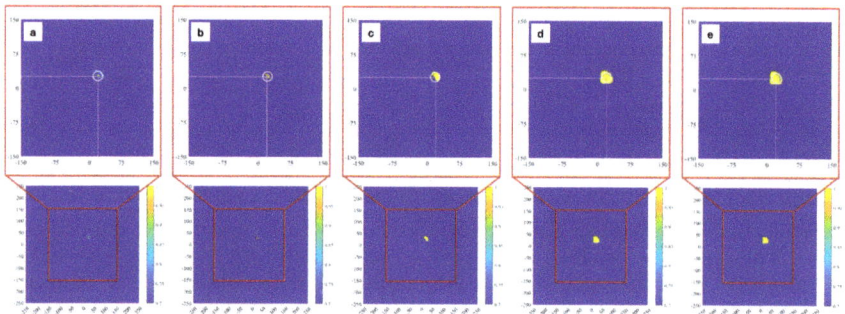

Fig. 5: Localized positions of damage using the algorithm exhibited by higher values of intensity (>0.9) compared with the actual position of damage represented by a white circle, with varying central frequencies of LW with dominant symmetric mode: (a) 100 kHz, (b) 200 kHz, (c) 300 kHz, (d) 400 kHz, and (e) 500 kHz. As the frequency increases, the localized spot increases, indicating the expansion of a higher-intensity area due to a slight decrease in the speed.

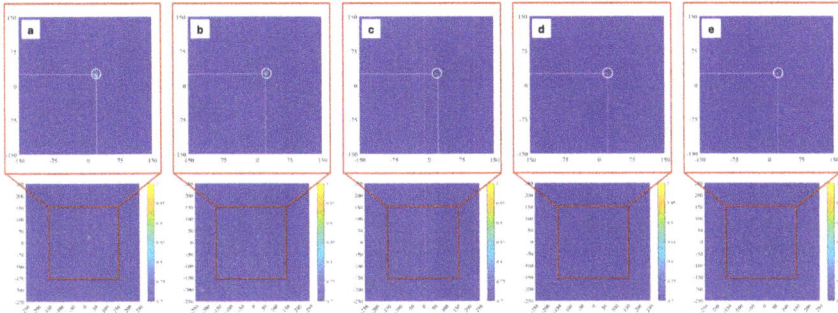

Fig. 6: Localized positions of damage using the algorithm exhibited by higher values of intensity (>0.9) compared with the actual position of damage represented by a white circle, with varying central frequencies of LW with dominant antisymmetric mode: (a) 100 kHz, (b) 200 kHz, (c) 300 kHz, (d) 400 kHz, and (e) 500 kHz. As the frequency increases, the localized spot decreases, indicating the reduction of a higher-intensity area due to a rapid increase in the speed.

Summary

This work examined the effect of Lamb wave frequency variations on the defect localization in metallic plates. For this investigation, a total of twenty FE models of 2 mm aluminum plates are modelled, out of which ten of them represent the defective state of the plate with a hole of 10 mm radius as the defect. Fundamental symmetric Lamb waves are actuated on the five pristine and five defective cases, with five different frequencies ranging from 100 kHz to 500 kHz. The same is repeated for the fundamental antisymmetric mode. The localized defect result showed that the frequency selection greatly affects the defect detection sensitivity and resolution. It is determined that the lower frequency values give better results using the symmetric mode for a significant defect.

Acknowledgement

The authors are grateful for the research project grant from the Defence Research and Development Organisation (DRDO) under the project number ARDB/01/1082021/M, DRDO, Govt. of India.

References

[1] H. Lamb, Proc. Roy. Soc. (London) A90, 111, 114 (1916).

[2] D. C. Worlton; Experimental Confirmation of Lamb Waves at Megacycle Frequencies. J. Appl. Phys. 1 June 1961; 32 (6): 967–971.

[3] Kumar, Deepak, Sahil Kalra, and Mayank Shekhar Jha. "Recent advancements on structural health monitoring using lamb waves." Computational and Experimental Methods in Mechanical Engineering: Proceedings of ICCEMME 2021. Springer Singapore, 2022. https://doi.org/10.1007/978-981-16-2857-3_15

[4] Zhang, G., Kundu, T., Deymier, P. A., & Runge, K. (2024). Defect localization in plate structures using the geometric phase of Lamb waves. Ultrasonics, 107492. https://doi.org/10.1016/j.ultras.2024.107492

[5] Zheng, S., Luo, Y., Xu, C., & Xu, G. (2023). A review of laser ultrasonic lamb wave damage detection methods for thin-walled structures. Sensors, 23(6), 3183. https://doi.org/10.3390/s23063183

[6] Ling, F., Chen, H., Lang, Y., Yang, Z., Xu, K., & Ta, D. (2023). Lamb wave tomography for defect localization using wideband dispersion reversal method. Measurement, 216, 112965. https://doi.org/10.1016/j.measurement.2023.112965

[7] Huang, L., Luo, Z., Zeng, L., & Lin, J. (2024). Detection and localization of corrosion using the combination information of multiple Lamb wave modes. Ultrasonics, 138, 107246. https://doi.org/10.1016/j.ultras.2024.107246

[8] Giurgiutiu, Victor. "Tuned Lamb wave excitation and detection with piezoelectric wafer active sensors for structural health monitoring." Journal of intelligent material systems and structures 16.4 (2005): 291-305. https://doi.org/10.1177/1045389X05050106

[9] Rabbi, M. S., Teramoto, K., Ishibashi, H., & Roshid, M. M. (2023). Imaging of sub-surface defect in CFRP laminate using A0-mode Lamb wave: Analytical, numerical and experimental studies. Ultrasonics, 127, 106849. https://doi.org/10.1016/j.ultras.2022.106849

[10]Zhang, N., Zhai, M., Zeng, L., Huang, L., & Lin, J. (2023). Damage assessment using the Lamb wave factorization method. Mechanical Systems and Signal Processing, 190, 110128. https://doi.org/10.1016/j.ymssp.2023.110128

[11]Xu, H., Liu, L., Xu, J., Xiang, Y., & Xuan, F. Z. (2024). Deep learning enables nonlinear Lamb waves for precise location of fatigue crack. Structural Health Monitoring, 23(1), 77-93. https://doi.org/10.1177/14759217231167076

[12]Ruzzene, Massimo. "Frequency–wavenumber domain filtering for improved damage visualization." Smart materials and structures 16.6 (2007): 2116. https://doi.org/10.1088/0964-1726/16/6/014

[13]Michaels, Thomas E., Jennifer E. Michaels, and Massimo Ruzzene. "Frequency–wavenumber domain analysis of guided wavefields." Ultrasonics 51.4 (2011): 452-466. https://doi.org/10.1016/j.ultras.2010.11.011

[14]SIMULIA, Abaqus/CAE 2021, Version 6.22. Dassault Syst`emes Simulia Corp, Johnston, RI, United States.

[15]MathWorks, MATLAB (R2022b). The MathWorks Inc., Natick, Massachusetts, United States.

Structural Health Monitoring: 10APWSHM
Materials Research Proceedings 50 (2025) 105-112

Materials Research Forum LLC
https://doi.org/10.21741/9781644903513-12

Structural damage identification method based on transfer learning and heterogeneous data alignment

Liu Mei[1,a *], Ying Zhou[1,b] and Wujian Long[1,c]

[1]State Key Laboratory of Intelligent Construction and Healthy Operation and Maintenance of Deep Underground Engineering, Guangdong Provincial Key Laboratory of Durability for Marine Civil Engineering, College of Civil and Transportation Engineering, Shenzhen University, Shenzhen, 518060, Guangdong, PR China

[a]meiliu@szu.edu.cn, [b]2200471020@email.szu.edu.cn, [c]longwj@szu.edu.cn

Keywords: Structural Damage Identification, Transfer Learning, Heterogeneous Data Alignment, JS Divergence

Abstract. Structural damage identification is crucial for ensuring building safety, as it helps safeguard both lives and property. Recently, deep learning has become a prominent approach for damage detection. However, effective training of deep learning models requires abundant structural response data with accurate damage labels, which are often scarce in practical engineering applications. While existing models based on simulated data perform well in simulation environments, they typically struggle to achieve satisfactory performance on real-world measured data. To address this issue, this paper proposes a novel method for structural damage identification that combines transfer learning with the alignment of heterogeneous data (simulated and measured data). This method enables the transfer of damage detection capabilities from simulated data to measured data. First, finite element software is used to simulate various structural damage scenarios, generating a large volume of simulated data with precise damage labels. This data is then used to pre-train an initial model for damage identification. Next, transfer learning is applied, where the model is further trained using a combination of a small amount of measured data and the large simulated dataset. To address the distributional differences between the heterogeneous data sources, the Jensen-Shannon (JS) divergence is employed to quantify the discrepancy in the high-dimensional feature space. This is combined with the cross-entropy classification loss to form a composite loss function for model training. The resulting model is capable of accurately identifying structural damage in measured data. Experimental results show that this approach improves damage identification accuracy by 6% compared to traditional methods, demonstrating its effectiveness.

1. Introduction

With the rapid development of modern urban infrastructure, various large-scale civil engineering buildings are continuously being constructed. These buildings, designed to last for decades or even centuries, must endure multiple factors such as environmental erosion, material aging, sustained loading, and fatigue under the prolonged influence of complex and harsh conditions. Over time, these factors contribute to the gradual accumulation of structural damage and deterioration. This is particularly concerning when buildings are subjected to sudden disasters like typhoons, earthquakes, or fires, as such damage can lead to major safety accidents, significant economic losses, and serious social impacts, ultimately threatening public safety. A structural damage identification and health monitoring system can detect damage early, offering targeted maintenance and reinforcement recommendations for the operation and maintenance of engineering facilities, thus significantly reducing long-term maintenance costs. Therefore, implementing timely and accurate health monitoring for large civil engineering structures is of paramount importance.

Structural Health Monitoring: 10APWSHM   Materials Research Forum LLC
Materials Research Proceedings 50 (2025) 105-112   https://doi.org/10.21741/9781644903513-12

In recent years, deep learning techniques have been widely applied in structural damage identification. These techniques are capable of automatically learning abstract and complex features, enabling them to effectively handle complex data and environments for structural damage identification [1]. Cha et al. [2] used convolutional neural networks (CNNs) to detect cracks on concrete surfaces, categorizing them into various types such as delamination, voids, and spalling. Kim et al. [3] applied the U-net architecture for concrete crack detection. Compared to traditional CNNs, U-net requires a smaller dataset for model training while achieving higher accuracy in damage identification. However, effective training of deep learning models necessitates large amounts of accurate, labeled data [4], which is often difficult to obtain in practical engineering scenarios, limiting the availability of high-quality data for large-scale damage identification. Furthermore, existing methods primarily use simulated data for training, which results in models that perform well on simulated datasets but struggle to accurately identify damage in measured datasets.

To address the aforementioned challenges, this study proposes a structural damage identification method based on transfer learning and heterogeneous data alignment, enabling the transfer of damage identification capabilities from simulated to measured data sources. First, a large dataset of simulated data with accurate damage labels is used to pre-train an initial model for structural damage identification. Next, the model undergoes two stages of training using transfer learning, combining a small amount of measured data with the large simulated dataset containing precise damage labels. To minimize the impact of feature distribution differences between the heterogeneous data sources on the transfer learning process, a heterogeneous data alignment strategy is employed. This results in a final structural damage identification model capable of accurately identifying structural damage in measured data.

2. Methodology

2.1 Feature Extraction

In this study, to mitigate the impact of different excitations applied to heterogeneous data on the effectiveness of transfer learning, the acceleration sequence is modeled using the Autoregressive (AR) Model [5]. The AR model is a linear predictive model that uses historical signal data to construct regression equations for predicting future signal values, making it well-suited for modeling and forecasting time-series data. This method not only effectively captures the dynamic features within the acceleration sequence but also enhances the model's generalization capability. The mathematical expression of the AR model is given in Eq. 1.

$$x(t) + \sum_{i=1}^{q} a_i x(t-i) = e(t) \tag{1}$$

At time t, $x(t)$ represents the output of the model, while $e(t)$ is the model residual, which is a white noise series with zero mean and variance σ^2. The parameter q denotes the order of the model, and a_i represents the i-th order coefficient of the model.

2.2 Pre-training of Initial Model for Structural Damage Identification

Different damage states of the simulated structure are generated to create structural response simulation data, which includes damage condition labels. This data is then used for the pre-training of the initial model for structural damage identification. Since structural damage identification is a multi-class classification problem, the cross-entropy loss function is employed to effectively measure the difference between the predicted labels and the true labels [6], thereby enhancing classification performance and ensuring accurate damage identification. Consequently, the cross-entropy loss function is selected to pre-train the convolutional neural network during the model training process, facilitating the construction of the initial model for structural damage

Structural Health Monitoring: 10APWSHM   Materials Research Forum LLC
Materials Research Proceedings 50 (2025) 105-112   https://doi.org/10.21741/9781644903513-12

identification and ensuring that the model can accurately identify structural damage in the simulated data.

2.3 Final Model Training for Structural Damage Identification

The acceleration response signals from key nodes of real engineering structures, subjected to different damage conditions, are collected using acceleration sensors to obtain measured structural response data with damage condition labels. A large volume of structural response simulation data and a small amount of structural response measurement data are used as input, and the initial pre-trained model for structural damage recognition is trained twice using the composite loss function and transfer learning techniques.

To mitigate the impact of feature distribution differences between heterogeneous data on transfer learning, a heterogeneous data alignment module is introduced. This module facilitates the effective transfer of the model's damage recognition capability from simulated to measured data, resulting in a final model capable of accurately recognizing structural damage in measured data.

In this study, Jensen-Shannon (JS) divergence is used to measure the difference between the distributions of heterogeneous data in the high-dimensional feature space. JS divergence is a commonly employed method for quantifying the difference between two probability distributions and can effectively assess the distributional differences between simulated and measured data. Compared to Kullback-Leibler (KL) divergence, JS divergence is symmetric and more stable, as its value is always finite. When evaluating feature distributions between simulated and measured data, JS divergence provides a more accurate reflection of the distributional differences in the high-dimensional feature space, thereby enhancing the model's generalization ability for cross-data source tasks. The formulation of the loss function for heterogeneous data alignment based on JS divergence is shown in Eq. 2.

$$JS(P\|Q) = \frac{1}{2}KL(P\|M) + \frac{1}{2}KL(Q\|M) \tag{2}$$

where P and Q denote the probability distributions of simulated and measured data features in the high-dimensional feature space, respectively. M denotes the average distribution of P and Q, and KL refers to the Kullback-Leibler (KL) divergence, as expressed in Eq. 3.

$$KL(P\|Q) = \sum P \cdot \ln \frac{P}{Q} \tag{3}$$

In the process of training the final model for structural damage identification, a composite loss function is constructed by adding the heterogeneous data alignment loss to the original cross-entropy classification loss. The composite loss function is shown in Eq. 4.

$$L = L_c(X_L, Y) + \lambda \cdot JS(P\|Q) \tag{4}$$

Where L denotes the composite loss function, L_c represents the cross-entropy classification loss function, X_L and Y denote the predicted and real values of the structural damage in the structural response data, respectively, and λ are the weight hyperparameters of the heterogenous data alignment loss function.

During the training process, the model simultaneously optimizes both the classification loss and the heterogeneous data alignment loss. This approach enables the distribution of the heterogeneous data in the feature space to be aligned, ultimately achieving high-accuracy recognition of structural damage information in the measured data.

3. Case Study: IASC-ASCE Benchmark Structural Model

To demonstrate the feasibility and advantages of the proposed method, this study uses the IASC-ASCE benchmark structural model as a case study [7]. The IASC-ASCE benchmark model is a 4-

Structural Health Monitoring: 10APWSHM   Materials Research Forum LLC
Materials Research Proceedings 50 (2025) 105-112   https://doi.org/10.21741/9781644903513-12

story, 2-span × 2-span steel frame, with a floor height of 0.9 m. The structure is equipped with eight diagonal braces on each floor. Each floor consists of four floor slabs, each measuring 2.5 m × 2.5 m. The numerical simulation model is shown in Fig. 1, and the experimental model is presented in Fig. 2. The test was conducted with three acceleration sensors placed on the ground floor and on each of the floors (1 through 4) of the structure.

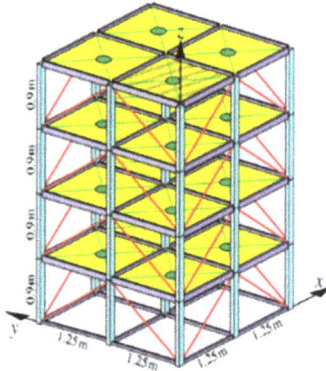

Fig. 1. The numerical simulation model

Fig. 2. The experimental model

Different structural damage conditions were simulated by removing the diagonal braces during the test. These included one non-damage condition and five distinct damage scenarios, each with varying damage locations and severity, resulting in a total of six damage conditions. The specific details of these conditions are provided in Table 1.

Structural Health Monitoring: 10APWSHM   Materials Research Forum LLC
Materials Research Proceedings 50 (2025) 105-112   https://doi.org/10.21741/9781644903513-12

Table 1. Damage Scenarios

Table 1. Case	Table 2. Condition of damage
Table 3. damage 0	Table 4. Safe
Table 5. damage 1	Table 6. Remove east side diagonal bracing on all floors
Table 7. damage 2	Table 8. Remove south-east side diagonal bracing on all floors
Table 9. damage 3	Table 10. Remove south-east side diagonal bracing on the first and fourth floors
Table 11. damage 4	Table 12. Remove southeast side diagonal bracing on ground floor
Table 13. damage 5	Table 14. Remove north diagonal braces on the ground floor and east diagonal brace on all floors.

The order of the AR model was set to 10, and the data size of each sample was 10×12, where 10 represents the number of AR orders and 12 denotes the number of accelerometers. For the structural response simulation data, 200 samples were collected for each damage condition, resulting in a total of 1200 simulation data samples used for model pre-training across the six conditions. The dataset was then split into a training set (840 samples), a validation set (112 samples), and a test set (240 samples) in a 7:2:1 ratio. For the structural response measured data, 120 samples per damage condition were used, yielding a total of 720 measured data samples. These were divided into a training set (504 samples), a validation set (72 samples), and a test set (144 samples), also in a 7:2:1 ratio.

The t-SNE algorithm is applied to downscale the heterogeneous data from the high-dimensional feature space to a two-dimensional feature space before and after implementing the heterogeneous data alignment strategy. Fig. 3 and 4 illustrate the feature distributions of the two data sources, both before and after applying the heterogeneous data alignment strategy, respectively. A comparative analysis clearly demonstrates that after applying the heterogeneous data alignment strategy, the simulated structural response data are more closely aligned with the measured structural response data in the feature space. This effectively reduces the distance between the two, achieving accurate alignment of the heterogeneous data.

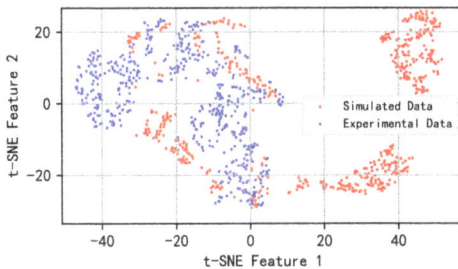

Fig. 3. Distribution of data features after applying the heterogeneous data alignment strategy

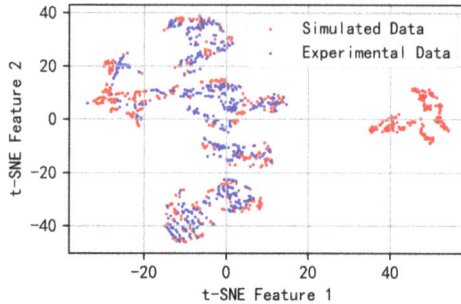

Fig. 4. Distribution of data features without applying the heterogeneous data alignment strategy

The damage identification performance of the model was evaluated using precision and recall, with the results presented in Table 2. The model without the heterogeneous data alignment strategy achieved a precision rate of 93.14% and a recall rate of 93.30%. The corresponding confusion matrix for the classification results on the test set is shown in Fig. 5. In contrast, the model utilizing the heterogeneous data alignment strategy showed significant improvement, with a precision rate of 99.31% and a recall rate of 99.33%. The confusion matrix for the classification results on the test set is shown in Fig. 6. These results indicate that the introduction of the heterogeneous data alignment strategy significantly enhances the model's damage identification performance on the structural response test data, further confirming the effectiveness of the proposed technical solution.

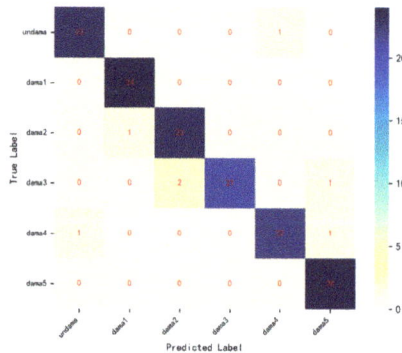

Fig. 5. Classification results without heterogeneous data alignment strategy

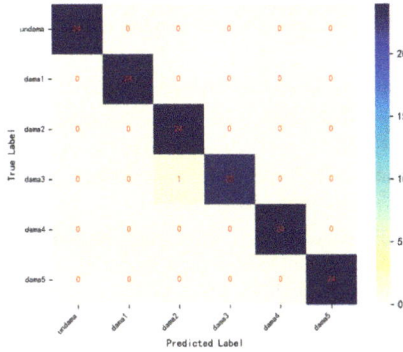

Fig. 6. Classification results using heterogeneous data alignment strategy

Table 2. Comparison of model performance before and after applying heterogenous data alignment strategy

Table 15. Model	Table 16. accuracy	Table 17. recall rate
Table 18. Before applying heterogenous data alignment strategy	Table 19. 93.14%	Table 20. 93.30%
Table 21. After applying heterogenous data alignment strategy	Table 22. 99.31%	Table 23. 99.33%

Conclusion

In this study, a structural damage identification method based on transfer learning and heterogeneous data alignment is proposed. This method uses physically simulated damage to generate a large dataset of structural response data that closely resembles real damage, which is then used to design and train the initial model for damage identification. The model is subsequently adapted to a smaller amount of actual structural response data through the heterogeneous data alignment technique, followed by further training to create the final model. This approach ensures accurate damage recognition for real-world scenarios. The proposed method overcomes the limitations of traditional deep learning-based damage identification methods, significantly enhances the generalization capability of the model, and makes a valuable contribution to the damage identification of real engineering structures.

Acknowledgements

The authors are grateful for the financial support from the Shenzhen Municipal Natural Science Foundation Basic Research General Project (Grant No. JCYJ20240813142818025), Guangdong Provincial Natural Science Foundation General Project (Grant No. 2023A1515011376).

References

[1] Dong C Z, Catbas F N. A review of computer vision-based structural health monitoring at local and global levels[J]. Structural Health Monitoring, 2021, 20(2): 692-743. https://doi.org/10.1177/1475921720935585

[2] Cha Y J, Choi W, Büyüköztürk O. Deep learning-based crack damage detection using convolutional neural networks[J]. Computer-Aided Civil and Infrastructure Engineering, 2017, 32(5): 361-378. https://doi.org/10.1111/mice.12263

[3] Kim B, Cho S. Image-based concrete crack assessment using mask and region-based convolutional neural network[J]. Structural Control and Health Monitoring, 2019, 26(8): e2381. https://doi.org/10.1002/stc.2381

[4] LeCun Y, Bengio Y, Hinton G. Deep learning[J]. nature, 2015, 521(7553): 436-444. https://doi.org/10.1038/nature14539

[5] Figueiredo E, Figueiras J, Park G, et al. Influence of the autoregressive model order on damage detection[J]. Computer-Aided Civil and Infrastructure Engineering, 2011, 26(3): 225-238. https://doi.org/10.1111/j.1467-8667.2010.00685.x

[6] Atha D J, Jahanshahi M R. Evaluation of deep learning approaches based on convolutional neural networks for corrosion detection[J]. Structural Health Monitoring, 2018, 17(5): 1110-1128. https://doi.org/10.1177/1475921717737051

[7] Johnson E A, Lam H F, Katafygiotis L S, et al. Phase I IASC-ASCE structural health monitoring benchmark problem using simulated data[J]. Journal of engineering mechanics, 2004, 130(1): 3-15. https://doi.org/10.1061/(ASCE)0733-9399(2004)130:1(3)

Structural Health Monitoring: 10APWSHM
Materials Research Proceedings 50 (2025) 113-118

Materials Research Forum LLC
https://doi.org/10.21741/9781644903513-13

Application of embedded fiber-optic distributed strain sensing for detection of matrix cracks in CFRP laminates

Shin-ichi Takeda[1,a] *, Yukino Ikeda[2,b], Shinsaku Hisada[1,c] and Toshio Ogasawara[2,d]

[1]6-13-1 Osawa, Mitaka-shi, Tokyo 181-0015, JAPAN, Aviation Technology Directorate, Japan Aerospace Exploration Agency

[2]2-24-16 Naka-cho, Koganei-shi, Tokyo 184-8588 JAPAN, Department of Mechanical Systems Engineering, Tokyo University of Agriculture and Technology

[a]takeda.shinichi@jaxa.jp, [b]ikeda-y@st.go.tuat.ac.jp, [c]hisada.shinsaku@jaxa.jp, [d]ogasat@cc.tuat.ac.jp

Keywords: Damage Detection, Fiber-Optic Sensors, Distributed Strain Sensing, CFRP Laminates

Abstract. FBG sensors with a gauge length of 100 mm were embedded in CFRP laminate specimens and monitored by the OFDR-FBG distributed strain measurement system. When a tensile load was applied to the specimen, matrix cracks were generated internally and their number increased. As the cracks developed, a peak appeared in the distributed strain. Soft X-ray inspection and optical microscopy of the specimen edge showed good agreement between the location of the cracks and the peaks. The finite element method was used to calculate the internal strain change during crack initiation, and its effect on the distributed strain was discussed. The peaks were clearly visible when the crack density was low, and when the crack density was high, the peaks were difficult to see due to peak overlap. With the resolution of the system and sensor used in this study, it is practically possible to distinguish areas where cracks have accumulated from areas where they have not accumulated. The proposed matrix crack detection method can also be applied to cases using other types of fiber optic distributed strain measurement systems. Given the widespread use of machine learning and other analytical techniques for measurements, the use of distributed strain is not limited to determining the number of peaks and can have many possibilities.

Introduction

CFRP is expected to be used in the construction of liquid hydrogen (LH2) tanks for hydrogen-powered aircraft and reusable space transportation systems due to its superior strength, stiffness, and thus lightweight properties. At liquid hydrogen temperatures (around -253°C), the difference in thermal expansion coefficients of each layer causes large thermal stresses, and microscopic damage such as matrix cracking occurs in the low strain region [1]. As matrix cracks can become a pathway for gas leakage, it is important to monitor their occurrence and evolution.

Fiber Bragg Grating (FBG) sensors, a type of optical fiber sensor, can measure strain and temperature with high accuracy, and their application in SHM is being considered [2]. As a method of distributed strain measurement using FBG sensors, distributions in gauges with high spatial resolution have been developed [3]. In earlier studies using OFDR-FBG, the application has been limited to the measurement of spatial strain distributions on the order of tens of millimeters in actual aircraft structures [4], and there are no reports on the measurement of matrix crack initiation and propagation in CFRP laminates. Therefore, the aim of this study was to verify the possibility of detecting matrix crack initiation and propagation by measuring the strain distribution in an orthogonal CFRP laminate with an FBG sensor using OFDR-FBG under room temperature, high temperature and low temperature environments.

Structural Health Monitoring: 10APWSHM   Materials Research Forum LLC
Materials Research Proceedings 50 (2025) 113-118   https://doi.org/10.21741/9781644903513-13

Distributed measurement using an OFDR system [5]

A schematic of the OFDR system used for the distributed measurement is presented in Fig. 1. The OFDR system consisted of a tunable laser source and two interferometers. Swept light entered each interferometer, with each returning a different interference signal. The reference interferometer returned an interference signal of the light reflected from Reflectors 1 and 2. The interference signal I_{PD1} detected by Photodiode 1 is expressed as follows:

$$I_{PD1} \propto \cos(2n_{eff}L_r k), \tag{1}$$

where L_r and k represent the distance between Reflectors 1 and 2 and the wavenumber, respectively. L_r was set to 80 m in the system adopted in this study. This interference signal triggered the acquisition of the main interferometer signal. The main interferometer returned an interference signal of the reflected light from Reflector 3 and from each position of the FBG sensor. The interference signal I_{PD2} detected by Photodiode 2 is expressed as follows:

$$I_{PD2} \propto \sum R_g(z_i, k) \cos(2n_{eff}z_i k), \tag{1}$$

where z_i and R_g represent the position of the i-th grating and the reflection spectrum of the grating, respectively. By applying a short-time Fourier transform (STFT) to this signal, the intensity at each position and wavelength could be obtained. Therefore, the strain and temperature distributions could be calculated from the Bragg wavelength shift at each position.

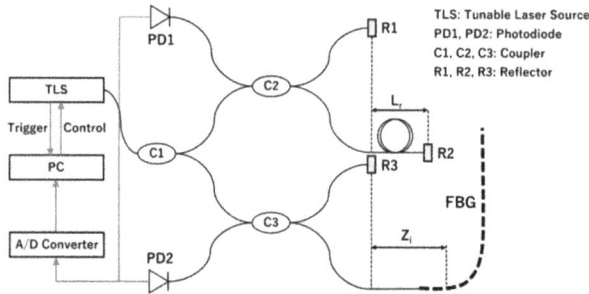

Fig. 1 Schematic of the OFDR system used for distribution measurement.

Experiment

The material used was a 130 °C curing unidirectional carbon fiber/epoxy prepreg (T700G / #2510, Toray). The specimen dimensions were 250 mm x 20 mm, and the laminate configuration was a $[0_4/OF/90_8/0_4]$ orthogonal laminate (16 plies). As shown in Figure 2, an FBG sensor (gauge length 100 mm, Fujikura Electric Wire) was embedded in the 0° ply so that it was in contact with the 90° ply. The thickness of the test plate after molding was 2.5 mm. Load/unload tests at room temperature (25°C), high temperature (80°C), and low temperature (-50°C) were performed on the fabricated test plate, and matrix cracks were generated in the 90° layer. A hydraulic fatigue testing machine (Model 8804, INSTRON) was used, and the test was displacement-controlled (loading 0.5 mm/min, unloading -1.0 mm/min). A strain gauge (5 mm gauge length, Kyowa Electronic Instruments) and a K-type thermocouple were attached to the center of the specimen. A tunable laser light source (TLB-8800, Venturi) and an OFDR interferometer (LAOFDR1500C1, LAZOC) were used for measurement. The wavelength resolution of the OFDR interferometer is 0.05 nm,

Structural Health Monitoring: 10APWSHM
Materials Research Proceedings 50 (2025) 113-118

Materials Research Forum LLC
https://doi.org/10.21741/9781644903513-13

and the position resolution is 1.0 mm. The occurrence and development of matrix cracks were also evaluated by lateral optical microscopy and X-ray transmission.

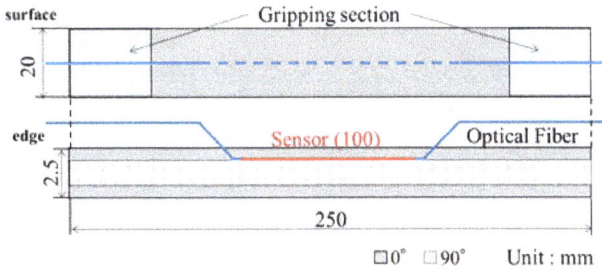

Fig. 2 Schematic of specimen with embedded FBG sensor.

Results and discussion

Figure 3(a) shows a comparison between the distribution strain obtained from the FBG sensor and the crack position when the strain in the evaluation section reaches 6,000 µε at room temperature and the load is then released to zero. The red line shows the position of the matrix crack identified by optical microscope observation from the side. The crack location and the strain data peak are in good agreement. The reason for the positive strain peak at the crack location is believed to be that the residual strain from the compression applied to the 90° layer of the sample during molding was released due to the occurrence of the crack.

Figure 3(b) shows the results of loading the specimen until the strain reached 12,000 µε, then unloading and measuring. As the crack density increases, the spatial distribution of the strain measured by the FBG sensor becomes flatter, making it difficult to identify the crack positions. This is because the strain distribution between cracks becomes smooth as the number of cracks increases.

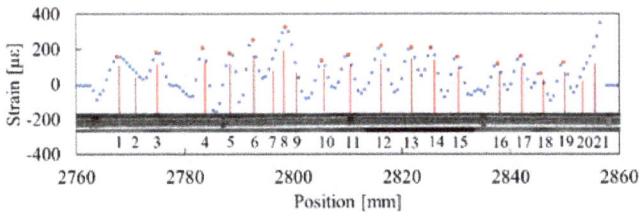

(a) 6,000 µε

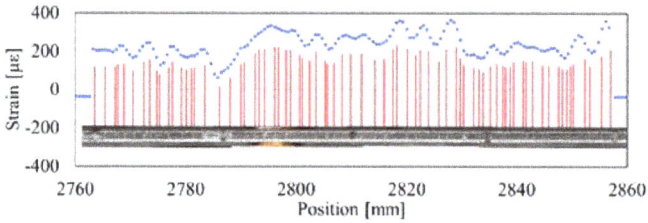

(b) 12,000 με

Fig. 3 Comparison of distributed strain and crack location at 25 °C (unloading).

The temperature of the specimen was raised to 80 °C or lowered to -50 °C, and the strain in the evaluation section was loaded until it reached 10,500 με and 6,800 με, respectively, and then the distribution strain obtained from the FBG sensor with the load removed is shown in Fig. 4(a) and (b), respectively. The reference (zero value) for the strain is the state at room temperature and before loading. From these results, the peak of the strain data corresponds well with the position of the matrix crack even at 80 °C and -50 °C.

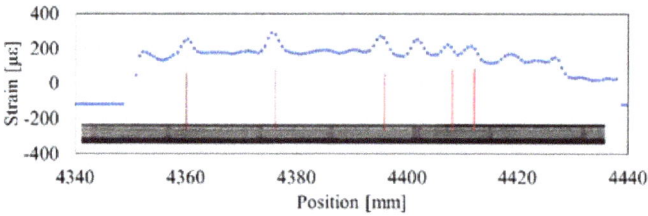

(a) 10,500 με, 80°C

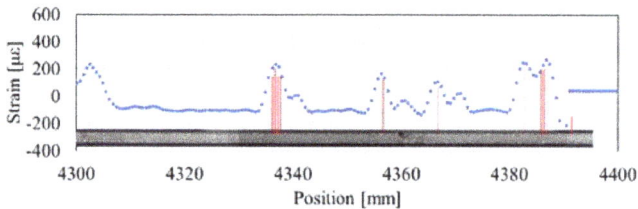

(b) 6,800 με, -50 °C

Fig. 4 Comparison of distributed strain and crack location at 80 °C and -50 °C (unloading).

Optical microscope observation showed that multiple cracks occurred in a specific location at -50 °C. The area where the cracks were concentrated was detected as a single peak because the largest crack spacing was 0.78 mm and the cracks could not be detected as individual cracks due

Structural Health Monitoring: 10APWSHM
Materials Research Proceedings 50 (2025) 113-118

Materials Research Forum LLC
https://doi.org/10.21741/9781644903513-13

to the distributed strain. Focusing on the strain values of the peaks, they are about 300 με at room temperature, about 100 με at 80 °C, and about 300-400 με at -50 °C, and vary with the test temperature. This is because the greater the difference between the molding temperature (130 °C) and the test temperature, the greater the thermal residual stress at the time of molding and the greater the amount of compressive stress released by the occurrence of cracks.

Fig. 5 shows the comparison of the crack position obtained from the microscopic observation image with the peak position of the distributed strain. Crack 1 indicates the origin; the distance from the origin is plotted as the position (circular symbols), and the errors are plotted as triangular symbols. Cracks 3, 5, 8, and 21 have measurement errors greater than 2%; however, most of the errors are less than 2%, showing reasonable accuracy.

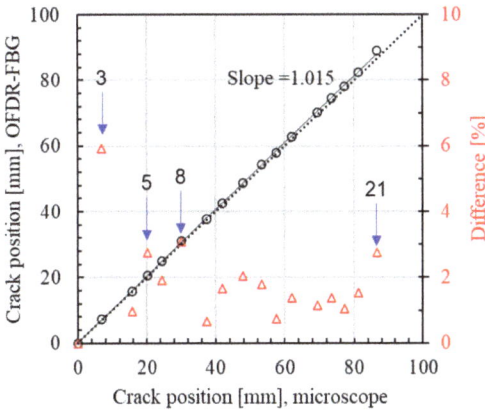

Fig. 5 Comparison of crack positions at 25 °C.

In the LH2 tank structure, the aim is to monitor the occurrence and evolution of matrix cracks at extremely low temperatures, so OFDR-FBG would be a promising sensing method. In the future, we plan to further verify the possibility of measurement at extremely low temperatures and under biaxial loading.

Conclusions

The possibility of detecting the occurrence and evolution of matrix cracks was verified by measuring the strain distribution at room temperature, 80 °C, and -50 °C using OFDR-FBG for CFRP cross-ply laminates. The results showed that the occurrence of matrix cracks could be detected as strain peaks in regions of low crack density. The accuracy of crack location was also sufficient for practical use.

References

[1] H. Kumazawa, T. Aoki and I. Susuki, Analysis and Experiment of Gas Leakage through Composite Laminates for Propellant Tanks, AIAA Journal 41 (2003) 2037-2044. https://doi.org/10.2514/2.1895

[2] S. Minakuchi and N. Takeda, Recent Advancement in Optical Fiber Sensing for Aerospace Composite Structures, Photonic Sensors Vol. 3 No. 4 (2013) 345–354. https://doi.org/10.1007/s13320-013-0133-4

[3] H. Igawa, K. Ohta, T. Kasai, I. Yamaguchi, H. Murayama, K. Kageyama, Distributed Measurements with a Long Gauge FBG Sensor Using Optical Frequency Domain Reflectometry, Journal of Solid Mechanics and Materials Engineering 2 (2008) 1242-1252. https://doi.org/10.1299/jmmp.2.1242

[4] D. Wada, H. Igawa, M. Tamayama, T. Kasai, H. Arizono and H. Murayama, Flight demonstration of aircraft wing monitoring using optical fiber distributed sensing system, Smart Materials and Structures 28 (2019) 055007. https://doi.org/10.1088/1361-665X/aae411

[5] S. Hisada, U. Kodakamine, D. Wada, H. Murayama and H. Igawa, Simultaneous Measurement of Strain and Temperature Distributions Using Optical Fibers with Different GeO_2 and B_2O_3 Doping, Sensors Vol. 23 No. 3 (2023) 1156. https://doi.org/10.3390/s23031156

Structural Health Monitoring: 10APWSHM
Materials Research Proceedings 50 (2025) 119-126

Materials Research Forum LLC
https://doi.org/10.21741/9781644903513-14

Online inverse solution for deep learning-based prognostics

Tianzhi Li[1,a] *, Morteza Moradi[3,b], Ming Xiao[2,c], Lihui Wang[1,d]

[1]Department of Production Engineering, KTH Royal Institute of Technology, 10044 Stockholm, Sweden

[2]Division of Information Science and Engineering, KTH Royal Institute of Technology, 10044 Stockholm, Sweden

[3]Center of Excellence in Artificial Intelligence for structures, prognostics & health management, Aerospace Engineering Faculty, Delft University of Technology, Delft, 2629 HS, the Netherlands

[a]tianzhil@kth.se, [b]M.Moradi-1@tudelft.nl, [c]mingx@kth.se, [d]lihuiw@kth.se

Keywords: Structural Health Monitoring, Remaining Useful Life Prediction, Inverse Solution, State and Parameter Estimation, Multiple Local Particle Filter

Abstract. Data-driven prognostic models have been extensively utilized in current structural health monitoring (SHM) practices. They are designed to provide the health indicator (HI) - a representation of the system's current health state - through sensor data. To enhance performance, online learning is often used to take care of uncertainties that arise from the run-to-failure process. The inverse solution, though demonstrated in online uncertainty quantification applications, remains unexplored in the context of online data-driven prognostics. Therefore, this work proposes a generic inverse solution for a deep prognostic model to online address uncertainties. The proposed method is tested using the open-access XJTU-SY bearing datasets, showcasing its capacity to online enhance the performance of a given model.

Introduction

Structural health monitoring (SHM) has been extensively investigated across various engineering application scenarios, such as in metals [1], composites [2], or rotating machinery [3-5]. SHM encompasses four key levels: damage detection [6], isolation or localization [7, 8], identification [9], and prognostics [1], efficiently ensuring the integrity and safety of engineering structures. Regarding prognostics, it often consists of three main sequential steps [1]: defining the damage state or health indicator (HI) to assess potential failure, building the prognostic models to calculate the RUL, and (iii) refining these models to improve the RUL prediction accuracy.

The first task typically involves characterizing the structure's health status by either a physics-based damage state or a data-driven HI. The definition of a physics-based state varies: for metal, it can be crack length [1, 10-12] or crack shape [13]; for composites, matrix cracking density [14], delamination length [15], or delamination shape [16, 17]; and for gears, crack length, pitting level, or wear depth [18]. On the other hand, a data-driven HI is often obtained by online structural health monitoring (SHM) data, such as acceleration [19, 20], strain [21], guided waves [22], or acoustic emission [23]. Deep models are often designed to provide advanced HI through sensor data, involving two steps. First, a function that involves the service time and end-of-life (EOL), such as the square ratio of the current time to EOL, is designed as a HI label simulator. Second, a deep model is trained by simulated labels and sensor data [22, 23]. During the testing phase, the model output DHI can directly produce EOL or RUL, typically without future HI projections [24, 25]. Therefore, this DHI construction model is often considered the prognostic model, though it does not describe the HI evolution with time.

Given the uncertainties arising from factors such as complex structure degradation and environmental influences [26], when the same model is applied to several identical specimens undergoing identical run-to-failure tests, significant variations in prognostic performance will occur. Therefore, online updating is often necessary to take care of the uncertainties arising from the degradation process. Regarding the HI-based model, since the EOL cannot be obtained during the run-to-failure process, it is rarely possible to acquire the true label for each data stream in real-time. As a result, updating this deep model with the latest data represents a typical unsupervised online learning problem, which only received a few investigations, i.e., online transfer learning [27-29] and online incremental learning [30].

These investigations [27-30] have provided successful online prognostic solutions, and they all consider data-driven prognostics as a mapping problem - progressing sequentially from sensor measurement, through the prognostic model, and finally resulting in the HI or RUL prediction. On the other hand, inverse solution has been extensively demonstrated in online uncertainty quantification [16], with the potential to be applied to a deep prognostic model for addressing degradation uncertainties. This, however, has not been explored.

In this context, this work proposes the a generic inverse solution for a given deep prognostic model to online address uncertainties. First, a prognostic model is constructed using a user-defined modeling strategy. Next, a state-space model is developed by incorporating the prognostic model and prior information. Finally, state and parameter estimation is performed to generate the HI posterior. The proposed method is tested using the open-access XJTU-SY bearing datasets, showcasing its capacity to enhance the performance of a given prognostic model.

The rest of this paper is organized as follows: Section 2 introduces the proposed method. Section 3 provides the results of the proposed method applied to the XJTU-SY bearing datasets, respectively. Finally, Section 4 concludes this paper.

Proposed method

The proposed inverse solution with both the offline and online phases. The offline phase involves the development of four models: prognostic model (PM), feature extraction model (FEM), measurement model (MM), and state space model (SSM). The online phase focuses on the state and parameter estimation with the latest measurement from the FEM model.

This section introduces the development of four models: training the PM, defining the FEM, training the MM, and formulating the SSM. To train the first model, a function needs to generate the HI labels. For example, the HI can be defined by a linear function as the ratio of current time to EOL:

$$x_k = \frac{t_k}{EOL} \tag{1}$$

where x, k, and t denote the HI, time step, and service time, respectively. Then, given the simulated labels, a data-driven model as PM can be built to link the sensor data u and the HI x as:

$$x_k = f\left(u_k\right) \tag{2}$$

Then, a FEM model is firstly split from the PM excluding certain last layers, as follows:

$$Y_k = f_s\left(u_k\right) \tag{3}$$

In this study, FEM is considered as PM without only the last layer, i.e., the HI output layer. As a result, the FEM's output Y is a vector of the neurons of PM last hidden layer. By using the HI labels as input and the FEM's output as output, the MM can be constructed as follows:

$$Y_k = g\left(x_k\right) \tag{4}$$

Then, by combining Eqs. (1) and (4), the SSM can be formulated as:

Structural Health Monitoring: 10APWSHM   Materials Research Forum LLC
Materials Research Proceedings 50 (2025) 119-126   https://doi.org/10.21741/9781644903513-14

$$\begin{cases} X_k = \begin{bmatrix} \theta_k \\ EOL_k \\ x_k \end{bmatrix} = \begin{bmatrix} \theta_{k-1} + \omega_{\theta,k} \\ EOL_{k-1} + \omega_{e,k} \\ t_k / EOL_k + \omega_{x,k} \end{bmatrix} \\ Y_k = g(x_k, \theta_k) + v_k \end{cases} \tag{5}$$

where θ is a vector of the model parameters within the measurement equation, ω_θ, ω_e, and ω are the process noises for the model parameters, EOL and HI, respectively, and v is the measurement noise. Moreover, certain prior information can be included, such as:

$$k \le EOL_k \le EOL_{max} \tag{6}$$

which means EOL should always lie within the range between the current service time and the maximum EOL. The EOL should be adjusted to the nearest boundary when falling out of the range.

The SSM Eq. (5) is developed based on the PM Eq. (2), while it can provide more accurate prognostic performance, as it incorporates additional prior knowledge, and takes care of the uncertainties arising from the degradation process.

The online phase includes the feature extraction and the state and parameter estimation. The latest sensor data should be processed through the aforementioned FEM to extract specific features, which serve as measurements in the state space model. Finally, a state and parameter estimation algorithm will be used to provide the HI posterior.

Given that Eq. (5) is often high-dimensional and nonlinear, because of the parameter vector and the prognostic and measurement models, respectively. A high-dimensional system identification method, i.e., multiple local particle filter [16], is used in this study for online state and parameter estimation.

Testing with XJTU-SY Bearing Datasets

The XJTU-SY bearing datasets [19] are used in this study. Figure 1 depicts one bearing under the run-to-failure test. Two PCB 352C33 accelerometers are placed on the bearing housing to collect the vertical and horizontal accelerations at 60-second intervals. The sampling frequency and duration are 25.6 kHz and 1.28 seconds, respectively. Table 1 presents one operating condition from the datasets, involving tests on five bearings. For safety reasons, testing for each bearing is stopped once the vibration amplitude exceeds 20 g. Consequently, the time when the vibration amplitude exceeds the threshold is considered the EOL. Due to the uncertainties inherent in the degradation process, bearings may have different failure modes, including inner race wear, outer race wear, and outer race fractures, which result in varying EOLs. One can refer to [19] for further experimental details.

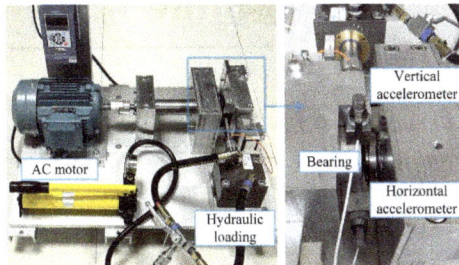

Figure 1: XJTU-SY Bearing run-to-failure test.

Structural Health Monitoring: 10APWSHM | Materials Research Forum LLC
Materials Research Proceedings 50 (2025) 119-126 | https://doi.org/10.21741/9781644903513-14

Table 1: Operating condition.

Condition	Rotating speed [rpm]	Radial force [kN]	Bearing specimen	EOL [Minute]
I	2100	12	1_1, 1_2, 1_3, 1_4, 1_5	123, 161, 158, 122, 52

Cross-validation will be performed for each bearing under Condition I. Specifically, one bearing will be designated for testing, while the remaining four bearings will be utilized for modeling. Table 2 lists the prognostic and measurement models used for each testing scenario. For each, two CNN-based PMs are separately used for testing, and their layouts are defined as 'Layout 1' and 'Layout 2', respectively. Then, the FEM, MM, and SSM have to be developed based on each PM.

Table 2: Prognostic and measurement models used for XJTU-SY datasets.

	Input	Output	Layout
Prognostic model 1 (PM1)	Raw data	HI	CNN 1
Prognostic model 2 (PM2)	Raw data	HI	CNN 2
Measurement model 1 (MM1)	HI	FEM 1 output	MLP
Measurement model 2 (MM2)	HI	FEM 2 output	MLP

The performance of the proposed method is first tested by applying PM1 to the bearings under Condition I. Three separate routines are compared:

- PM1: PM1 is used to provide HI prediction results.
- PM1 (P): The PM1 results are modified by the prior knowledge. Specifically, at each time step, the HI is set as the ratio of service time to the maximum EOL when it is lower than the ratio, and it is set as one when it is above the value of one.
- New: The proposed method is developed based on the given PM.

The parameter estimation results using the new approach are shown in Figures 2 (a) - (e). As the estimation process progresses, the spread of parameter samples decreases and converges around final values, indicating satisfactory convergence. The HI predictions using both the PM1 and the new approach are presented in Figures 2 (f) - (j). The new approach yields significantly smoother and more accurate results, highlighting its superiority in improving prognostic performance. Then, the results from the above three routines are evaluated by three performance metrics: root-mean-square error (RMSE), mean absolute percentage error (MAPE), and mean absolute error (MAE). Results from the three above routines are given in Table 3. The proposed method consistently achieves lower RMSE, MAPE, and MAE values, demonstrating its ability to enhance the prognostic performance of the given model PM1.

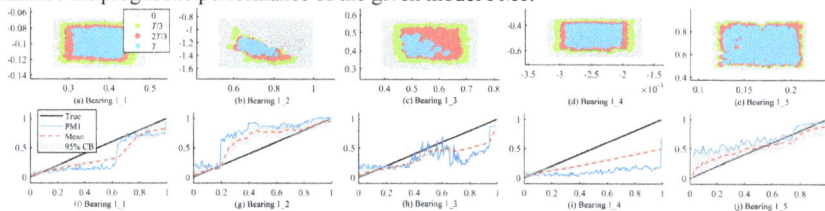

Figure 2: Results of PM1 and new methods for bearings under Condition I.

Note: (i) 'T' means the number of total time steps, (ii) the x and y axes for parameter estimation are two model parameters, while those for HI prediction are 'Service time / True EOL', and 'HI', respectively, and (iii) 'Mean' and '95% confidence boundary (CB)' are derived from the results of new method.

Structural Health Monitoring: 10APWSHM   Materials Research Forum LLC
Materials Research Proceedings 50 (2025) 119-126   https://doi.org/10.21741/9781644903513-14

Table 3: Prognostic performances of PM1 and new methods for bearings under Condition I.

	Bearing	1_1	1_2	1_3	1_4	1_5
RMSE	PM1	0.216	0.285	0.256	0.466	0.204
	PM1 (P)	0.173	0.284	0.188	0.301	0.204
	New	0.172*	0.171*	0.186*	0.291*	0.097*
MAPE	PM1	46.6	112.7	82.6	86.3	130.2
	PM1 (P)	41.0	112.6	77.0	61.7	130.2
	New	31.3*	37.4*	33.6*	45.2*	42.6*
MAE	PM1	0.174	0.237	0.196	0.402	0.163
	PM1 (P)	0.147*	0.235	0.155	0.261	0.163
	New	0.147*	0.125*	0.153*	0.247*	0.088*

Note: The symbol '*' denotes 'best performance among the three routines'.

The same validation is conducted on the five bearings using PM2. Three routines are included: PM2, PM2 (P), and the new method. The estimation and prognostic results are given in Figure 3 and Table 4. Although a different PM can yield slightly different prognostic performance for the same bearing, the proposed method can always have the capacity to improve the performance of the given model.

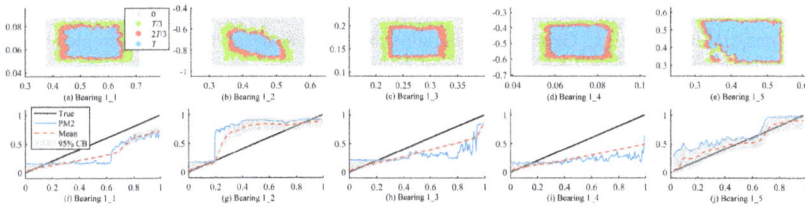

Figure 3: Results of PM2 and new methods for bearings under Condition I.

Note: (i) '*T*' means the number of total time steps, (ii) the *x* and *y* axes for parameter estimation are two model parameters, while those for HI prediction are 'Service time / True EOL', and 'HI', respectively, and (iii) 'Mean' and '95% confidence boundary (CB)' are derived from the results of new method.

Table 4: Prognostic performances of PM2 and new methods for bearings under Condition I.

	Bearing	1_1	1_2	1_3	1_4	1_5
RMSE	PM2	0.244	0.301	0.285	0.379	0.184
	PM2 (P)	0.213	0.301	0.206	0.299	0.184
	New	0.212*	0.211*	0.198*	0.295*	0.068*
MAPE	PM2	75.0	108.0	97.9	86.4	107.6
	PM2 (P)	71.0	108.0	90.6	79.0	107.6
	New	38.7*	44.2*	34.4*	47.5*	23.0*
MAE	PM2	0.214	0.240	0.235	0.314	0.154
	PM2 (P)	0.192	0.240	0.182	0.258	0.154
	New	0.188*	0.160*	0.168*	0.250*	0.059*

Note: The symbol '*' denotes 'best performance among the three routines'.

Conclusion
Data-driven prognostic models must be continuously updated to account for uncertainties in the degradation process. This work introduces a novel inverse solution to address these uncertainties in real time. Based on prognostic results from the open-access XJTU-SY bearing datasets, the following conclusions can be drawn: Incorporating prior knowledge, such as the maximum EOL

of certain specimens, can significantly enhance prognostic performance, even with simple online adjustments. For further improvements, it is essential to employ prognostic and measurement models to construct a state-space model, integrating prior knowledge for real-time state and parameter estimation. The effectiveness of the proposed method has been shown to rely on the proper utilization of prior information and measurement models. This work has only adopted a very simple prior, i.e., a constant maximum EOL. One may consider leveraging more advanced priors, such as the degradation-related physical constraints, or incorporating a physics-informed neural network into the proposed method.

References

[1] T. Li, Particle filter-based fatigue damage prognosis using prognostic-aided model updating, Mechanical Systems and Signal Processing, 211 (2024) 111244. https://doi.org/10.1016/j.ymssp.2024.111244

[2] L. Lomazzi, R. Junges, M. Giglio, F. Cadini, Unsupervised data-driven method for damage localization using guided waves, Mechanical Systems and Signal Processing, 208 (2024) 111038. https://doi.org/10.1016/j.ymssp.2023.111038

[3] J. Tian, D. Han, H.R. Karimi, Y. Zhang, P. Shi, A universal multi-source domain adaptation method with unsupervised clustering for mechanical fault diagnosis under incomplete data, Neural Networks, (2024) 106167. https://doi.org/10.1016/j.neunet.2024.106167

[4] D. Yang, H.R. Karimi, L. Gelman, An explainable intelligence fault diagnosis framework for rotating machinery, Neurocomputing, 541 (2023) 126257. https://doi.org/10.1016/j.neucom.2023.126257

[5] R. Zhong, Y. Feng, P. Li, X. Wu, A. Guo, A. Zhang, C. Li, Uncertainty-aware nuclear power turbine vibration fault diagnosis method integrating machine learning and heuristic algorithm, IET Collaborative Intelligent Manufacturing, 6 (2024) e12108. https://doi.org/10.1049/cim2.12108

[6] C. Lai, P. Baraldi, E. Zio, Physics-Informed deep Autoencoder for fault detection in New-Design systems, Mechanical Systems and Signal Processing, 215 (2024) 111420. https://doi.org/10.1016/j.ymssp.2024.111420

[7] J. Tian, D. Han, M. Li, P. Shi, A multi-source information transfer learning method with subdomain adaptation for cross-domain fault diagnosis, Knowledge-Based Systems, 243 (2022) 108466. https://doi.org/10.1016/j.knosys.2022.108466

[8] R. Junges, Z. Rastin, L. Lomazzi, M. Giglio, F. Cadini, Convolutional autoencoders and CGANs for unsupervised structural damage localization, Mechanical Systems and Signal Processing, 220 (2024) 111645. https://doi.org/10.1016/j.ymssp.2024.111645

[9] X. Zhou, C. Sbarufatti, M. Giglio, L. Dong, A fuzzy-set-based joint distribution adaptation method for regression and its application to online damage quantification for structural digital twin, Mechanical Systems and Signal Processing, 191 (2023) 110164. https://doi.org/10.1016/j.ymssp.2023.110164

[10] J. Chen, S. Yuan, C. Sbarufatti, X. Jin, Dual crack growth prognosis by using a mixture proposal particle filter and on-line crack monitoring, Reliability Engineering & System Safety, 215 (2021) 107758. https://doi.org/10.1016/j.ress.2021.107758

[11] T. Li, L. Lomazzi, F. Cadini, C. Sbarufatti, J. Chen, S. Yuan, Numerical simulation-aided particle filter-based damage prognosis using Lamb waves, Mechanical Systems and Signal Processing, 178 (2022) 109326. https://doi.org/10.1016/j.ymssp.2022.109326

[12] T. Li, J. Chen, S. Yuan, F. Cadini, C. Sbarufatti, Particle filter-based damage prognosis using online feature fusion and selection, Mechanical Systems and Signal Processing, 203 (2023) 110713. https://doi.org/10.1016/j.ymssp.2023.110713

[13] X. Zhou, S. He, L. Dong, S.N. Atluri, Real-Time Prediction of Probabilistic Crack Growth with a Helicopter Component Digital Twin, AIAA Journal, 60 (2022) 2555-2567. https://doi.org/10.2514/1.J060890

[14] M. Chiachío, J. Chiachío, S. Sankararaman, K. Goebel, J. Andrews, A new algorithm for prognostics using Subset Simulation, Reliability Engineering & System Safety, 168 (2017) 189-199. https://doi.org/10.1016/j.ress.2017.05.042

[15] D. Cristiani, C. Sbarufatti, M. Giglio, Damage diagnosis and prognosis in composite double cantilever beam coupons by particle filtering and surrogate modelling, Structural Health Monitoring, (2020) 1475921720960067. https://doi.org/10.12783/shm2019/32281

[16] T. Li, C. Sbarufatti, F. Cadini, Multiple local particle filter for high-dimensional system identification, Mechanical Systems and Signal Processing, 209 (2024) 111060. https://doi.org/10.1016/j.ymssp.2023.111060

[17] T. Li, F. Cadini, M. Chiachío, J. Chiachío, C. Sbarufatti, Particle filter-based delamination shape prediction in composites subjected to fatigue loading, Structural Health Monitoring, 22 (2023) 1844-1862. https://doi.org/10.1177/14759217221116041

[18] P. Kundu, A.K. Darpe, M.S. Kulkarni, A review on diagnostic and prognostic approaches for gears, Structural Health Monitoring, 20 (2020) 2853-2893. https://doi.org/10.1177/1475921720972926

[19] B. Wang, Y. Lei, N. Li, N. Li, A Hybrid Prognostics Approach for Estimating Remaining Useful Life of Rolling Element Bearings, IEEE Transactions on Reliability, 69 (2020) 401-412. https://doi.org/10.1109/TR.2018.2882682

[20] B. Wang, Y. Lei, T. Yan, N. Li, L. Guo, Recurrent convolutional neural network: A new framework for remaining useful life prediction of machinery, Neurocomputing, 379 (2020) 117-129. https://doi.org/10.1016/j.neucom.2019.10.064

[21] G. Galanopoulos, N. Eleftheroglou, D. Milanoski, A. Broer, D. Zarouchas, T. Loutas, A novel strain-based health indicator for the remaining useful life estimation of degrading composite structures, Composite Structures, 306 (2023) 116579. https://doi.org/10.1016/j.compstruct.2022.116579

[22] M. Moradi, F.C. Gul, D. Zarouchas, A novel machine learning model to design historical-independent health indicators for composite structures, Composites Part B: Engineering, 275 (2024) 111328. https://doi.org/10.1016/j.compositesb.2024.111328

[23] M. Moradi, A. Broer, J. Chiachío, R. Benedictus, T.H. Loutas, D. Zarouchas, Intelligent health indicator construction for prognostics of composite structures utilizing a semi-supervised deep neural network and SHM data, Engineering Applications of Artificial Intelligence, 117 (2023) 105502. https://doi.org/10.1016/j.engappai.2022.105502

[24] X. Liu, Y. Lei, N. Li, X. Si, X. Li, RUL prediction of machinery using convolutional-vector fusion network through multi-feature dynamic weighting, Mechanical Systems and Signal Processing, 185 (2023) 109788. https://doi.org/10.1016/j.ymssp.2022.109788

[25] Q. Ni, J.C. Ji, K. Feng, Data-Driven Prognostic Scheme for Bearings Based on a Novel Health Indicator and Gated Recurrent Unit Network, IEEE Transactions on Industrial Informatics, 19 (2023) 1301-1311. https://doi.org/10.1109/TII.2022.3169465

[26] R. Gao, L. Wang, R. Teti, D. Dornfeld, S. Kumara, M. Mori, M. Helu, Cloud-enabled prognosis for manufacturing, CIRP Annals, 64 (2015) 749-772. https://doi.org/10.1016/j.cirp.2015.05.011

[27] F. Zeng, Y. Li, Y. Jiang, G. Song, An online transfer learning-based remaining useful life prediction method of ball bearings, Measurement, 176 (2021) 109201. https://doi.org/10.1016/j.measurement.2021.109201

[28] J. Zhuang, Y. Cao, M. Jia, X. Zhao, Q. Peng, Remaining useful life prediction of bearings using multi-source adversarial online regression under online unknown conditions, Expert Systems with Applications, 227 (2023) 120276. https://doi.org/10.1016/j.eswa.2023.120276

[29] W. Mao, K. Liu, Y. Zhang, X. Liang, Z. Wang, Self-Supervised Deep Tensor Domain-Adversarial Regression Adaptation for Online Remaining Useful Life Prediction Across Machines, IEEE Transactions on Instrumentation and Measurement, 72 (2023) 1-16. https://doi.org/10.1109/TIM.2023.3265109

[30] C. Liu, L. Zhang, Y. Zheng, Z. Jiang, J. Zheng, C. Wu, Online industrial fault prognosis in dynamic environments via task-free continual learning, Neurocomputing, 598 (2024) 127930. https://doi.org/10.1016/j.neucom.2024.127930

Structural Health Monitoring: 10APWSHM   Materials Research Forum LLC
Materials Research Proceedings 50 (2025) 127-132   https://doi.org/10.21741/9781644903513-15

In-situ modal decomposition of acoustic emission events arising from low-velocity impacts

Jaslyn Gray[1,2,a] *, Cedric Rosalie[1,b], Ben Vien[2,c], Wing Kong Chiu[2,d] and Nik Rajic[1]

[1]Defence Science and Technology Group, 506 Lorimer Street, Fishermans Bend, VIC 3207, Australia

[2]Department of Mechanical and Aerospace Engineering, Monash University, Wellington Rd, Clayton, VIC 3800, Australia

[a]jaslyn.gray@defence.gov.au, [b]cedric.rosalie1@defence.gov.au, [c]ben.vien@monash.edu, [d]wing.kong.chiu@monash.edu

Keywords: Modal Decomposition, Acoustic Emission, Damage Detection, Piezo-Electric Sensing, Low-Velocity Impact

Abstract. In the field of structural health monitoring (SHM), it is common to employ the use of sensors to provide information about the state of a platform, particularly when concerned with whether damage has occurred as a result of impact. Acoustic emission (AE) sensing using piezoelectric elements is well suited to the detection of guided plate waves arising from impact events on thin-walled aerospace structures. This paper presents a systematic evaluation of impact induced AE with increasing velocity and explores changes in the resulting signals model content. A spring-loaded mechanism is used to accelerate ball bearings up to 7 m/s and the corresponding AE event is recorded using a LAMDA (Linear Array for Modal Decomposition and Analysis) sensor. This LAMDA sensor is an in-situ sensing capability that allows for frequency-wavenumber decomposition of the acquired acoustic signal into its constituent modes, which can then be mapped against theoretical dispersion curves for the material under investigation. To date, most research in the low velocity regime is concerned with the fundamental antisymmetric (A0) and symmetric (S0) wave modes from multiple sensing locations. Using LAMDA, it is possible to build a more complete picture of an impact signal, not currently achievable with conventional sensing methods.

Introduction

Impact damage poses a significant threat to the operational capability of aerospace platforms, particularly in thin structural components, which are common in air and space applications. In the field of structural health monitoring (SHM), methods of non-destructive testing and evaluation (NDT&E) allow information about the structural state of a platform to be determined without the need for it to be deconstructed or removed from operation. Such evaluation can include the integration of sensors in order to acquire, transmit and process data [1]. The importance of this data is only growing as the industry shifts to integrate machine learning, artificial intelligence and digital twins into the operational environment in order to predict structural damage and improve platform survivability [2].

AE methods excite stress waves as a result of loading, and when a thin-walled structure is impacted, an acoustic disturbance is excited in the form of Lamb waves. These waves propagate as two basic types: symmetric (axial) and antisymmetric (flexural) wave modes [3]. Lamb waves are dispersive in nature and often more than one mode can be excited for any given frequency [4]. Therefore, the wave field produced as a result of impact typically consists of multiple overlapping modes which makes AE analysis challenging [5].

Structural Health Monitoring: 10APWSHM   Materials Research Forum LLC
Materials Research Proceedings 50 (2025) 127-132   https://doi.org/10.21741/9781644903513-15

Alleyne and Cawley [4] first introduced the use of a two-dimensional fast Fourier transform (2DFFT) to statically resolve the static wave field in terms of its wavenumber-frequency decomposition (k-ω) and map the wavenumber dispersion curves theoretically for a given material. In this approach, the modes are distinguishable from each other due to their dispersion characteristics and it is possible to analyse individual modes without interference. To achieve k-ω decomposition experimentally using the 2DFFT method, a measurement of the wave field is required that is of high spatial resolution and includes time domain readings taken at multiple points. Array-based piezoelectric sensing enables AE signals to be experimentally acquired with sufficient resolution to apply this method [6].

An array based polymer piezoelectric sensor developed by the Material State Awareness Group at DSTG and suited to this application is called LAMDA (Linear Array for Modal Decomposition and Analysis) [7]. It is a flexible PVDF (polyvinylidene fluoride) sensor with multiple sensing elements arranged in a linear array. As the AE signal travels through each element, an individual measurement of the travelling wave is recorded. The wave dispersion can then be calculated using the described 2DFFT method. Previously, studies have analysed the AE response to ball-drop impacts with ceramic PZT (Lead zirconate titanate) sensors [8-11]. However, these examples only consider velocities under 4 m/s. Additionally, these studies require multiple sensors at different locations to seperate the modes [8-9]. An array-based approach such as LAMDA simplifies the process of acquiring AE events and allows each unique impact event to be mapped visually, so that the wave modes may be compared.

Methodology
Experimental Setup
A square aluminium panel (AA5005) with side dimensions of 600 mm and 1.6 mm thickness was used as the target panel for impact experiments with 10 mm diameter steel ball bearings. The panel was placed on a damped optical table to minimise external vibration and separated from the surface with a plastic sheet. For the freehand ball "drop" experiments, the bearings were released from a predetermined height to impact the panel at a prescribed velocity. In the "pinball" experiments, a spring loaded mechanism with a pin release accelerated the ball bearing and its velocity was measured using a chronograph and outputted to the display. Target velocities of 2.0 m/s through to 6.0 m/s were achieved with 1.0 m/s increments. 2.0 m/s and 3.0 m/s experiments were obtained from freehand drops with a drop height estimated by equating the potential and kinetic energy equations for $h = v^2/2g$ where $g = 9.81 m/s^2$. Higher velocities were recorded using the "pinball" mechanism and values read from the velocity display. For all velocity cases, a sample size of 20 was taken to understand the statistical spread of the data.

For data acquisition, a 16-chiplet PVDF LAMDA sensor was bonded directly to the panel with M-Bond 200 adhesive and connected to four AUSAM+ (Acousto Ultrasonic Structural health monitoring Array Module) units. Each unit is comprised of four channels of bandwidth range 50kHz - 5MHz. Together, the four AUSAM+ setup is capable of simultaneously acquiring 16 AE signals as voltage-time measurements over the array, one for each chiplet, for each unique impact event. Technical details relating to the LAMDA sensor and AUSAM+ units can be found in [6]. The described experimental setup can be seen in Figure 1.

Data Processing
The spectral decomposition algorithm was first run on the original data set and the output stored in an array. This algorithm uses a 2DFFT as described in [4] to transform experimental space-time wave data and generate corresponding wavenumber-frequency data. A series of filters were applied to the dataset in order to highlight the low amplitude, high frequency signals which were overshadowed by the dominate high amplitude, low frequency A0 (antisymmetric) component of

Structural Health Monitoring: 10APWSHM   Materials Research Forum LLC
Materials Research Proceedings 50 (2025) 127-132   https://doi.org/10.21741/9781644903513-15

the impact wave response. Two separate high-pass filters with cut-off frequencies of 250 kHz and 1 MHz were used. In each filter case, logical comparison is performed between the initial data and the newly filtered data with the higher value for each frequency stored. The resulting array was then plotted against the fundamental wave modes for the plate, with the latter generated using the commercial package DISPERSE [12].

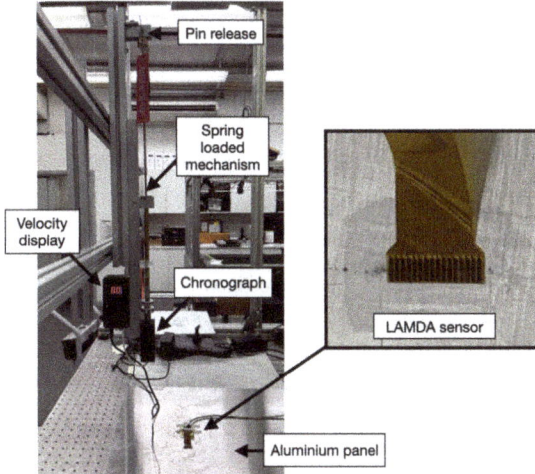

Figure 1: AE impact experiment setup with spring loaded "pinball" mechanism, aluminium test panel and PVDF LAMDA sensor

Results and Discussion

Representative AE waveforms for the lowest and highest recorded impact velocities can be seen in Figure 2. It is important to note that while each impact event is intrinsically unique, the samples presented reflect the overall data sets for each velocity, and as such, conclusions can be drawn comparatively. The AE time signature for the ball drops are expectedly similar across the time axis while the higher velocity impact results in a higher amplitude signal. It is difficult to make any further statements about the content of the AE resulting from impact in the time domain.

Figure 3 contains the spectral decomposition maps for each investigated velocity, obtained through the spectral decomposition data processing methodology and multi-stage filtering as previously outlined. In each case the fundamental A0 and S0 modes are clearly visible and align well with the theoretical predictions obtained from DISPERSE. For each impact velocity case, A0 content is present up to 500 kHz and the signature aligns to the S0 band from approximately 900 kHz though to just below 1.5 MHz. The overall strength of the S0 mode also appears to increase as the velocity increases. The highest velocity investigated was 6.1 m/s with the result shown in Figure 4. In this case, there is evidence that the A1 mode is present in the AE signal, along with the A0 and S0 modes, highlighted by the red ovals.

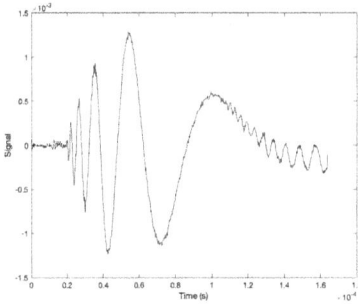

2.0 m/s impact AE waveform   6.0 m/s impact AE waveform

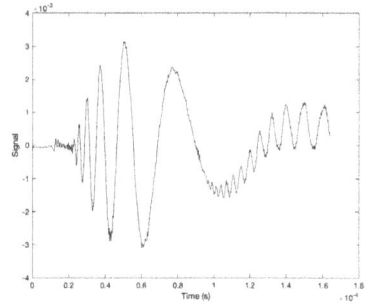

Figure 2: Acoustic emission waveforms as recorded by 16th chiplet of LAMDA sensor, closest to impact location

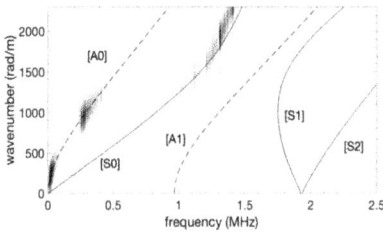

20 cm height ball drop at 2.0 m/s   50 cm height ball drop at 3.1 m/s

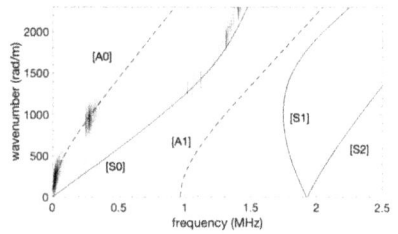

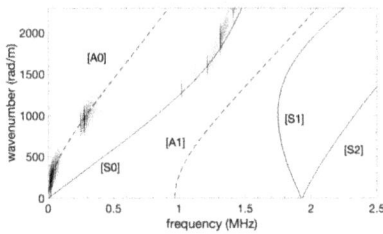

"Pinball impact" at 4.1 m/s   "Pinball impact" at 5.2 m/s

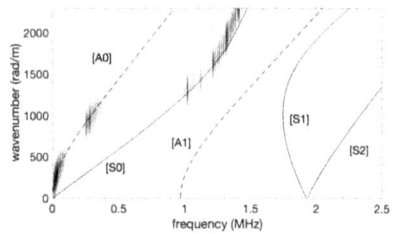

Figure 3: Comparison of acoustic emission for various velocities from approximately 2.0 m/s to 5.0 m/s plotted using spectral decomposition algorithm

The practical meaning of these modes in relation to the impact event is being investigated but it is evident that this spectral decomposition approach to interpreting AE signals arising from impact provides clarity in separating modal content. This, is turn, allows future characterisation of the modal content relevant to the impact type. For example, if individual modes can be linked to particular types of damage, then the modal content could be used as a way of identifying the type of failure that has occurred due to impact. It is hypothesised that as impact velocity increases, the frequency bandwidth of the AE signal will expand, with evidence of higher-order modes becoming visible in the spectral decomposition maps. Furthermore, this array based sensing methodology presented (LAMDA) appeals as a practical implementation of lightweight in-situ sensing from a single sensing location.

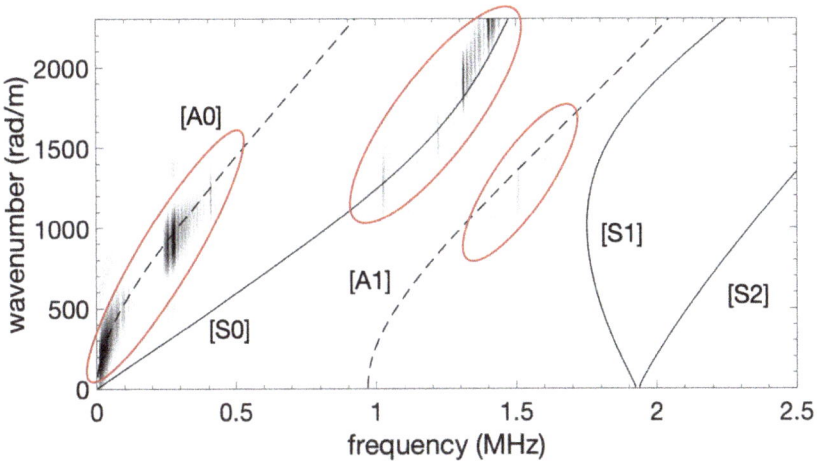

Figure 4: Acoustic emission for highest recorded "pinball" ball bearing impact at 6.1 m/s plotted using spectral decomposition algorithm

Conclusion
This study has demonstrated a distributed sensing approach to AE using PVDF LAMDA for detecting ball drop impact events up to 6 m/s. Spectral decomposition of the acquired signal using the 2DFFT method allows the resulting frequencies to be mapped against the wave modes for the given material. As the velocity of impact increases, the amplitudes of signals also increases and richer frequency content is observed. All experimental cases contained the fundamental A0 and S0 mode. In the highest tested velocity, the A1 mode is also evident. Work to understand the practical significance of these modes in terms of structural damage is underway. Additional work is planned to investigate higher velocity impacts using this in-situ sensing approach in the near future.

Acknowledgements
The authors would like to acknowledge and thank: Matthew Ibrahim, Francis Rose, Daniel Bitton, Stephen van der Velden, George Jung, Joel Smithard, Pooia Lalbakhsh, Patrick Norman and Scott Moss from the Material State Awareness Group at Defence Science Technology Group for their support on this project.

This paper includes research that was supported by the Advanced Piezoelectric Materials and Applications Program, through DMTC Limited (Australia), and funded by Defence under the Next

Structural Health Monitoring: 10APWSHM
Materials Research Proceedings 50 (2025) 127-132

Materials Research Forum LLC
https://doi.org/10.21741/9781644903513-15

Generation Technologies Fund, which is now administered through the Advanced Strategic Capabilities Accelerator. The authors have prepared this paper in accordance with the intellectual property rights granted by a DMTC Project Agreement.

References

[1] M. Senthilkumar, T.G. Sreekanth and S.M Reddy. "Nondestructive health monitoring techniques for composite materials: A review," Polymers and Polymer Composites, vol. 29 (5), pp. 528-540, 2021. https://doi.org/10.1177/0967391120921701

[2] S. Khalid, J. Song, M.M. Azad, M.U. Elahi, J. Lee, S. Jo, H.S. Kim, "A comprehensive Review of Emerging Trends in Aircraft Structural Prognostics and Health Management," Mathematics, vol. 11 (18) 3837, 2023. https://doi.org/10.3390/math11183837

[3] V. Giurgiutiu, "6.4 Lamb waves," in Structual health monitoring: with piezoelectric wafer active sensors, Academic Press. Elsevier, 2008, pp. 198-228.

[4] D. Alleyne and P. Cawley. 'A two-dimensional Fourier transform method for the measurements of propagating multimode signals.' The Journal of the Acoustical Society of America, vol. 89 (3), pp. 1159-1168, 1991. https://doi.org/10.1121/1.400530

[5] C. Rosalie, N. Rajic, S. van der Velden, F. Rose, J. Smithard and W.K. Chiu, "Towards composite damage classification using in situ wavenumber-frequency modal decomposition of Acoustic Emissions," in European Workshop on Structural Health Monitoring. EWSHM 2020. P. Rizzo, A. Milazzo (eds). Lecture Notes in Civil Engineering, vol 127. Springer, Cham, 2021. https://doi.org/10.1007/978-3-030-64594-6_63

[6] N. Rajic, C. Rosalie, BS. Vien, S. van der Velden, LRF. Rose, J. Smithard and WK. Chiu. "In situ wavenumber-frequency modal decomposition of acoustic emission," Structural Health Monitoring, vol. 19 (6), pp. 2033-2050, 2020. https://doi.org/10.1177/1475921719885324

[7] N. Rajic, C. Rosalie, S. van der Velden, et al. "A flexible high density piezoelectric sensor array for in situ modal decomposition," in Proceedings of the 9th European Workshop on Structural Health Monitoring, Manchester, 2018.

[8] J. Zhu, S.M. Parvasi, S.C.M. Ho, D. Patil, M. Ge, H. Li and G. Song, "An innovative method for automatic determination of time of arrival for lamb wave excited by impact events," Smart Materials and Structures, vol. 26, 2017. https://doi.org/10.1088/1361-665X/aa63e1

[9] L. Capineri and A. Bulletti, "A versatile analog electronic interface for piezoelectric sensors used for impacts detection and positioning in Structural Health Monitoring (SHM) systems," Electronics, vol. 10, 1047, 2021. https://doi.org/10.3390/electronics10091047

[10] S. Guo, H. Ding, Y. Li, H. Feng, X. Xiong, Z. Su and W. Feng, "A hierarchical deep convolutional regression framework with sensor network fail-safe adaption for acoustic-emission- based structural monitoring," Mechanical Systems and Signal Processing, vol. 181, 2022. https://doi.org/10.1016/j.ymssp.2022.109508

[11] N. Daniele Boffa, N. Area, E. Manaco, M. Viscardi, F. Ricci and T. Kundu, "About the combination of high and low frequency methods for impact detection on aerospace components," Progress in Aerospace Sciences, vol. 129, 2022. https://doi.org/10.1016/j.paerosci.2021.100789

[12] Pavlakovic, B., M.J.S. Lowe, D. Alleyne, and P. Cawley, Disperse: A General Purpose Program for Creating Dispersion Curves, in Review of Progress in Quantitative Nondestructive Evaluation, D. Thompson and D. Chimenti, Editors. 1997, Springer: US. pp. 185-192. https://doi.org/10.1007/978-1-4615-5947-4_24

Structural Health Monitoring: 10APWSHM
Materials Research Proceedings 50 (2025) 133-140

Materials Research Forum LLC
https://doi.org/10.21741/9781644903513-16

Zero group velocity mode enhanced electro-mechanical impedance spectroscopy and nonlinear ultrasonics

Runye Lu[1,a] and Yanfeng Shen[1,b] *

[1]University of Michigan-Shanghai Jiao Tong University Joint Institute, Shanghai Jiao Tong University, Shanghai, 200240, China

[a]runye.lu@sjtu.edu.cn [b]yanfeng.shen@sjtu.edu.cn

Keywords: Structural Health Monitoring; Electro-Mechanical Impedance Spectroscopy; Zero Group Velocity Mode; Local Resonances; Nonlinear Ultrasonics

Abstract. The zero group velocity (ZGV) mode possesses the distinctive attribute of an elapsed group velocity with a finite wavenumber, indicating a spatially propagating wave under a motionless package. This stationary mode engenders a localized resonance, confining the wave energy in the vicinity of actuation. Researchers have utilized ZGV modes for structural health monitoring (SHM) scenarios mostly in transient analysis and linear ultrasonic regime. Nevertheless, it remains an uncharted frontier that how ZGV mode, especially via its peculiar characteristics, can empower SHM methodologies regarding harmonics analysis and nonlinear ultrasonics. Therefore, this paper, on one hand, explores the trembling feature of ZGV resonances in steady-state electro-mechanical impedance spectroscopy (EMIS) and utilizes it for structural sensing. On the other hand, this paper leverage nonlinear generation of ZGV modes to amplify the higher harmonics signal to enhance conventional nonlinear ultrasonic methodology. This paper culminates in summary, concluding remarks, and suggestions for future work.

Introduction

The zero group velocity (ZGV) mode, as a peculiar non-propagating Lamb wave mode, has ignited enormous research interests. ZGV modes are endowed with special characteristics of a vanishing group velocity and simultaneously a non-zero phase velocity, resulting in a spatially propagating phase information under a motionless wave package [1]. Such a stationary mode can trigger intense local resonances in the thickness direction, confining the wave energy within the vicinity of the actuation site [2, 3].

The existence of ZGV modes has been validated in thin plates and membranes [1, 4-8], thin films [9], isotropic [10]/hollow cylinders [11], rigid/soft strips [12], concrete materials [13], and multi-layered structures [14, 15]. Furthermore, Li et al. validated the presence of ZGV features in a welded joint subjected to a force source [16]. Zhu et al. demonstrated the ZGV modes in free rails by numerical modelling and experimental verification [17]. ZGV modes have also been applied for various structural health monitoring (SHM) purposes. The potential application of ZGV modes for material properties determination were demonstrated, encompassing elastic constant [18], Poisson's ratio [19], interfacial stiffness [5], and bulk acoustic wave velocity [20]. Regarding various damage types, the ZGV modes enabled the diagnosis of fatigue damage [8], laminate disbond [14], delamination in composites [13]. To sum, the current SHM techinques mostly rely on the direct use of ZGV modes in transient analyses and linear ultrasonic regimes. It still remains an uncharted frontier that how ZGV resonances manifest themselves in steady-state harmonic anlaysis and whether ZGV modes can be utlized for nonlinear ultrasonics. It is worth mentioning that the research on nonlinear generation of ZGV modes have been pioneered by Li et al. and Mora et al. [21, 22].

The paper commences with the identification and extraction of ZGV modes in aluminum plate by analytical modelling, obtaining the ZGV frequencies as references. Following this, the paper

Structural Health Monitoring: 10APWSHM   Materials Research Forum LLC
Materials Research Proceedings 50 (2025) 133-140   https://doi.org/10.21741/9781644903513-16

delves into the trembling feature following the ZGV resonant peaks in steady-state frequency spectra. The mechanism and nature behind the trembling features are investigated from various aspects. Subsequently, such feature is leveraged for the efficient ZGV identification and structural sensing in harmonic analysis. When it comes to the nonlinear ultrasonics, the paper conducts the comparative finite element analyses to demonstrate the enhancement performance of ZGV modes. Different from the normal frequency actuation, a half ZGV frequency actuation is employed so that the second harmonics component, which is exactly the ZGV frequency, can be generated at the fatigue crack location. The nonlinear ZGV generation triggers local resonance and thus intensive interactions, therefore enhancing the second harmonic spectral amplitude. Afterwards, some bizarre phenomena accompanying the nonlinear ZGV generation are explored and elucidated. This paper culminates in summary, concluding remarks and suggestions for future work.

Analytical Investigation on ZGV Modes in Al Plates
This section conducts the analytical investigation on the ZGV mode frequencies in the target aluminum plate as depicted in Figure 1. The dispersion curve of the aluminum plate can be obtained by solving the Rayleigh Lamb equation as depicted below:

$$\frac{\tan \eta_P d}{\tan \eta_S d} = -\left[\frac{\left(\xi^2 - \eta_S^2 \right)^2}{4\xi^2 \eta_P \eta_S} \right]^{\pm 1} \tag{1}$$

where d denotes the half plate thickness; $\eta_P^2 = \omega^2/c_P^2 - \xi^2, \eta_S^2 = \omega^2/c_S^2 - \xi^2$; ξ signifies the wavenumber; the power of +1 refers to symmetric modes, while -1 corresponds to anti-symmetric modes. Numerical solution of Rayleigh-Lamb equation yields the eigenvalues, $\xi_0^S, \xi_1^S, \xi_2^S, ..., \xi_0^A, \xi_1^A, \xi_2^A, ...,$ representing the wavenumbers of the lamb wave modes. The relationship $c = \omega/\xi$ engenders the dispersive phase velocity, as a function of the the frequency and half thickness product, fd. At a given fd product, several modes exist simultaneously, designating the multimodal feature of Lamb waves.

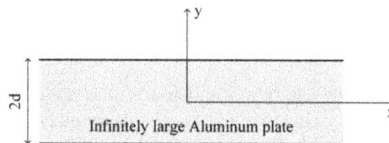

Figure 1: Diagram of the target infinitely large aluminum plate.

The aluminum plates to be utilized in the following sections take 4 mm and 1 mm in thickness respectively. The dispersion curves of the two target Al plates and the corresponding S₁-ZGV mode shape are displayed in Figure 2. The ZGV frequencies can be determined to be 707 kHz and 2828 kHz by tracing the points with zero slope in dispersion curves. The ZGV frequencies function as the reference frequency for the subsequent analyses on the two ZGV enhanced SHM methodologies.

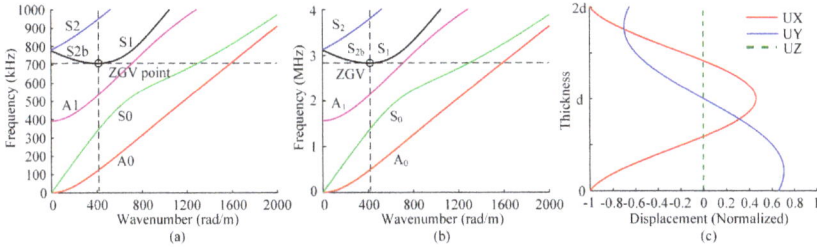

Figure 2: dispersion curves for (a) 4 mm Al plate; (b) 1 mm Al plate; (c) S_1-ZGV mode shape

Trembling Features Following ZGV Resonances in EMIS Framework

This section presents the FE analyses and experimental verification of the distinctive trembling features following ZGV resonances and corresponding application for structural sensing. The target structure is a 4 mm thick aluminum plate, the FE model of which is displayed in Figure 3. A piezoelectric wafer active sensor (PWAS) was bonded on the top center of the plate, implemented by the 2-D multi-physics element. The 100 mm non-reflective boundary at both sides was achieved by absorbing layers with increasing damping to simulate the infinitely large structure. The harmonic analysis was conducted with the frequency sweeping from 700 kHz to 720 kHz. The steady-state displacement frequency spectra were to be captured at all points in the sampling region with a spatial resolution of 0.25 mm.

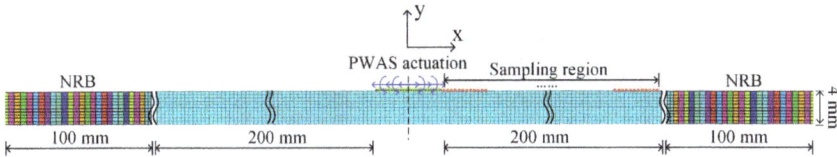

Figure 3: The FE model under EMIS configuration.

The steady-state displacement frequency spectrum captured at the first sampling point was illustrated in Figure 4. The steep ZGV resonance peak at 707 kHz can be obviously observed. The intriguing trembling features appeared right after the resonant peak and gradually faded away with the frequency increasing, akin to the accompanying local resonances.

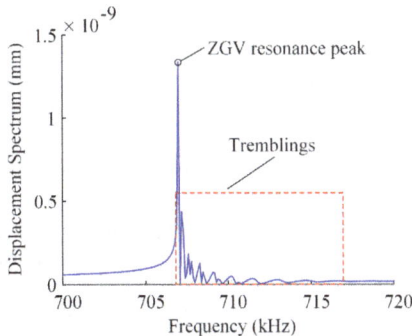

Figure 4: The steady-state displacement frequency spectrum manifesting trembling features.

Afterwards, the ZGV resonances and the trembling features were examined in frequency-wavenumber domain. Since the steady-state frequency spectra were obtained, only 1-D Fast Fourier Transformation (FFT) was required to calculate the dispersion curves. It is worth mentioning that the dispersion curves obtained here were steady-state dispersion curve, different from the conventional dispersion curves implemented by 2-D FFT from transient analysis as illustrated in Figure 5.

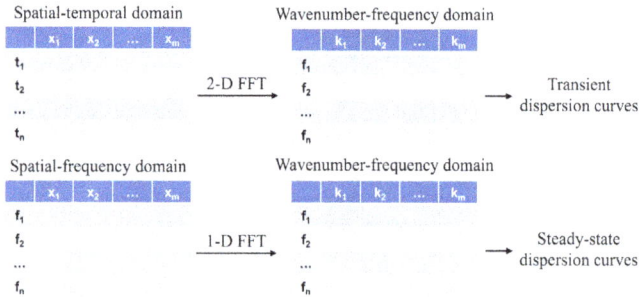

Figure 5: Comparison on the signal processing procedures.

The steady-state dispersion curve and the corresponding 3-D version were displayed in Figure 6. The dispersion curves clearly illustrated the existence of ZGV resonance occurring at 707 kHz and the alternatively dark-shallow color indicated the trembling features right after the ZGV frequency. The 3-D surface plots in Figure 6 elucidated the alternating peaks and valleys of the subsequent resonances more clearly. That is to say, following the generation of ZGV resonance, the participation level of each mode revealed drastic fluctuations and gradually settled down.

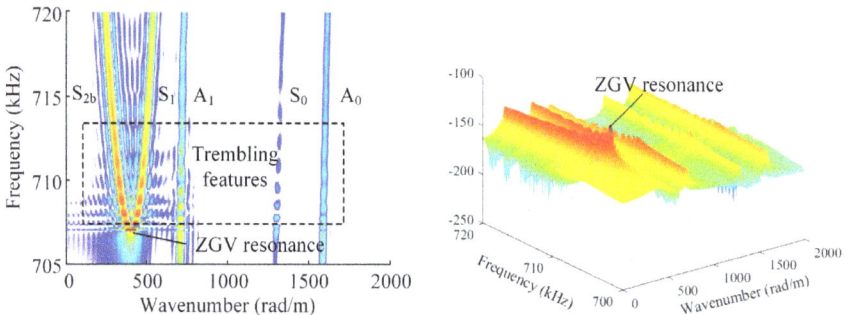

Figure 6: Steady-state dispersion curves ellucidating ZGV resonance and trembling features.

Furthermore, the distinctive trembling features were utilized for structural sensing via electromechanical impedance spectroscopy (EMIS), targeting at the multi-layer aluminum laminate. As depicted in Figure 7(a), the laminate was composed of two aluminum plates (4 mm and 6 mm in thickness) bonded by the epoxy adhesive layer. A PWAS was mounted at the top center of the plate, functioning as both the actuator and receiver. Similarly, NRB was utilized to simulate the infinitely large structure. The EMI spectrum was obtained and displayed in Figure 7(b), where the ZGV resonances were identified and extracted by the unique trembling features. The group velocity dispersion curve further validated the observation.

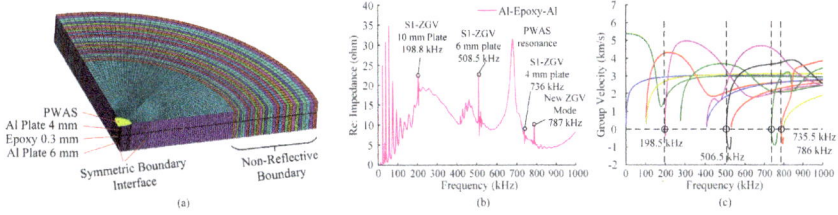

Figure 7: Structural sensing application on Al laminate (a) FE model configuraion; (b) electromechanical impedance spectra; (c) frequency-wavenumber dispersion curves.

ZGV Resonance Enhanced Nonlinear Ultrasonics for Fatigue Crack Monitoring

This section conducts comparative finite element (FE) analyses on the ZGV enhanced nonlinear ultrasonic method. The FE model of the 1 mm thick Al plate is displayed in Figure 8. The symmetric pin-force actuation was applied to output the tone burst signal at certain frequencies. The non-reflective boundaries (NRB) were employed to absorb the wave reflections to simulate the infinitely large structure. The fatigue crack was simulated by the contact element to integrate the contact acoustic nonlinearity (CAN) effect. The actuation frequencies were determined to be 1000 kHz and 1383 kHz, corresponding to normal excitation and half ZGV frequency (2776 kHz /2) excitation respectively. The temporal displacement responses of the sampling points in the denoted sampling region were captured.

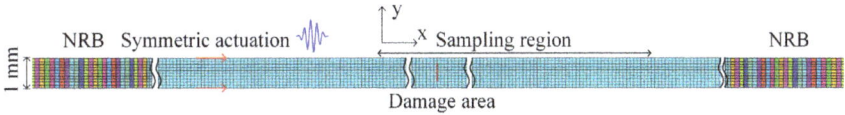

Figure 8: FE model configurations for the ZGV enhanced nonlinear ultrasonic method.

The temporal domain displacement responses three points (0 mm, 5 mm, 10 mm at the surface) are displayed in Figure 9. The displacement amplitude of the half ZGV actuation overwhelms the normal frequency actuation and the ringing effect of the ZGV mode with a longer time duration is observed.

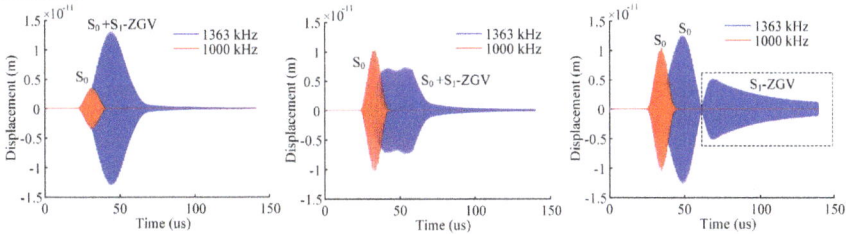

Figure 9: Temporal displacement responses for the two actuation frequencies at three selective points.

Afterwards, FFT was applied on the temporal signal at the point top of the crack to obtain the frequency spectra as depicted in Figure 10. The results from the half ZGV excitation manifested a drastic amplification of the second harmonic component, even larger than the fundamental frequency component. Following the frequency spectra, 2-D FFT was implemented on the signals at all sampling points to obtain the frequency-wavenumber dispersion curves as showcased in Figure 11. The nonlinear ZGV generation can be further validated via the partially marked curves.

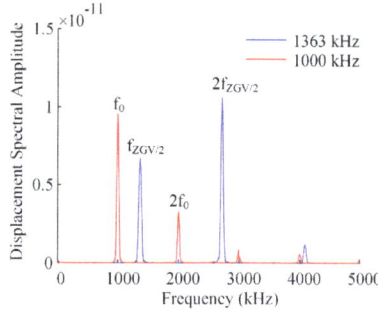

Figure 10: The comparative displacement frequency spectra for the two actuation frequencies.

To sum, the half ZGV frequency actuation can be utilized to generate the ZGV mode at the vicinity of damage site, triggering drastic local resonance and thus intense interactions of damage interfaces. Such interactions could largely amplify the nonlinear ultrasonic response against the external disturbances to enhance the performance of nonlinear ultrasonic methodologies.

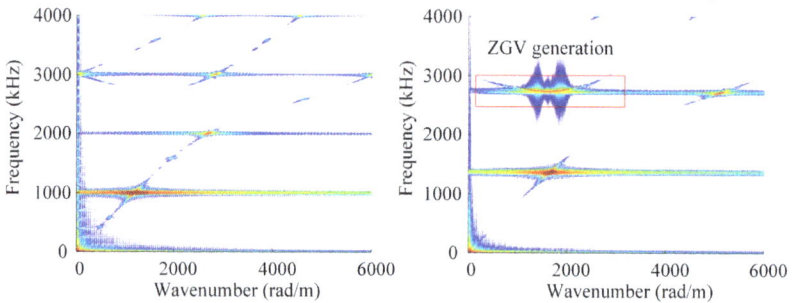

Figure 11: The comparative dispersion curve for the two actuation frequencies.

Summary
This paper proposed the zero group velocity (ZGV) mode enhanced structural health monitoring (SHM) strategies to enhance the performance of EMIS and nonlinear ultrasonics methodologies. From the FE analyses, the distinctive trembling features following ZGV resonances were visualized and validated, and subsequently leveraged for structural sensing of multi-layer laminate. Further, the ZGV resonances were utilized for the amplification of second harmonics generation. It demonstrated that the nonlinear ZGV generation can drastically enlarge the second harmonic component, thus promoting the effectiveness of nonlinear ultrasonics.

Acknowledgement
The support from the National Natural Science Foundation of China (contract number 52475161) is thankfully acknowledged; Funding of John Wu and Jane Sun Endowed Professorship is gratefully acknowledged.

Structural Health Monitoring: 10APWSHM
Materials Research Proceedings 50 (2025) 133-140

Materials Research Forum LLC
https://doi.org/10.21741/9781644903513-16

References

[1] C. Prada, D. Clorennec, and D. Royer, "Local vibration of an elastic plate and zero-group velocity Lamb modes," *J Acoust Soc Am,* vol. 124, no. 1, pp. 203-12, Jul 2008. https://doi.org/10.1121/1.2918543

[2] S. D. Holland and D. E. Chimenti, "Air-coupled acoustic imaging with zero-group-velocity Lamb modes," *Applied Physics Letters,* vol. 83, no. 13, pp. 2704-2706, 2003. https://doi.org/10.1063/1.1613046

[3] C. Prada, O. Balogun, and T. W. Murray, "Laser-based ultrasonic generation and detection of zero-group velocity Lamb waves in thin plates," *Applied Physics Letters,* vol. 87, p. 194109, November 01, 2005 2005. https://doi.org/10.1063/1.2128063

[4] C. Prada, O. Balogun, and T. W. Murray, "Laser-based ultrasonic generation and detection of zero-group velocity Lamb waves in thin plates," *Applied Physics Letters,* vol. 87, no. 19, p. 194109, 2005. https://doi.org/10.1063/1.2128063

[5] D. Clorennec, C. Prada, and D. Royer, "Laser ultrasonic inspection of plates using zero-group velocity lamb modes," *IEEE Transactions on Ultrasonics, Ferroelectrics, and Frequency Control,* vol. 57, no. 5, pp. 1125-1132, 2010. https://doi.org/10.1109/TUFFC.2010.1523

[6] C. M. Grunsteidl, I. A. Veres, and T. W. Murray, "Experimental and numerical study of the excitability of zero group velocity Lamb waves by laser-ultrasound," *J Acoust Soc Am,* vol. 138, no. 1, pp. 242-50, Jul 2015. https://doi.org/10.1121/1.4922701

[7] O. Tofeldt and N. Ryden, "Zero-group velocity modes in plates with continuous material variation through the thickness," *J Acoust Soc Am,* vol. 141, no. 5, p. 3302, May 2017. https://doi.org/10.1121/1.4983296

[8] G. Yan, S. Raetz, N. Chigarev, V. E. Gusev, and V. Tournat, "Characterization of Progressive Fatigue Damage in Solid Plates by Laser Ultrasonic Monitoring of Zero-Group-Velocity Lamb Modes," *Physical Review Applied,* vol. 9, no. 6, 2018. https://doi.org/10.1103/PhysRevApplied.9.061001

[9] V. Yantchev, L. Arapan, I. Katardjiev, and V. Plessky, "Thin-film zero-group-velocity Lamb wave resonator," *Applied Physics Letters,* vol. 99, no. 3, 2011. https://doi.org/10.1063/1.3614559

[10] J. Laurent, D. Royer, T. Hussain, F. Ahmad, and C. Prada, "Laser induced zero-group velocity resonances in transversely isotropic cylinder," *J Acoust Soc Am,* vol. 137, no. 6, pp. 3325-34, Jun 2015. https://doi.org/10.1121/1.4921608

[11] M. Ces, D. Royer, and C. Prada, "Characterization of mechanical properties of a hollow cylinder with zero group velocity Lamb modes," *J Acoust Soc Am,* vol. 132, no. 1, pp. 180-5, Jul 2012. https://doi.org/10.1121/1.4726033

[12] J. Laurent, D. Royer, and C. Prada, "In-plane backward and zero group velocity guided modes in rigid and soft strips," *J Acoust Soc Am,* vol. 147, no. 2, p. 1302, Feb 2020. https://doi.org/10.1121/10.0000760

[13] Y.-T. Tsai and J. Zhu, "Simulation and Experiments of Airborne Zero-Group-Velocity Lamb Waves in Concrete Plate," *Journal of Nondestructive Evaluation,* vol. 31, no. 4, pp. 373-382, 2012/12/01 2012. https://doi.org/10.1007/s10921-012-0148-6

[14] J. Spytek, A. Ziaja-Sujdak, K. Dziedziech, L. Pieczonka, I. Pelivanov, and L. Ambrozinski, "Evaluation of disbonds at various interfaces of adhesively bonded aluminum plates using all-

optical excitation and detection of zero-group velocity Lamb waves," *NDT & E International,* vol. 112, 2020. https://doi.org/10.1016/j.ndteint.2020.102249

[15] Q. Liu *et al.,* "Advancing measurement of zero-group-velocity Lamb waves using PVDF-TrFE transducers: first data and application to in situ health monitoring of multilayer bonded structures," *Structural Health Monitoring,* 2022. https://doi.org/10.1177/14759217221126812

[16] X. Meng, M. Deng, and W. Li, "Validation of zero-group-velocity feature guided waves in a welded joint," *Ultrasonics,* vol. 136, p. 107173, Jan 2024. https://doi.org/10.1016/j.ultras.2023.107173

[17] Y. Wu, R. Cui, K. Zhang, X. Zhu, and J. S. Popovics, "On the existence of zero-group velocity modes in free rails: Modeling and experiments," *NDT & E International,* vol. 132, 2022. https://doi.org/10.1016/j.ndteint.2022.102727

[18] B. Ji, R. R. Mehdi, G.-W. Jang, and S. H. Cho, "Determination of third-order elastic constants using change of cross-sectional resonance frequencies by acoustoelastic effect," *Journal of Applied Physics,* vol. 130, no. 23, p. 235105, 2021. https://doi.org/10.1063/5.0069579

[19] C. Grunsteidl, T. W. Murray, T. Berer, and I. A. Veres, "Inverse characterization of plates using zero group velocity Lamb modes," *Ultrasonics,* vol. 65, pp. 1-4, Feb 2016. https://doi.org/10.1016/j.ultras.2015.10.015

[20] D. Clorennec, C. Prada, and D. Royer, "Local and noncontact measurements of bulk acoustic wave velocities in thin isotropic plates and shells using zero group velocity Lamb modes," *Journal of Applied Physics,* vol. 101, no. 3, p. 034908, 2007. https://doi.org/10.1063/1.2434824

[21] P. Mora, M. Chekroun, S. Raetz, and V. Tournat, "Nonlinear generation of a zero group velocity mode in an elastic plate by non-collinear mixing," *Ultrasonics,* vol. 119, p. 106589, Feb 2022. https://doi.org/10.1016/j.ultras.2021.106589

[22] W. Li, C. Zhang, and M. Deng, "Modeling and simulation of zero-group velocity combined harmonic generated by guided waves mixing," *Ultrasonics,* vol. 132, p. 106996, Jul 2023. https://doi.org/10.1016/j.ultras.2023.106996

Structural Health Monitoring: 10APWSHM
Materials Research Proceedings 50 (2025) 141-146

Materials Research Forum LLC
https://doi.org/10.21741/9781644903513-17

Changes in wave propagation due to damages and automatic recognition using a CNN model

Xin Wang[1,a] *, Aijia Zhang[2,b*], Yoshihiro Nitta[1,c], Ji Dang[2,d]

[1]Ashikaga University, Omae Cho, Ashikaga, Tochigi, Japan

[2]Saitama University, Shimo-Okubo Sakuraku, Saitama, Japan

[a]wang.xin@g.ashikaga.ac.jp, [b]zhang.a.968@ms.saitama-u.ac.jp,
[c]nitta.yoshihiro@g.ashikaga.ac.jp, [d]dangji@mail.saitama-u.ac.jp

Keywords: Wave Propagation, Interference, Impulse Response, CNN Model

Abstract. The vibration of buildings is generally considered as a superposition of natural vibration modes, but it can also be interpreted from the perspective of wave propagation in the vertical direction of SH wave. When the virtual source is placed on the roof, the impulse response consists of one non-causal wave and one causal wave, making the wave propagation path clearer. In this study, using vibration table experiments with building models, we visualized the changes in the overall wave field from the foundation to the roof when interlayer stiffness decreases due to damage and temporal changes of the impulse response in the damaged and adjacent layers. These changes were automatically recognized using a trained CNN model.

Introduction

System identification, such as natural frequencies and mode shapes, can indicate the possibility of structural damage or deterioration. Conventional methods of building system identification in the frequency domain may not be feasible when natural frequencies are affected by the soil-structure interaction. On the other hand, when building horizontal vibration is viewed from the perspective of shear wave propagation in the vertical direction, it is possible to evaluate the properties of the superstructure only, as the boundary conditions are changed by wave field reconstruction by setting the virtual impulse source on the roof using the deconvolution analysis. The reconstructed new wavefield consist of one acausal up-going wave and one causal down-going wave, and the propagation path of the waves is clearer.

When a building vibrates in an elastic state, the interlayer stiffness is constant, so the speed of the wave propagating in the vertical direction remains steady, and the waves propagate orderly. However, once damage occur in a certain layer, the interlayer stiffness of that layer decreases, causing the velocity within the layer to slow down, and interference or dispersion occurs at the layer boundaries. In this case, the wave field of the waves propagating vertically becomes unstable. By using the temporal changes in this wave field, it is possible to understand the inter-story damage of the building.

In this study, shake-table tests of a six-story building model with weak stories at the second and the third story are used to examine the changes of the wave field to examine the anomaly in the wave field. Then using a trained CNN model to automatically recognize the anomaly from the figure of temporal the impulse response.

Time-Space Wavefield Reconstruction

Figure 1(a) shows a 1-D layered model of a building and an image of the propagation of pulses incident from the foundation of the building between the foundation and the top in the vertical direction. The propagation of the SV waves incident from the foundation through the layers of the super structure is identical to the motion of a semi-infinite layered structure excited by vertically incident SH waves. Reflections at the building foundation and the top, as well as transmissions

Structural Health Monitoring: 10APWSHM   Materials Research Forum LLC
Materials Research Proceedings 50 (2025) 141-146   https://doi.org/10.21741/9781644903513-17

and reflections at internal stories, cause vibrations in the superstructure. Figure 1(b) shows the time history waveforms at the boundaries of the layers, i.e., at each floor. We call it the original time-space wave field.

To separate the waves, we construct a new wave field (impulse response) when the virtual source is placed on a top using the deconvolution method shown in Equation (1)[1]. The deconvolved wave composed of one up-going wave (negative time side) and one down-going wave (positive time side). The wave propagation path can be more easily read from the deconvolved waves.

$$D_i(\omega) = \frac{u_i(\omega)u_t^*(\omega)}{|u_t(\omega)|^2 + \varepsilon} \qquad (1)$$

where $u_t(\omega)$ is the Fourier transform of the vibration on the top, $u_i(\omega)$ is the Fourier transform of the vibration on the i-th floor, $u_t^*(\omega)$ is the conjugate of $u_t(\omega)$, and ε is a constant to avoid instability near the notch of $|u_t(\omega)|$, set as 10% of the average of $|u_t(\omega)|$.

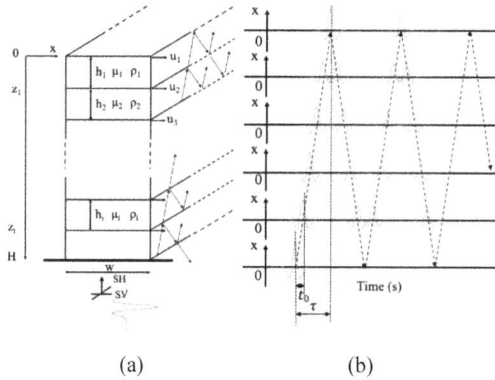

(a)   (b)

Figure 1. (a) Imaged diagram of wave propagation in a 1-D layered structure; (b) Imaged diagram of time history of waveforms at each floor (original wavefield)

Outline of the Building Model and the Shake-table Test

In this study, a six-story wall-type building model as shown in Figure 2 is used. The floor material of the building model is 1-mm thick bonded steel plate, 300 mm x 120 mm (folded 10 mm on both sides). The columns are 1-mm thick aluminum plates with two types of size (i) 70 mm x 250 mm for the normal stories and (ii) 60 mm x 250 mm for the damaged stories. The joints are 1 mm thick aluminum equal-sided angles, 25 mm x 25 mm x 70 mm. Assuming that the aluminum angles are not deformed, the effective height of the column is 200 mm. The mass of each floor is 0.45 kg. The damaged stories are set to be the second and the third story.

High-sensitivity single-axis horizontal accelerometer ARS-10A manufactured by Tokyo Sokki Research Laboratory was used to observe the response at each floor. The observed acceleration waveforms are shown in Figure 3. The sampling rate fs was set to 1000 Hz to accurately read the propagation time of the new wave field reconstructed from the acceleration response. The direction of excitation in the white noise excitation experiment is in the direction of the weak axis of the building model, which is the direction of the arrow in Figure 2(b).

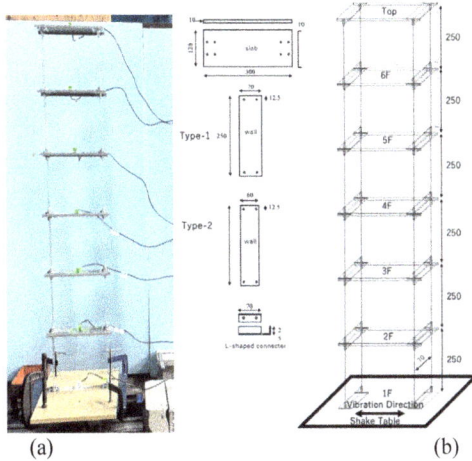

Figure 2. (a) Photo of the six-story building model; (b) the size of the model.

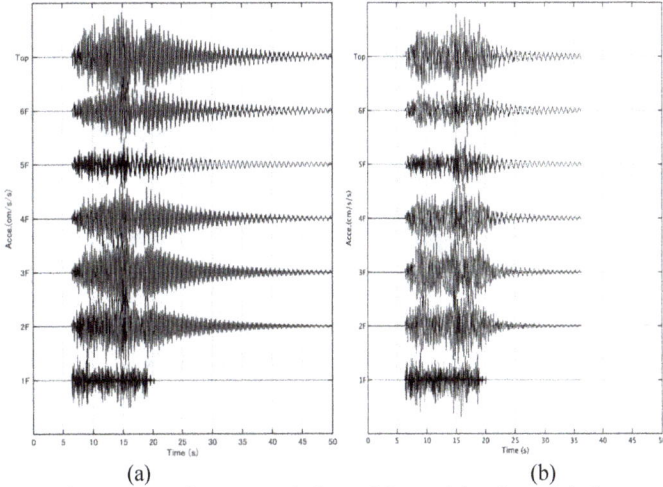

Figure 3. (a) Acceleration waveforms at each floor of the model with weak 2ed story and (b) that of the model with weak 3rd story.

Wavefiled Reconstruction

Figure 4 shows the deconvolved waveforms (new wavefield) when the virtual source is located on the top, reconstructed from the acceleration response waveforms (original wavefield in Figure 3) of the white noise shake-table tests. The non-causal up-going and causal down-going waves of the deconvolved waves at each floor are clearly visible. Reflected waves with reversed phase due to the high impedance at the damaged story are shown in green broken lines. To show the path of impulse more clearly, the figures of the deconvolved waveforms are shown in Figure 5.

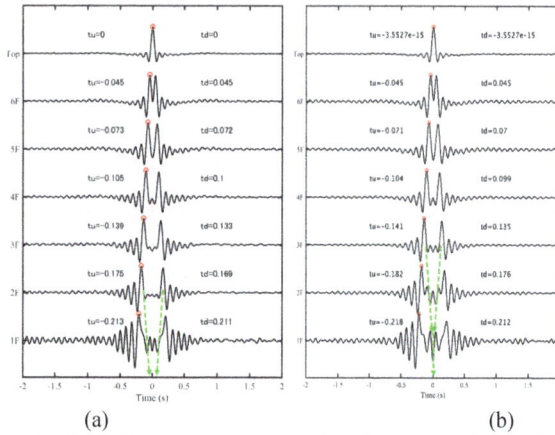

Figure 4. Deconvolved waveforms respect to the virtual source on the top. (a) is for the model with weak story at the 2ed story and (b) is for the model with weak story at the 3rd story.

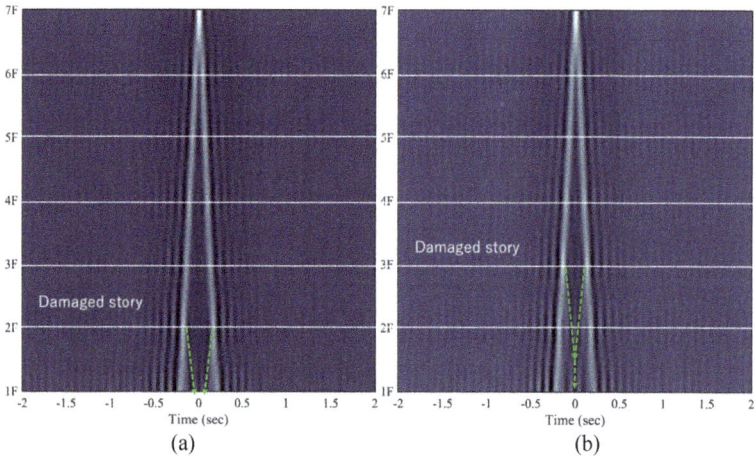

Figure 5. Bone coloured figures of the waveforms in the figure 4 to show the path of impulse.

Time-sequential Impulse Response

Time-sequential impulse responses within the duration of vibration at all floors have been obtained using a 10-second sliding window. The results are shown in Figure 6. From these figure it can be seen that the reflected wave between the two main impulse can be seen at the floors under the damaged story.

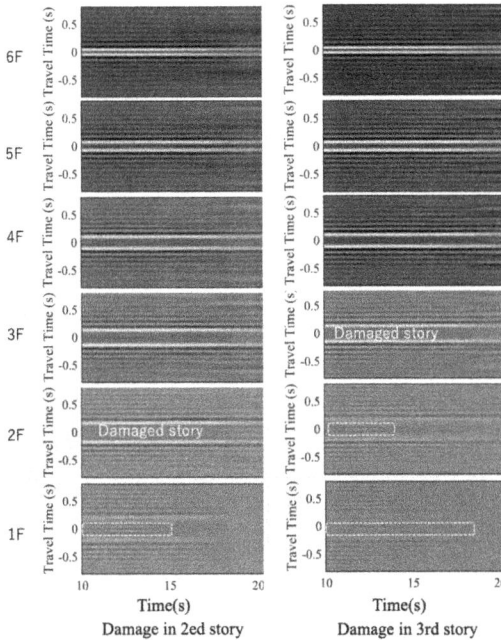

Figure 5. Time-sequential impulse responses within the duration of vibration at all floors of the six-story building model.

Automatic Recognition Using a Trained CNN Model

In previous study, we have constructed a CNN model to automatically recognize the change of the visualized wavefield[2]. The trained CNN model is used to recognize figures of linear and non-linear cases of structures. The accuracy of proposed method is satisfactory. In this study the trained CNN model is used to recognize the anomaly from the figure of time-sequential impulse responses of the damaged story, the stories under them and the stories with no damages.

The confusion matrix of the total 12 figures shown in the Figure 5 are shown in the Figure 6. The accuracy is 50%. The miss recognition happened to the classification of the time-sequential impulse responses of the stories with no damaged. On the other hand, the time-sequential impulse responses of the stories with damages and the stories under the damaged stories are all correctly classified.

The Grad-CAM results of the time-sequential impulse responses of the stories under the damaged stories are shown in Figure 7. It can be seen that the reflected waves are not recognized.

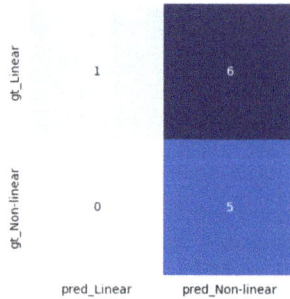

Figure 6. Confusion matrix of the classification results of the total 12 figures in figure 5.

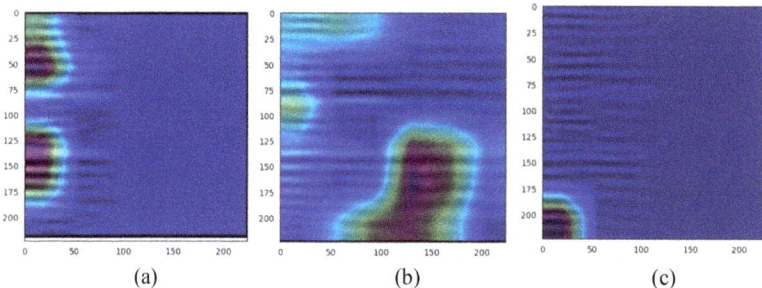

(a)   (b)   (c)

Figure 7. Grad-CAM figures of the time-sequential impulse responses of the (a) the first floor of the model with weak 2ed story, (b) the first and (c) the 2ed floor of that with weak 3rd story.

Summary

In this study, we examined how wave propagation in a building's vertical direction (shear wave) changes when structural damage occurs using shake-table test of six-story building models combining with a trained CNN model to recognize structural anomalies automatically. Visualizing wave propagation paths by creating a virtual source on the roof and employing deconvolution allows clearer identification of causal and non-causal wave components. The trained CNN model automatically identifies visual patterns of damage within time-sequential impulse response figures. While the model accurately identifies damaged and adjacent layers, there is some misclassification with undamaged stories. Grad-CAM analysis further identifies critical areas in the impulse response where reflections are detected but not fully recognized by the model, suggesting a direction for improving model sensitivity.

References

[1] Snieder R, Safak E. Extracting the building response using seismic interferometry: Theory and application to the Millikan Library in Pasadena, California. Bulletin of the Seismological Society of America, 2006; 96(2): 586-598. https://doi.org/10.1785/0120050109

[2] Aijia Zhang, Xin Wang, Ji Dang: CNN‐based damage detection of buildings from wave propagation between two adjacent floors, Earthquake Engineering and Resilience, Volume2, Issue4:479-492,2023. https://doi.org/10.1002/eer2.62

Structural Health Monitoring: 10APWSHM
Materials Research Proceedings 50 (2025) 147-154

Materials Research Forum LLC
https://doi.org/10.21741/9781644903513-18

Analysis of the mechanical properties of surrounding rock in large span tunnel with variable cross-section under different excavation methods

Changze Li[1,a] *, Dongming Zhang[1,b], Yuxuan Xia[1,c]

[1]Key Laboratory of Geotechnical Engineering, Dept. of Geotechnical Engineering, Tongji Univ., Shanghai 200092, China

[a]2232580@tongji.edu.cn, [b]09zhang@tongji.edu.cn, [c]2232581@tongji.edu.cn

Keywords: Variable Cross-Section Tunnel, Numerical Simulation, Distribution Expansion Excavation, Step-By-Step Excavation

Abstract. In order to explore the impact of variable cross-section tunnel excavation on the surrounding rock, this paper takes the pump room section of the second Qingdao Jiaozhou Bay submarine tunnel as the engineering background. Using the FLAC3D finite difference software, a three-dimensional numerical simulation of the entire excavation and support process for variable cross-section tunnels with different excavation methods was conducted. The displacement field, stress field, and the variation in the forces acting on the support structure were analyzed to select a reasonable excavation method. The results show that adopting the distribution expansion excavation method combined with a three-step excavation method for large span variable cross-section tunnels can effectively control stress concentration and deformation. Compared with the direct use of the step-by-step excavation method, both the surrounding rock deformation and the force on the support structure are smaller. This study provides theoretical support and guidance for the construction design of large-span variable cross-section tunnels.

1. Introduction

With the continuous development of infrastructure construction in China, tunnel engineering plays an increasingly important role in urban transportation, railway, and highway construction. However, large-span tunnels with variable cross-sections face challenges due to their complex cross-sectional changes and the varying mechanical properties of surrounding rock. During the construction process, these tunnels are prone to stress concentration and ground deformation issues, which significantly increase construction risks and difficulty. These problems not only affect the stability of the tunnel structure but also pose potential threats to surrounding buildings and surface facilities. Therefore, it is particularly necessary to conduct in-depth research on the mechanical characteristics during construction.

In recent years, researchers have explored the stress and deformation characteristics of large-span tunnels with variable cross-sections through numerical simulations and field measurements, proposing various construction control and optimization schemes. For example, regarding the surface settlement problem during the construction of variable cross-section tunnels, Li Tao et al.[1] proposed an effective prediction model based on random medium theory and verified the applicability of the model with measured data. Fang Gang et al.[2] used the elastoplastic finite element method to simulate the mechanical characteristics of variable cross-section tunnels under different construction methods, showing that a reasonable excavation method can significantly improve the stability of the surrounding rock. Additionally, He Minghua et al.[3] studied the construction risks of variable cross-section tunnels passing under existing structures and proposed effective ground reinforcement and construction control measures.

Structural Health Monitoring: 10APWSHM   Materials Research Forum LLC
Materials Research Proceedings 50 (2025) 147-154   https://doi.org/10.21741/9781644903513-18

This study is based on the second Qingdao Jiaozhou Bay submarine tunnel pump room section, which mainly crosses slightly weathered granite formations with relatively hard and stable geological conditions. The FLAC3D numerical simulation was used to analyze the mechanical properties of surrounding rock in large-span tunnels with variable cross-sections under different excavation methods, aiming to provide theoretical support and construction guidance for similar projects.

2. Engineering Overview

The pump room section of the Qingdao Jiaozhou Bay Second Submarine Tunnel passes through granite and other igneous rocks. The surrounding rock is intact and strong, with drilling and blasting methods used for excavation. Located at SK14+707 (NK14+658), the pump room lies within slightly weathered granite, classified as Grade III2 rock. The overburden is at least 30 meters thick, with a permeability coefficient of 0.03 m/d and a calculated seepage flow of 6.36 m^3/d*m. The groundwater is primarily fissure water from the bedrock, divided into highly weathered pore-fissure water and moderately weathered bedrock fissure water.

The foundation rests on stable bedrock, which can fully support the required load, with no exposed cavities or weak layers that could cause instability. The strata below the foundation consist of slightly weathered medium-grained granite with some diabase dykes, making the foundation relatively uneven but stable.

The strata at the top and sidewalls consist of slightly weathered granite, with well-developed joints and good rock mass integrity, rated Grade III2. During excavation, a combination of bolt-spraying and reinforced mesh support is recommended, using a step-by-step excavation method. For areas with fractures or abundant groundwater, pre-grouting should be applied. The bedrock foundation has a high bearing capacity, much greater than the load of the superstructure, so settlement is not a concern.

The submarine pump room section will be excavated using the distribution expansion method, which is complex and presents significant support challenges. The presence of seawater adds difficulty, with high water pressure at the junctions making reinforcement difficult. Tunnel instability could lead to severe water inflow hazards, posing substantial construction risks.

3. Excavation Scheme for Variable Cross-Section Subsea Pump Room

The excavation of the variable cross-section submarine pump room adopts the distribution expansion method (as shown in Fig.1). When the tunnel cross-section changes, the distribution expansion method effectively prevents stress concentration. This method involves not directly excavating the entire variable cross-section at once, but instead, excavating in stages or segments. This allows for a gradual reduction in the sudden increase of local stress at each stage. By incrementally increasing the excavation volume, the method effectively mitigates the abrupt changes in ground stress caused by the cross-section transition.

The basic process of the pump room expansion excavation is as follows: First, demolish the original left-side shotcrete and anchor lining of the service tunnel, in excavation area ①. Perform initial support for the tunnel structure, spraying 4 cm thick concrete. Next, demolish the shotcrete and anchor lining of the original service tunnel arch in excavation area ②, and apply initial support for the tunnel structure. Then, demolish the shotcrete and anchor lining of the original service tunnel arch again in excavation area ③, and apply initial support for the tunnel structure. Finally, excavate areas ④, ⑤, and ⑥, and apply initial support for the tunnel structure.

Considering different excavation methods, the excavation advance is 2 meters, and the distance between the faces of the step-by-step excavation method is 10 meters. The simulation working conditions are shown in Table 1.

Structural Health Monitoring: 10APWSHM
Materials Research Proceedings 50 (2025) 147-154

Materials Research Forum LLC
https://doi.org/10.21741/9781644903513-18

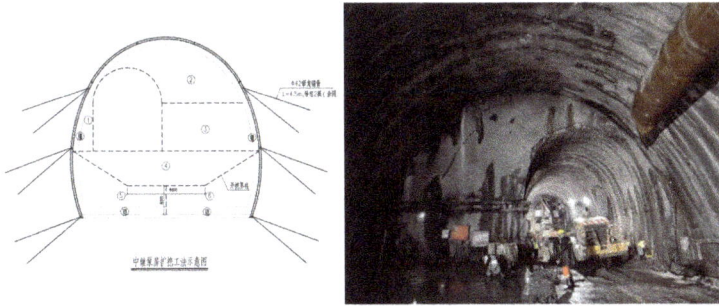

Fig.1 Submarine Pump Room Construction Plan Diagram

Table 1 Tunnel Excavation Conditions for Two Excavation Methods

Excavation Method	Excavation Advance	Distance Between Steps
Distribution Expansion Method	2m	10m
Direct Step Method	2m	10m

4. Numerical Modeling and Results Analysis

4.1 Introduction to FLAC 3D FLAC3D (Fast Lagrangian Analysis of Continua in 3 Dimensions) is a three-dimensional numerical simulation software based on the finite difference method, widely used in geotechnical engineering, underground structure analysis, and mining engineering. The software employs the Lagrangian method, which can simulate the complex nonlinear behaviors of materials under external forces, including deformation, stress distribution, and instability failure processes. Due to its powerful features and flexible computational capabilities, FLAC3D is extensively applied in both academic research and engineering practice.It has the following key features: Constitutive Models: FLAC3D provides a variety of constitutive models, such as the elastoplastic model, Mohr-Coulomb model, and Duncan-Chang model, which can be used to simulate the different mechanical behaviors of geotechnical materials. Users can also customize material models according to engineering requirements, increasing the accuracy and flexibility of calculations[4]. Lagrangian Method: FLAC3D uses the Lagrangian method, which has unique advantages in simulating large deformation problems, such as the surrounding rock deformation and failure processes caused by tunnel excavation. The software dynamically tracks the movement of nodes and elements, making large deformation analysis more accurate and stable[5]. Stepwise Loading and Real-Time Adjustments: The software supports stepwise loading and real-time adjustment of excavation and support sequences. Users can simulate complex construction processes and analyze the impact of each step on the surrounding rock and support structures. Through these simulations, engineers can optimize construction plans, improving both safety and efficiency[6].

4.2 Model Establishment Based on the actual excavation conditions of the service tunnel pump room section of the Qingdao Jiaozhou Bay Second Submarine Tunnel project, a three-dimensional complex model was established using the finite element software ABAQUS, combined with large-scale finite difference software FLAC3D for computational analysis. According to the Saint-Venant principle, the transverse calculation range for tunnel excavation is taken as 3 to 5 times the tunnel diameter. In this model, the left and right boundaries are set at 3 times the tunnel diameter

Structural Health Monitoring: 10APWSHM   Materials Research Forum LLC
Materials Research Proceedings 50 (2025) 147-154   https://doi.org/10.21741/9781644903513-18

to eliminate boundary effects. The specific model dimensions are defined as follows: the transverse width is 100 m, the longitudinal length is 60 m (in the direction of tunnel excavation, with small and large spans of 30 m each), and the model height is 100 m. The left and right boundary directions are constrained, as well as the front and back boundary directions, with constraints applied to the bottom surface, while the upper surface is free. The three-dimensional computational model is shown in Fig.2.

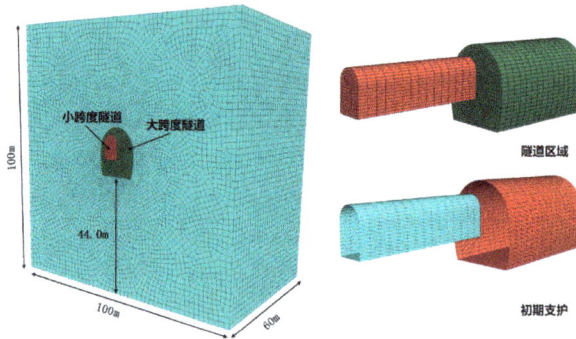

Fig.2 Three-Dimensional Model of the Subsea Pump Room

4.3 Ground Reinforcement and Tunnel Support Simulation Advance Support Simulation: Advance support in the mined section primarily uses small forepoling pipes and pre-grouting. To reflect the support effect, the small forepoling pipes are simulated using circumferentially inclined anchor elements (Cable), and the pre-grouting effect is modeled by increasing the mechanical parameters of the surrounding rock at the working face. The elastic modulus of the reinforcement zone is increased by four times, while the internal friction angle remains unchanged, and cohesion is increased by 30%[7][8].

Primary Support: The primary support consists of shotcrete and steel arch frames, simplified based on equivalent stiffness principles. The elastic modulus of the shotcrete is increased, and steel arch frames are equivalently modeled. The support structure is simulated using shell elements, and the equivalent elastic modulus is calculated as follows[9]:

$$E_{\text{equivalent}} = \frac{E_{\text{concrete}} A_{\text{concrete}} + E_{\text{steel}} A_{\text{steel}}}{A_{\text{total}}} \tag{1}$$

Where $E_{\text{equivalent}}$ is the equivalent elastic modulus, E_{concrete} and E_{steel} are the elastic moduli of shotcrete and steel, and A_{concrete} and A_{steel} are their respective cross-sectional areas.

Table 2 and Table 3 show the physical and mechanical parameters of the model soil layer and the supporting structure parameters after equivalence.

Table 2 Physical and Mechanical Parameters of Model Rock Layers

Material	Density (kg/m³)	Elastic Modulus (GPa)	Poisson's Ratio	Cohesion (kPa)	Friction Angle(°)
Silt	1600	12	0.3	30	10
Silty Clay	2000	19	0.4	19	18
Medium-Coarse Sand	2000	25	0.26	20	35
Slightly Weathered Granite	2500	30	0.31	26	36.4

Table 3 Support Structure Parameters

Material	Density (kg/m³)	Elastic Modulus (GPa)	Poisson's Ratio	Thickness (m)
Primary Support (Equivalent)	2500	35	0.2	0.1
Secondary Lining	2200	30	0.2	0.4

4.4 Numerical Model Basic Assumptions The following assumptions are made for the FLAC3D analysis of the distribution expansion method for submarine pump room excavation:

The soil is considered a homogeneous, isotropic continuous medium, and the Mohr-Coulomb constitutive model is applied. The excavation is carried out in cycles, with immediate support installed after each excavation step. Ground reinforcement is simulated according to the design requirements. Given that the rock layer at the cavern is slightly weathered granite, with good lithology and low permeability, classified as Grade III surrounding rock, the influence of groundwater seepage during tunnel excavation is ignored. The overlying water body is simplified as a uniformly distributed load, with tidal effects ignored. The focus is on the tunnel excavation and initial support processes.

4.5. Analysis of Numerical Simulation Results This study primarily explores the impact of different excavation methods on the surrounding rock stress and deformation, as well as the stress on the support structure in large-span variable cross-section tunnels. The focus of the numerical calculation is twofold: one is the disturbance to the surrounding rock caused by different excavation methods during construction, analyzed by monitoring the change in the crown settlement of a typical cross-section with each excavation step. The cross-section is selected from the middle of the large-span tunnel (i.e., 15 meters from the variable cross-section). The second focus is the effect of different excavation methods on the surrounding rock displacement field, stress field, and support structure stress after the completion of excavation, used to evaluate the rationality of the two excavation methods.

The crown settlement values of typical cross-sections for both excavation methods are extracted, and the curve showing the crown settlement variation with excavation steps is plotted, as shown in Fig.3. The horizontal axis represents the computation steps of the numerical simulation.

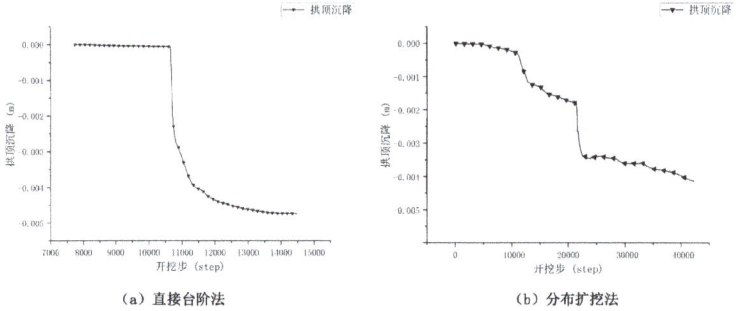

（a）直接台阶法 　　　　　　　　　　　（b）分布扩挖法

Fig.3 Crown Settlement Variation with Time Steps for the Monitored Section under Different Excavation Methods

As shown in Fig.3, regardless of the excavation method, vertical settlement has already occurred at the monitored section when the tunnel excavation reaches the front of the variable cross-section. However, this value is very small, accounting for about 5% of the final deformation. The direct step method, due to the use of the step-by-step excavation for the large span directly after completing the small span, without a transition section or the use of the distribution expansion method, causes greater disturbance to the surrounding rock, resulting in a more abrupt increase in vertical settlement. In contrast, the distribution expansion method causes less disturbance to the surrounding rock because of multiple small-scale excavations, which balances the excavation process. The overall crown settlement curve shows a stepped pattern. After the initial lining is completed, the crown settlement in both methods quickly stabilizes.

The final vertical displacement variation of the surrounding rock for the two excavation methods is shown in Fig.4.

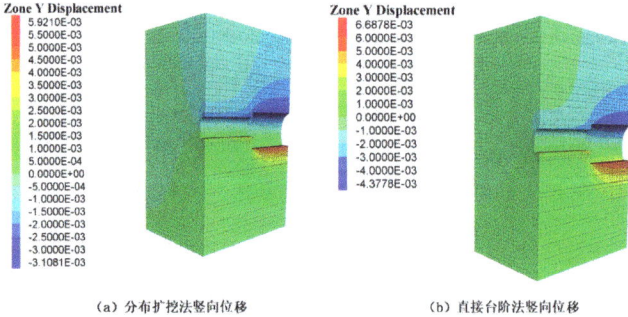

（a）分布扩挖法竖向位移 　　　　　　　（b）直接台阶法竖向位移

Fig.4 Displacement Contour Map for Different Excavation Methods

From the displacement contour map, it is clearly observed that the range of crown settlement for the surrounding rock is larger in the direct step method compared to the distribution expansion method. In both excavation methods, a significant displacement change is observed at the location of the variable cross-section. To compare the final crown settlement values in the longitudinal direction for the two methods, the data from the contour map is extracted and the corresponding displacement variation curve is plotted, as shown in Fig.5.

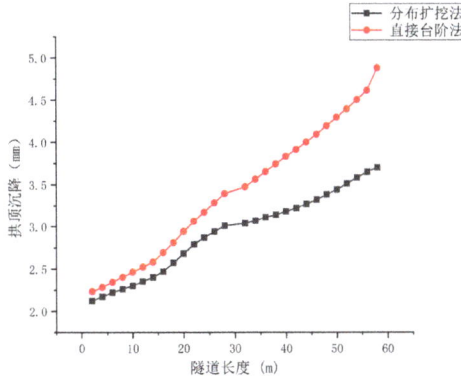

Fig.5 Vertical Displacement Comparison Chart for Different Excavation Methods

From Fig.5, it can be concluded that the crown settlement under both excavation methods exhibits similar fluctuating patterns. Specifically, the vertical settlement trends for the distribution expansion method and the direct step method are quite similar, but there are significant numerical differences. The maximum settlement for the distribution expansion method is 3.10 mm, while the maximum settlement for the direct step method is 4.38 mm, which is 29% higher than that of the distribution expansion method. The maximum heave for the direct step method is 6.88 mm, which is 14% higher than the 5.92 mm observed for the distribution expansion method.

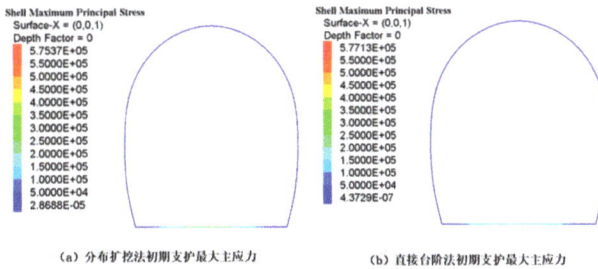

(a) 分布扩挖法初期支护最大主应力　　　(b) 直接台阶法初期支护最大主应力

Fig.6 Initial Support Stress Contour Map for the Monitored Section under Different Excavation Methods

From the analysis of Fig.6, it can be observed that under both excavation methods, the maximum principal stress of the initial support at the monitored section is compressive stress, which occurs at the arch bottom of the initial support. The maximum principal stress for the direct step method is higher than that of the distribution expansion method, with a compressive stress value of 0.668 MPa for the direct step method and 0.616 MPa for the distribution expansion method. The stress variation patterns for the tunnel initial support in both conditions are similar, with the stress decreasing from the crown to the arch bottom.

Summary
1. This study aims to explore the impact of different excavation methods on the surrounding rock stress and deformation, as well as the stress on the support structure, in large-span variable cross-section tunnels. Using FLAC3D for 3D numerical simulations, the effects of distribution

Structural Health Monitoring: 10APWSHM   Materials Research Forum LLC
Materials Research Proceedings 50 (2025) 147-154   https://doi.org/10.21741/9781644903513-18

expansion method and direct step method were compared. The simulation results indicate that the distribution expansion method is more effective in controlling tunnel surrounding rock stress concentration and deformation, reducing rock disturbance and alleviating support structure stress.

2. Analysis of Numerical Simulation Results

Crown Settlement: The simulations show that the distribution expansion method effectively controls crown settlement, with a maximum settlement of 3.10 mm, significantly lower than the 4.38 mm of the direct step method. This indicates that the distribution expansion method causes less disturbance to the surrounding rock and results in more stable settlement.

Surrounding Rock Stress and Displacement: The surrounding rock displacement and stress patterns for both methods are similar, but the direct step method causes slightly higher stresses and a larger disturbance range. The distribution expansion method, by using step-by-step excavation, effectively controls the surrounding rock disturbance and stress concentration.

3. This study provides theoretical support for the design and construction of large-span variable cross-section tunnels. The distribution expansion method performs better, especially in complex geological conditions, by reducing surrounding rock deformation and stress concentration. It is recommended to prioritize this method in similar tunnel projects, particularly in high water pressure or complex geological environments, to ensure tunnel stability and construction safety. Furthermore, future research should focus on the behavior of surrounding rock under complex geological conditions, offering more reliable technical support for similar projects.

References

[1] He Minghua. Analysis of the Impact of Boring a Varied-Section Tunnel Underground Passage. Railway Construction Technology, 2019.

[2] Fang Gang, Ding Chunlin, Jia Runzhi, Wu Chao. Analysis of Construction Mechanical Properties of a Large-span Varied-section Tunnel Excavation. Modern Tunneling Technology, 2018.

[3] Li Tao, Wang Yibo, Yu Zhiwei, Liu Bo. Prediction of Surface Soil Movement and Settlement in Varied-section Tunnel Excavation. Journal of Central South University (Natural Science Edition), 2020.

[4] Itasca Consulting Group. (2019). FLAC3D Manual. Minneapolis: Itasca.

[5] Potts, D. M., & Zdravković, L. (2001). Finite Element Analysis in Geotechnical Engineering: Theory and Application. Thomas Telford Publishing. https://doi.org/10.1680/feaigea.27831

[6] Cundall, P. A. (2000). Numerical modeling of jointed and faulted rock. International Journal of Rock Mechanics and Mining Sciences, 37(1-2), 165-187.

[7] Hou Chaojiong, Gou Panfeng. Research on the Mechanism of Strengthening Rock Mass by Anchored Bolts in Roadway Support[J]. Acta Geotechnica et Engineering Geophysica, 2000(03):342-345.

[8] He Xiangang, Bi Ran. Numerical Analysis of Mechanical Properties of Continuous Super-ahead Pipe Shielding Support in Fully Weathered Granite Tunnel[J]. Journal of Highway Engineering, 2019, 44(03):216-221+252.

[9] Zhu Zhengguo, Li Wenjiang, Song Yuxiang. 3D Numerical Simulation Analysis of Super Large-span Railway Station Tunnel[J]. Rock and Soil Mechanics, 2008, 29(S1):277-282. https://doi.org/10.16285/j.rsm.2008.s1.036

Structural Health Monitoring: 10APWSHM
Materials Research Proceedings 50 (2025) 155-162

Materials Research Forum LLC
https://doi.org/10.21741/9781644903513-19

Enhancing displacement control performance of base-isolated structures using TVMD and SIS

Shotaro Kezuka[1,a*] and Kohju Ikago[1,b]

[1]Earthquake Engineering Lab., International Research Institute of Disaster Science, Tohoku University, 468-1, Aramaki-Aza Aoba, Aoba-ku, Sendai 980-8572 Japan

[a]shotaro.kezuka.q8@dc.tohoku.ac.jp, [b]koju.ikago.e8@tohoku.ac.jp

Keywords: Displacement Control Design, Seismically Isolated Structures, Inerter, Long-Period Ground Motion

Abstract. Base isolation of structures allows for large isolator displacements during strong ground motions, which reduces the seismic forces acting on a superstructure. However, recent large-scale earthquakes have triggered long-period ground motions, raising concerns regarding excessive isolator displacement beyond design limits. This study proposes a method to selectively increase damping around the isolation period using an inerter-based dynamic vibration absorber, thereby reducing excessive deformation in the isolation layer without compromising the isolation effect.

Introduction

Since the 2000s, a growing concern in Japan has emerged regarding the potential adverse effects of long-period seismic motion—often associated with large-amplitude seismic motion—on long-period structures, including base-isolated structures [1]. Base-isolation structures effectively reduce the absolute acceleration responses of their superstructures at the expense of isolator displacement. A critical research question in enhancing seismic isolation performance involves improving displacement control while maintaining or enhancing acceleration control performance. In conventional velocity-dependent dampers, an increase in the damping ratio for the first mode, which predominantly influences the displacement response, simultaneously results in an increased higher-order damping ratio, thereby compromising acceleration control performance.

Previous research on existing issues has adopted approaches such as increasing the damping coefficient of the energy-dissipating device to mitigate displacement at the expense of acceleration control performance when excessive isolator displacement is detected [2], and designing a mechanical network that achieves the desired frequency characteristics for dissipating energy [3,4]. The latter primarily aims to realize the benefits of rate-independent linear damping, also known as structural or linear hysteretic damping, which enhances acceleration control performance when incorporated into a long-period structure.

This study focuses on designing an inerter-based mechanical network according to the desired frequency characteristics to develop a novel isolation damper model that improves displacement control without compromising acceleration control performance. The proposed model is linear; thus, its effectiveness is verified within the linear range.

Damping Conditions Enhancing Seismic Isolation Performance

For this study, we considered base-isolated structures with sufficiently rigid superstructures. We examined a single-degree-of-freedom (SDOF) system incorporated with linear viscous damping (LVD) (hereafter referred to as SDOFBI-L, Fig. 1(b)) as a reduced model of an nDOF system incorporated with LVD (hereafter referred to as nDOFBI-L, Fig. 1(a)). In existing base-isolated structures, nDOFBI-L is considered a representative linear model.

In Fig. 1, m_j, c_j, and k_j ($j = 0,1,2,...,n-1$) denote the j-th mass, damping coefficient, and stiffness, respectively. Specifically, $j = 0$ denotes the isolation level. Following the Japanese conventional design practice of seismically isolated structures, we set the fundamental natural angular frequency ω_0 to $\pi/2$, implying that the period provided by the isolator and superstructure mass equaled 4 s. The damping ratio h_L at the fundamental natural frequency of the isolated system was considered to be 20% of critical damping, resulting in $\omega_0 = \sqrt{k_0/M} = \pi/2$, $M = \sum_{j=0}^{n-1} m_j$, $c_0 = c_L = 2h_L k_0/\omega_0$, and $h_L = 0.2$.

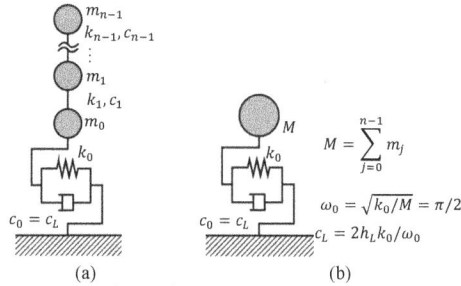

Figure 1: (a) nDOFBI-L as the model of a conventional base-isolated structure (b) SDOFBI-L

The relative transmissibility, defined as the absolute value of the transfer function from ground displacement to the displacement of SDOFBI-L relative to the ground, can be expressed as:

$$|H_{RT,L}| = \frac{\omega^2}{\sqrt{\left(-\omega^2+\omega_0^2\right)^2 + 4h_L^2\omega_0^2\omega^2}} . \tag{1}$$

where ω denotes the excitation angular frequency.

In Fig. 2(a), the relationship between $|H_{RT,L}|$ and h_L is illustrated. A higher damping ratio resulted in a smaller peak value for the relative transmissibility.

The absolute transmissibility, defined as the absolute value of the transfer function from the ground acceleration to the absolute acceleration response of the SDOFBI-L, can be expressed as:

$$|H_{AT,L}| = \sqrt{\frac{\omega_0^4 + 4h_L^2\omega_0^2\omega^2}{\left(-\omega^2+\omega_0^2\right)^2 + 4h_L^2\omega_0^2\omega^2}} , \tag{2}$$

The relationship between $|H_{AT,L}|$ and h_L is depicted in Fig. 2(b). A higher damping ratio resulted in a smaller peak value of $|H_{AT,L}|$ within the frequency below $\omega = \sqrt{2}\omega_0$. Conversely, within the frequency range beyond $\omega = \sqrt{2}\omega_0$, a lower damping ratio resulted in a smaller peak value of $|H_{AT,L}|$.

Further, we aimed to design the damping coefficient of the isolation layer for each frequency observing Figs. 2(a) and 2(b). The results are shown in Fig. 2(c). In the frequency region lower than $\sqrt{2}\omega_0$, increasing the damping coefficient improved both relative displacement and absolute acceleration. However, at the frequencies beyond $\sqrt{2}\omega_0$, only relative displacement improved with compromised absolute acceleration. Thus, this study proposes an energy-dissipating device to achieve the damping coefficient indicated by the green line in Fig. 2(c).

Structural Health Monitoring: 10APWSHM
Materials Research Proceedings 50 (2025) 155-162

Materials Research Forum LLC
https://doi.org/10.21741/9781644903513-19

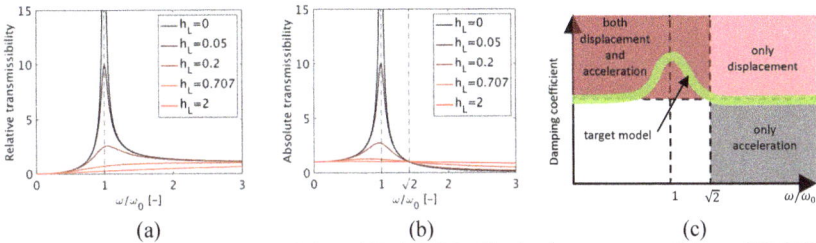

(a) (b) (c)

Figure 2: (a) Relative transmissibility of SDOFBI-L; (b) absolute transmissibility of SDOFBI-L; (c) the regions of the damping coefficient $c(\omega)$ of SDOF, where displacement response or acceleration response can be reduced compared to c_L, and the target model (green line).

Dividing LVD into Two Components

A parallel arrangement of a Maxwell element and a series inerter system (SIS) [5] reproduces LVD. Conversely, LVD can be divided into a Maxwell element and SIS. For verification, the following calculations were performed.

The dynamic stiffness of an LVD can be expressed as:

$$K_L = i\omega c_L. \tag{3}$$

The dynamic stiffness of a Maxwell element is given as:

$$K_M = \left[\frac{1}{k_M} + \frac{1}{i\omega c_M}\right]^{-1} = k_M \frac{i\omega}{i\omega + k_M/c_M}, \tag{4}$$

where k_M and c_M denote the spring stiffness and damping coefficient, respectively.

The dynamic stiffness of an SIS element is expressed as:

$$K_S = \left[\frac{1}{i\omega c_S} + \frac{1}{(i\omega)^2 m_S}\right]^{-1} = i\omega c_S \frac{i\omega}{i\omega + c_S/m_S}, \tag{5}$$

where c_S and m_S denote the damping coefficient and inertance, respectively.

Substituting $k_M = c_L\omega_0$, $c_M = c_S = c_L$, and $m_S = c_L/\omega_0$ into Eqs. 4 and 5 gives

$$K_M + K_S = c_L\omega_0 \frac{i\omega}{i\omega + \omega_0} + i\omega c_L \frac{i\omega}{i\omega + \omega_0} = i\omega c_L. \tag{6}$$

According to Eqs. 3 and 6, we have

$$K_L = K_M + K_S, \tag{7}$$

Thus, LVD was divided into a Maxwell element and SIS (Fig. 3).

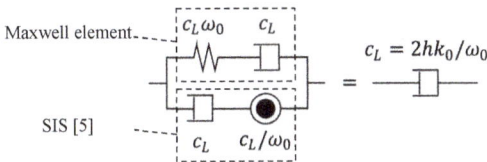

Figure 3: Dividing LVD into a Maxwell element and SIS [5]

Structural Health Monitoring: 10APWSHM | Materials Research Forum LLC
Materials Research Proceedings 50 (2025) 155-162 | https://doi.org/10.21741/9781644903513-19

The dynamic stiffness of the Maxwell element and SIS stemming from the LVD element with a damping coefficient of c_L is shown in Fig. 4. By enhancing the Maxwell element, the damping around the fundamental natural angular frequency ω_0 selectively increased to satisfy the criterion of the proposed damping device, as shown in Fig. 2(c) of Section 2. However, the undesirable storage stiffness increased in the high-frequency range, as shown in Fig. 4, which compromised the acceleration control performance compared with the LVD. Therefore, we propose replacing the Maxwell element with a tuned viscous mass damper (TVMD) [6].

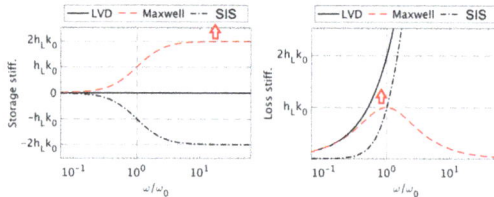

Figure 4: Dynamic stiffness of components of an LVD

Replacing the Maxwell Element of LVD with TVMD

Herein, we propose a hybrid viscous mass damper (HVMD), derived by replacing the Maxwell element in Fig. 3 with a TVMD, as shown in Fig. 5. The term "hybrid" indicates that the proposed model incorporated inerters in parallel and series with dampers.

The dynamic stiffness of the TVMD is expressed as:

$$K_T = \left[\frac{1}{k_T} + \frac{1}{(i\omega)^2 m_T + i\omega c_T}\right]^{-1} = k_T \frac{-\omega^2 + i\omega\epsilon_T}{-\omega^2 + i\omega\epsilon_T + \omega_0^2}, \tag{8}$$

where k_T, c_T, and m_T denote spring stiffness, damping coefficient, and apparent mass, respectively. In addition, ϵ_T denotes the ratio of c_T to m_T. For simplicity, we set the natural circular frequency to $\omega_0 = \sqrt{k_T/m_T}$), which resulted in a slight shift in the peak frequency of loss stiffness compared to the Maxwell element (Fig. 6 (a-2)).

Accordingly, the dynamic stiffness of HVMD is expressed as:

$$K_H = K_T + K_S. \tag{9}$$

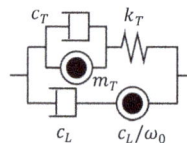

Figure 5: Hybrid VMD

The dynamic stiffnesses of the TVMD and Maxwell elements are shown in Fig. 6(a-1,a-2), respectively. The bandwidth of loss stiffness of the TVMD was narrower than that of the Maxwell element (Fig. 6(a-2)), and the storage stiffness of the TVMD in the high-frequency range was lower than that of the Maxwell element (Fig. 6(a-1)). Thus, the construction of the target model (Fig. 2(c)) could be approximately achieved using HVMD while eliminating the increase in storage stiffness in the high-frequency band (Fig. 6(b-1,b-2)).

Structural Health Monitoring: 10APWSHM
Materials Research Proceedings 50 (2025) 155-162

Materials Research Forum LLC
https://doi.org/10.21741/9781644903513-19

In Fig. 6, the parameters k_T, c_T, and m_T were set such that the loss stiffness of HVMD was a times as large as the loss stiffness of LVD at $\omega = \omega_0$ (Fig. 6(b-2)). Parameters k_T, c_T, and m_T were calculated using Eqs. 10, 11, and 12. The assignment of values to a and ϵ_T resulted in fixed parameters k_T, c_T, and m_T.

$$k_T/m_T = \omega_0^2. \tag{10}$$

$$\epsilon_T = c_T/m_T. \tag{11}$$

$$k_T = (2a-1)\frac{\epsilon_T}{\omega_0}h_L k_0, a > 1. \tag{12}$$

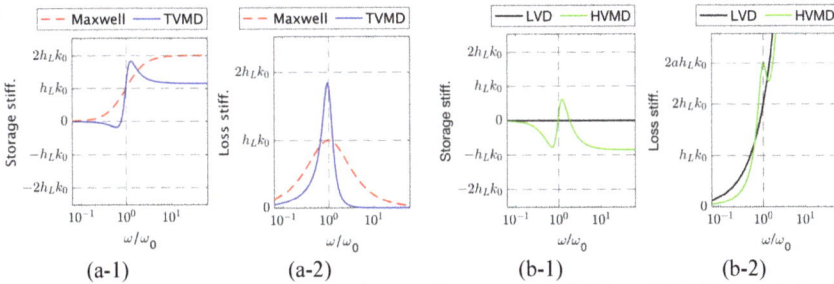

Figure 6: (a-1,2) dynamic stiffness of Maxwell component of LVD and TVMD ($a = 1.4, \epsilon_T = 1.0$) (b)dynamic stiffness of HVMD comprising SIS component of LVD, and TVMD($a = 1.4, \epsilon_T = 1.0$)

Optimal Design Method Based on SDOF system incorporated with HVMD

We propose an optimal design method for an SDOF system incorporating an HVMD and additional stiffness k_{adj} (hereafter referred to as SDOFBI-H, Fig.7). The objective of the optimal design is to minimize the infinity norm of the relative transmissibility of SDOFBI-H (H Infinity norm control problem (HINCP, Table. 1). The relative transmissibility of the SDOFBI-H is expressed as follows:

$$|H_{RT,H}| = \left|\frac{\omega^2}{-\omega^2+\omega_0^2+(K_H+k_{adj})/m}\right|. \tag{13}$$

In this study, we compared the control performance of damper models with their storage stiffness, equal to k_0 at $\omega = \omega_0$ (Eq. 14). In addition, we imposed the constraint in Eq. 15 to maintain the storage stiffness at $\omega = 0$ positive ensuring the static stability. The value of $k_{StaticMin}$ is provided by a structural designer without excessive reduction in its control performance against static or quasi-static forces, such as the average component of the wind force. To avoid the excessive effects of k_{adj}, Eq. 15 also demands maintaining the value $k_0 + k_{adj}$ below $2k_0$.

Structural Health Monitoring: 10APWSHM
Materials Research Proceedings 50 (2025) 155-162

Materials Research Forum LLC
https://doi.org/10.21741/9781644903513-19

Table 1

H Infinity Norm Control Problem (HINCP)	
find	a, ϵ_T, k_{adj}
to minimize	$\|H_{RT,H}\|_\infty$
subject to	$Re[k_0 + K_H + k_{adj}]\|_{\omega=\omega_0} = k_0,$ (14)
	$0 < k_{StaticMin} \le k_0 + k_{adj} < 2k_0.$ (15)

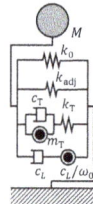

Fig. 7 SDOFBI-H

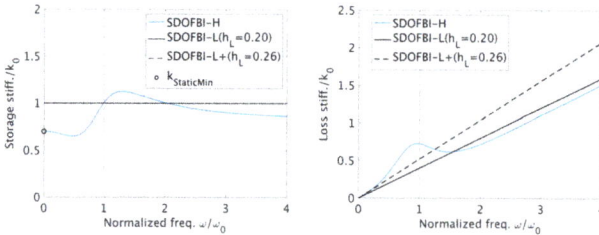

Figure 8: Dynamic stiffness of optimized SDOFBI-H
$(a = 1.81, \epsilon_T = 1.51, k_{adj} = -0.3k_0, k_{StaticMin} = 0.7k_0)$

Fig. 8 shows the optimized dynamic stiffness of SDOFBI-H with the following parameters: $a = 1.81, \epsilon_T = 1.51, k_{adj} = -0.3k_0$, which was obtained by solving HINCP with $k_{StaticMin} = 0.7k_0$. Fig. 8 shows the dynamic stiffness of the SDOFBI-L and SDOFBI-L+, where the damping ratio h_L of the SDOFBI-L increased from 0.2 to 0.26.

Figure 9(a) shows the relative transmissibility of the three models: SDOFBI-L, optimum SDOFBI-H, and SDOFBI-L+. Figure 9(b) shows the absolute transmissibilities of the aforementioned models. The absolute transmissibility of the SDOFBI-H is expressed as:

$$|H_{AT,H}| = \left| \frac{\omega_0^2 + (K_H + k_{adj})/m}{-\omega^2 + \omega_0^2 + (K_H + k_{adj})/m} \right|. \tag{16}$$

Figure 9(c) shows the absolute transmissibility normalized to that of the SDOFBI-L. In contrast to SDOFBI-L+, SDOFBI-H enhanced the acceleration control performance on the high-frequency side. This result was achieved by concentrating the range in which the acceleration control performance was inferior to that of the SDOFBI-L around the first natural frequency. Because the displacement response was dominant around the first natural frequency rather than acceleration, it can be posited that the SDOFBI-H enhances the displacement control performance while minimizing the deterioration of the acceleration control performance.

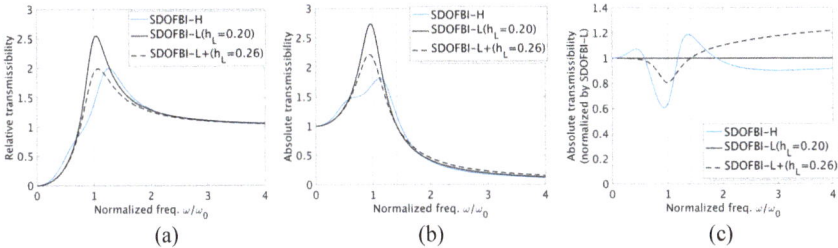

Figure 9
(a): Relative transmissibility, (b): Absolute transmissibility, (c): Normalized absolute transmissibility by SDOFBI-L ($h_L = 0.2$)

Summary

This study proposed a novel model of energy dissipation for seismic isolation comprising Maxwell and SIS elements in a parallel arrangement designated as HVMD, which was constructed using the decomposition theory of LVD originally discussed in this study as a starting point.

The optimum design for minimizing the infinite norm of the relative transmissibility of an SDOF system as a reduced model was examined. The frequency domain analysis of the optimized SDOFBI-H revealed that the HVMD increased the damping ratio compared to the LVD within a limited band around the fundamental natural frequency without increasing the transmissibility of the higher modes.

Therefore, the HVMD is considered more effective than the LVD in displacement control design. Future studies are planned to verify its effectiveness in the time domain by using a multi-degree-of-freedom system. Currently, we are conducting a nonlinear analysis that considers the influence of nonlinear elements in the seismic isolation layer; the results are expected to be presented shortly.

Acknowledgments

This study was supported by JST SPRING (grant number JPMJSP2114). We would like to thank Editage (www.editage.jp) for English language editing.

References

[1] I. Takewaki, S. Murakami, K. Fujita, S. Yoshitomi, M. Tsuji (2011) The 2011 off the Pacific coast of Tohoku earthquake and response of high-rise buildings under long-period ground motions, *Soil Dynamics and Earthquake Engineering*, 31(2011): 1511-1528. https://doi.org/10.1016/j.soildyn.2011.06.001

[2] K. Ikago, I. Fukuda, H. Luo, D. Li, H. Kanno, N. Hori (2021) Real-time hybrid testing of a passive variable orifice damper incorporated into base isolated building, *Proceedings of the 17th World Conference on Earthquake Engineering*, Paper No. C001728.

[3] H. Luo, K. Ikago, C. Chong, A. Keivan, B.M. Phillips (2019) Performance of low-frequency structure incorporated with rate-independent linear damping, *Engineering Structures*, 181: 324—335. https://doi.org/10.1016/j.engstruct.2018.12.022

[4] H. Luo, K. Ikago (2021) Unifying causal model of rate-independent linear damping for effectively reducing seismic response in low-frequency structures, *Earthquake Engineering and Structural Dynamics*, 50: 2355-2378. https://doi.org/10.1002/eqe.3450

[5] R. Zhang, Z. Zhao, C. Pan, (2018) Influence of mechanical layout of inerter systems on seismic mitigation of storage tanks, *Soil Dynamics and Earthquake Engineering*, 114: 639-649. https://doi.org/10.1016/j.soildyn.2018.07.036

[6] Ikago, K., Saito, K., and Inoue, N. (2012). Seismic control of single-degree-of-freedom structure using tuned viscous mass damper. *Earthq. Eng. Struct. Dyn.* 41, 453–474. https://doi.org/10.1002/eqe.1138

Structural Health Monitoring: 10APWSHM
Materials Research Proceedings 50 (2025) 163-171

Materials Research Forum LLC
https://doi.org/10.21741/9781644903513-20

Optimal design of a tuned viscous mass damper enhanced outrigger system

Dawei Li[1,2,a] *, and Xianghui Guo[1,b]

[1]School of civil engineering, Lanzhou University of Technology, Lanzhou, 730050, China

[2]China Electronics Engineering Design Institute Co., Ltd, Beijing, 100142, China

[a]ldw@lut.edu.cn, [b]1245715944@qq.com

Keywords: Damped Outrigger System; Tuned Viscous Mass Damper; Fixed-Point Method; Mode Superposition

Abstract. A damped outrigger system is presented to improve the seismic performance of the conventional outrigger system by incorporating a tuned viscous mass damper (TVMD) between the core tube and the perimeter columns. An optimal framework of the design parameters of the TVMD is developed, which includes mode superposition technique and fixed-point method. A simplified cantilever beam with distributed parameters represents the deformation characteristic of the core tube in the outrigger system. Mode superposition decouples the continuous system using a general single-degree-of-freedom system. Then, the fixed-point method is proposed to determine the optimal design parameters of the TVMD by minimizing the $H_2 -$ norm of the frequency response function of the displacement response of the outrigger system. A numerical study demonstrates that the designed TVMD effectively mitigates the roof displacement.

Introduction

The concept of a damped outrigger system, developed by Smith and co-author [1], has been widely implemented to enhance the resilience of super high-rise buildings. In comparison to the conventional outrigger system, the inserted damping achieves the cooperative work between the outrigger and perimeter columns, thereby reducing the displacement demand of the controlling building subjected to external excitation. An inerter element [2], which generates the apparent mass effect by utilizing the relative motion of two terminals, is employed in a damping network to enhance the energy dissipation capacity of a traditional damping device. The inerter damping system [3-4], which combines a complex configuration of a high-tech damping element, an inerter element, and a tuning spring element, has the potential to improve the seismic performance of a damped outrigger system.

For the implementation of the inerter damping system into the outrigger system, Asai and Watanabe [5] proposed the concept of a tuned inertial mass electromagnetic transducers-based outrigger system. Liu et al. [6] examined the seismic performance of the rotary inerter damping-enhanced outrigger system by using a set of typical ground motion records. Wang et al. [7] investigated the damping dissipation capacity of the inerter and negative stiffness-based damping system by considering the soil-structure-interaction. Xie et al. [8] proposed a cable-bracing-self-balanced inerter system to enhance the seismic performance of the outrigger system. While the contributions above effectively identify optimal solutions by combining the numerical eigen results from the finite element analysis and optimal design formulation for a single-degree-of-freedom system, it should be noted that the numerical measures cannot directly identify the optimal design parameters of the inerter damping in an outrigger system.

In this study, a tuned viscous mass damper (TVMD) is proposed to enhance the seismic performance of the outrigger system. A simplified cantilever beam with distributed mass is employed to represent the deformation of the core tube of the outrigger system. The designed

Materials Research Forum LLC

https://doi.org/10.21741/9781644903513-20

TVMD is positioned between the stiff outrigger and perimeter columns. The optimal frequency ratio of the TVMD is estimated by using the invariant point method of the displacement magnification factor (DMF) of the modal coordinates of the cantilever beam. The optimal damping ratio of the TVMD is obtained by minimizing the H_2- norm of the DMF under white noise. Finally, the efficacy of the proposed method is verified by using a 60-story benchmark model.

Motion equation of the analytical model

As illustrated in Figure 1, a high-rise building with outrigger system consists of a frame-core and damped outriggers. The analytical model of the undamped core tube can be simplified to a cantilever beam with elastic modulus E, cross section aera A and inertial moment I. The differential equation of the cantilever beam can be expressed as,

$$m\frac{\partial^2 y(x,t)}{\partial t^2} + \frac{\partial^2}{\partial x^2}\left[EI\frac{\partial^2 y(x,t)}{\partial x^2}\right] = f(x,t) \tag{1}$$

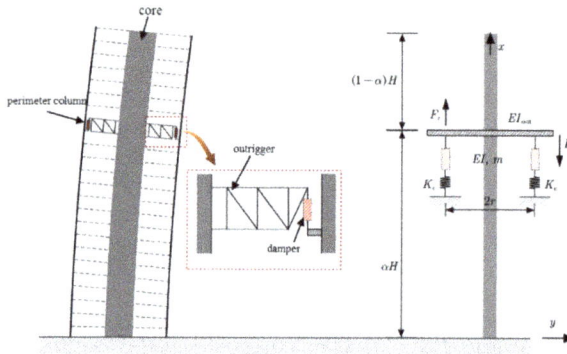

Figure 1 Schematics of cantilever beam with damped outrigger.

where, $y(x,t)$ denote the translational displacement response of cantilever beam with vertical coordinates x and time t, m denote the distributed mass along to vertical coordinate, $f(x,t)$ denote the external force. According to the modal superposition principle, the displacement response is expressed as

$$y(x,t) = \sum_{i=1}^{N}\phi_i(x)q_i(t) \tag{2}$$

where $\phi_i(x)$ and $q_i(t)$ denotes the eigen vector and general coordinates. By virtue of dynamic solution of the uniformed Euler-Bernoulli beam, the natural frequencies and eigen vector are given as follows,

$$\omega_i = \lambda_i^2\sqrt{\frac{EI}{mH^4}}, \quad \phi_i(x) = A_i\left[\sin\lambda_i\bar{x} - \sinh\lambda_i\bar{x} + B_i\left(\cosh\lambda_i\bar{x} - \cos\lambda_i\bar{x}\right)\right] \tag{3}$$

where the normalized coordinate the coefficient $\bar{x} = x/H$, H is the height of the cantilever beam, and coefficient B_i is expressed as,

Structural Health Monitoring: 10APWSHM
Materials Research Proceedings 50 (2025) 163-171

Materials Research Forum LLC
https://doi.org/10.21741/9781644903513-20

$$B_i = \left(\sin \lambda_i + \sinh \lambda_i\right) / \left(\cos \lambda_i + \cosh \lambda_i\right) \tag{4}$$

where the eigenvalue λ_i in Eq. (3) to Eq. (4) is approximated as $\lambda_1 = 1.875$, $\lambda_2 = 4.694$, $\lambda_3 = 7.855$, for $i \geq 4$, $\lambda_i \approx (i - 1/2)\pi$. The normalized coefficient A_i is determined by employing the normalized condition for modal mass M_i,

$$M_i = \int_0^H m\phi_i^2(x)\,dx = mH \tag{5}$$

where the coefficient $A_i = 2\sqrt{\lambda_i/\gamma_i}$, the coefficient γ_i is expressed as,

$$
\begin{aligned}
\gamma_i = {}& 4B_i^2 \lambda + 8B_i \sinh \lambda \sin \lambda + \left(B_i^2 - 1\right)\left(\sin 2\lambda - 4\sinh \lambda \cos \lambda\right) \\
& + \left(B_i^2 + 1\right)\left(\sinh 2\lambda - 4\cosh \lambda \sin \lambda\right) + 2B_i\left(\cos 2\lambda - \cosh 2\lambda\right)
\end{aligned} \tag{6}
$$

Substituting Eq. (2) into Eq. (1), and using modal superposition method, the motion equation of cantilever beam is rewritten as

$$\sum_{i=1}^{N} \ddot{q}_i(t) + \sum_{i=1}^{N} \omega_i^2 q_i(t) = \frac{1}{mH} \sum_{i=1}^{N} \int_0^H \phi_i(x) f(x,t)\,dx \tag{7}$$

On the other hand, the external force of the cantilever beam is expressed as the sum of the resistant moment of the outrigger and earthquake excitation.

$$\int_0^H \phi_i(x) f(x,t)\,dx = -ma_g(t)\int_0^H \phi_i(x)\,dx - \int_0^H \phi_i(x)\frac{\partial M_d(x,t)}{\partial x}\,dx \tag{8}$$

where, $a_g(t)$ is the acceleration the ground motions, $M_d(x,t)$ is the resistant moment provided by the outrigger. The first contribution of the right hand in Eq. (8) is expressed as,

$$-ma_g(t)\int_0^H \phi_i(x)\,dx = -s_i mH a_g(t) \tag{9}$$

where s_i the modal participation coefficient that is estimated as,

$$s_i = \frac{2}{\sqrt{\lambda_i \gamma_i}}\left[2 - \cos \lambda_i - \cosh \lambda_i + B_i\left(\sinh \lambda_i - \sin \lambda_i\right)\right] \tag{10}$$

The resistant moment in the right hand of Eq. (8) can be expressed as,

$$M_d(x,t) = -\delta(x - \alpha H)2rF_r(t) \tag{11}$$

where r and F_r denote the length of the outrigger and the resistant force provided by the outrigger. The Dirac function is expressed as,

$$\delta(x - \alpha H) = \begin{cases} \infty & x = \alpha H \\ 0 & x \neq \alpha H \end{cases} \tag{12}$$

By using the mathematical operation rule of Dirac function, the contribution of the resistant moment in outrigger can be expressed as,

$$-\int_0^H \phi_i(x)\frac{\partial M_d(x,t)}{\partial x}\,\mathrm{d}x = -2rF_r(t)\int_0^H \delta(x-\alpha H)\frac{\partial \phi_i(x)}{\partial x}\,\mathrm{d}x = -2rF_r(t)\frac{\partial \phi_i(x)}{\partial x}\Big|_{x=\alpha H} \qquad (13)$$

where,

$$\frac{\partial \phi_i(x)}{\partial x}\Big|_{x=\alpha H} = 2\frac{\lambda_i}{H}\sqrt{\frac{\lambda_i}{\gamma_i}}\left[\cos\lambda_i\bar{x} - \cosh\lambda_i\bar{x} + B_i\left(\sinh\lambda_i\bar{x} + \sin\lambda_i\bar{x}\right)\right] \qquad (14)$$

Thus, the motion equation of the generalized single-degree-of-freedom system for ith eigenvalue is written as,

$$\ddot{q}_i(t) + \omega_i^2 q_i(t) = -s_i a_g - \frac{2rF_r}{mH}\frac{\partial \phi_i(x)}{\partial x}\Big|_{x=\alpha H} \qquad (15)$$

The motion equation of the TVMD system is written as,

$$m_d(\ddot{x}_1 - \ddot{x}_d) + c_d(\dot{x}_1 - \dot{x}_d) - k_d x_d = 0 \qquad (16)$$

where, m_d, c_d, and k_d denote the apparent mass, damping coefficient, and equivalent spring stiffness of the TVMD, x_1 and x_d denote the deformation of the outside end of the outrigger and the equivalent spring in TVMD, respectively. The equivalent stiffness of the TVMD is express as,

$$k_d = \frac{k_s k_c k_o}{k_s k_c + k_c k_o + k_s k_o} \qquad (17)$$

where, k_s and k_c denote the stiffness of designed spring in TVMD and axial stiffness of perimeter columns. Additionally, k_o is the equivalent stiffness of outrigger system, that is approximate with following expression,

$$k_o = \frac{3E_o I_o}{r^3} \qquad (18)$$

where, E_o and I_o denote the equivalent elastic modulus and inertial moment of outrigger. The cumulative deformation of rotary outrigger is expressed as,

$$x_1 = \theta r = r\sum_{i=1}^{N}\frac{\partial \phi_i(x)}{\partial x}q_i(t) \qquad (19)$$

Based on above derivation, Eq. (15) can be expressed as,

$$\ddot{q}_i(t) + \omega_i^2 q_i(t) = -s_i a_g - \frac{2rk_d x_d}{mH}\frac{\partial \phi_i(x)}{\partial x}\Big|_{x=\alpha H} \qquad (20)$$

Let the external excitation $a_g(\omega, t) = A_g e^{j\omega t}$, and the Laplace transformation of the x_d can be expressed as,

$$X_d = \Gamma_{TVMD} X_1 = \Gamma_{TVMD} r \sum_{i=1}^{N} \frac{\partial \phi_i(x)}{\partial x} Q_i \tag{21}$$

where X_1 and Q_i is the Laplace transformation of x_1 and q_i, Γ_{TVMD} is expressed as,

$$\Gamma_{TVMD} = \frac{c_d j\omega - m_d \omega^2}{c_d j\omega - m_d \omega^2 + k_d} \tag{22}$$

where, $j = \sqrt{-1}$. Then, the Laplace transformation of Eq. (20) is obtained as,

$$-\omega^2 Q_i + \omega_i^2 Q_i = -s_i A_g - \frac{2r^2 k_d}{mH} \Gamma_{TVMD} \frac{\partial \phi_i(x)}{\partial x} \sum_i \frac{\partial \phi_i(x)}{\partial x} Q_i |_{x=\alpha H} \tag{23}$$

Let $\eta_i = 2r^2[\partial \phi_i(x)/\partial x]^2|_{x=\alpha H}$, Eq. (23) is further to simplified as,

$$\left(\omega_i^2 - \omega^2 + \frac{k_d}{mH} \Gamma_{TVMD} \eta_i \right) Q_i = -s_i A_g \tag{24}$$

Then, the displacement magnification factor (DMF) of the ESDOF can be expressed as,

$$\left| \frac{Q_i}{A_g} \right| = s_i \left| \frac{mH(c_d j\omega - m_d \omega^2 + k_d)}{k_d (c_d j\omega - m_d \omega^2) \eta_i + mH(\omega_i^2 - \omega^2)(c_d j\omega - m_d \omega^2 + k_d)} \right| \tag{25}$$

Closed-form formulation of optimal frequency ratio of the TVMD with DMF
Substituting the parameters in Eq. (26), the DMF in Eq. (25) is rewritten in Eq. (27),

$$\mu_d = \frac{m_d}{mH}, \omega_d = \sqrt{\frac{k_d}{m_d}}, \zeta_d = \frac{c_d}{2m_d \omega_d}, \beta_d = \frac{\omega_d}{\omega_i}, \beta_0 = \frac{\omega}{\omega_i}, \tag{26}$$

$$R_d = \frac{\omega_i^2}{s_i} \left| \frac{Q_i}{A_g} \right| = \left| \frac{\beta_d^2 - \beta_0^2 + 2j\zeta_d \beta_0 \beta_d}{(\beta_0^2 - 1)(\beta_0^2 - \beta_d^2) - \mu_d \eta_i \beta_0^2 \beta_d^2 + 2j\zeta_d \beta_0 \beta_d (1 + \beta_d^2 \mu_d \eta_i - \beta_0^2)} \right| \tag{27}$$

As indicated in fixed point, for case of $\zeta_d = 0$ and $\zeta_d = \infty$, Eq. (27) is rewritten as,

$$R_d \big|_{\zeta_d=0} = \left| \frac{\beta_d^2 - \beta_0^2}{(\beta_0^2 - 1)(\beta_0^2 - \beta_d^2) - \mu_d \eta_i \beta_0^2 \beta_d^2} \right| \tag{28}$$

$$R_d \big|_{\zeta_d=\infty} = \left| \frac{1}{1 + \beta_d^2 \mu_d \eta_i - \beta_0^2} \right| \tag{29}$$

As Brock indicated that Eq. (29) is opposite to Eq. (28) when the frequency ratio locates the fixed points P and Q,

$$\frac{\beta_{\mathrm{d}}^2 - \beta_0^2}{\left(\beta_{\mathrm{d}}^2 - \beta_0^2\right)\left(1 - \beta_0^2\right) - \beta_0^2 \beta_{\mathrm{d}}^2 \mu_{\mathrm{d}} \eta_i} = \frac{-1}{\beta_{\mathrm{d}}^2 \mu_{\mathrm{d}} \eta_i + \left(1 - \beta_0^2\right)} \tag{30}$$

Then, the equation of frequency ratio of excitation is expressed as,

$$2\beta_0^4 - 2\left(\beta_{\mathrm{d}}^2 \mu_{\mathrm{d}} \eta_i + \beta_{\mathrm{d}}^2 + 1\right)\beta_0^2 + \beta_{\mathrm{d}}^4 \mu_{\mathrm{d}} \eta + 2\beta_{\mathrm{d}}^2 = 0 \tag{31}$$

On the other hand, the equivalent amplitude of DMF at point P and Q can be used as,

$$\frac{1}{\beta_{\mathrm{d}}^2 \mu_{\mathrm{d}} \eta_i + \left(1 - \beta_P^2\right)} = -\frac{1}{\beta_{\mathrm{d}}^2 \mu_{\mathrm{d}} \eta_i + \left(1 - \beta_Q^2\right)} \tag{32}$$

$$\beta_Q^2 + \beta_P^2 = 2\left(1 + \beta_{\mathrm{d}}^2 \mu_{\mathrm{d}} \eta_i\right) \tag{33}$$

The Vieta's theorem of Eq. (31) gives that,

$$\beta_Q^2 + \beta_P^2 = \beta_{\mathrm{d}}^2 \mu_{\mathrm{d}} \eta_i + 1 + \beta_{\mathrm{d}}^2 \tag{34}$$

$$\beta_Q^2 \beta_P^2 = \frac{\beta_{\mathrm{d}}^4 \mu_{\mathrm{d}} \eta + 2\beta_{\mathrm{d}}^2}{2} \tag{35}$$

Then, the optimal frequency ratio is obtained by equalling Eq. (33) and Eq. (34),

$$\beta_{\mathrm{d}}^{\mathrm{opt}} = \sqrt{1 / \left(1 - \mu_{\mathrm{d}} \eta_i\right)} \tag{36}$$

Determination of the optimal damping ratio with H_2 norm
Consider the external excitation is white noise, the H_2- norm of the DMF is expressed as,

$$PI_1 = \frac{1}{\pi} \int_{-\infty}^{\infty} |R_{\mathrm{d}}|^2 \, \mathrm{d}\omega \tag{37}$$

The analytical formulation of PI_1 is obtained by using integral rules of the rational function,

$$PI_1 = \frac{\beta_{\mathrm{d}}^4 + \beta_{\mathrm{d}}^2 \mu_{\mathrm{d}} \eta_i + 4\beta_{\mathrm{d}}^2 \zeta_{\mathrm{d}}^2 - 2\beta_{\mathrm{d}}^2 + 1}{2\beta_{\mathrm{d}}^5 \mu_{\mathrm{d}} \eta_i \zeta_{\mathrm{d}}} \tag{38}$$

The minimization of the performance indicator is obtained by solving following equation,

$$\partial PI_1 / \partial \zeta_{\mathrm{d}} = 0 \tag{39}$$

Then, the square of damping ratio is approximated as,

$$\zeta_{\mathrm{d}}^2 = \frac{\beta_{\mathrm{d}}^2 \mu_{\mathrm{d}} \eta_i + \beta_{\mathrm{d}}^4 - 2\beta_{\mathrm{d}}^2 + 1}{4\beta_{\mathrm{d}}^2} \tag{40}$$

Substituting β_d^{opt} into Eq. (40), the optimized damping ratio is obtained as,

$$\zeta_{d,2}^{opt} = \frac{1}{2}\sqrt{\mu_d \eta_i / \left(1 - \mu_d \eta_i\right)} \tag{41}$$

And the corresponding performance indicator is estimated as,

$$PI_1^{opt} = 2\frac{1 - \mu_d \eta_i}{\sqrt{\mu_d \eta_i}} \tag{42}$$

Validation of the proposed design method with frequency response function
The St. Francis Shangri-La Palace Tower with 60-story and a height of 210 m is presented to illustrate the feasibility validation of the proposed optimal measures. An analytical model with parameters as listed in Table 1 was adopted in this paper.

Table 1 Detailed information of the benchmark model.

Parameters	Value	Parameters	Value
E	3.6×10^4MPa	A_c	$8.0m^2$
I	$507.91m^4$	n_s	60
H_i	3.5 m	m	90308.6kg/m

As indicated in Table 1, the first five natural frequencies of the core tube are estimated to be 0.18, 1.15, 3.14, 6.0 and 9.16 Hz, respectively. It is assumed that the Rayleigh model with an inherent damping ratio of 2 % applies to the first and third frequencies. A stiff outrigger with a length of 4m is adopted, and further detailed information regarding this building, can be found in Ref [1].

A TVMD with a mass ratio $\mu_d = 5$ is positioned between the stiff outrigger and the perimeter column at the 60th story. Based on the aforementioned information, the intermediate factor η_1 of the first-order mode is approximated to be 0.0055 and the optimal frequency ratio and damping ratio of the designed TVMD are $\beta_d^{opt} = 1.014$ and $\zeta_{d,2}^{opt} = 0.0084$, respectively. By substituting the optimal frequency ratio and damping ratio into Eq. (27), the DMF curve corresponding to the first order mode is plotted in Figure 2 (a) with a purple dashed line and the symbol $\zeta_{d,2}^{opt}$. In order to examine the feasibility and efficacy of the proposed optimal measure, the DMF obtained by using the invariant method and the minimization of H_∞- norm of the DMF is included in Figure 2 (a) with the yellow solid line and the symbol $\zeta_{d,1}^{opt}$. To illustrate the invariant points P and Q, two DMF curves of $\zeta_d = \infty$ and $\zeta_d = 0$ are supplemented in Figure 2 (a). Correspondingly, the peak value of the DMF curve corresponding to the invariant points is calculated as red solid line. To verify the accuracy of the proposed method, the counterplot of the performance indicator $\lg(PI_1)$ is approximated in the range of $\beta_d \in [0.1,10]$ and $\zeta_d \in [0.01,1]$ in Figure 2 (b).

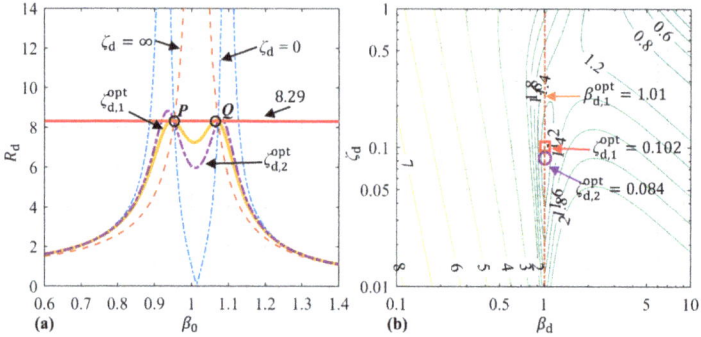

Figure 2 (a) DMF curves and (b) counterplot map of the performance indicators $\lg(PI_1)$ *of the designed TVMD outrigger at the 60th story with the first-order mode.*

As illustrated in Figure 2 (a), two fixed points, designated P and Q locates on the intersection of the DMF resonance curves with $\zeta_d = \infty$ and $\zeta_d = 0$. Based on the H_∞-norm minimization strategy, the peak value of the DMF is obtained as $R_d = 8.29$. The resonance curve obtained from the H_2 − norm minimization also crosses these two fixed points and exhibits a lower amplitude between them. Furthermore, it is evident that the envelope area of the DMF resonance curve of the H_2 − norm is smaller than the value of the H_∞ − norm strategy. As illustrated in Figure 2 (b), a global minimization value of the performance indicator is computed from the combination of $\beta_d = 10$ and $\zeta_d = 1$. However, the combination of optimal frequency ratio and damping ratio, obtained from H_∞ − norm and H_2 − norm, locates the local minimal value of the counterploted performance indicator $\lg(PI_1)$. This means that the proposed optimal method affords a viable solution for identifying the optimal values of $\lg(PI_1)$ other than the global minimization.

Conclusions

In order to enhance the seismic performance of the high-rise building, this work proposes a novel damped-outrigger system, which replaces the traditional viscous damper with a designed TVMD. By virtue of the hypothesis of the distributed system, a cantilever beam is employed to represent the motion characteristic of the core tube. The optimal strategy, which combines the invariant point and minimization of the H_2 − norm of the DMF is proposed to identify the optimal parameters of the TVMD. The frequency response function of the DMF and counterplot of the performance indicator $\lg(PI_1)$ demonstrate that the proposed optimal method affords a reliable measure for identifying the local minimization of the performance indicator. In conclusion, the proposed method is effective for mitigating the displacement response of the outrigger system.

Acknowledgements

The authors gratefully acknowledge financial support from the National Natural Science Foundation of China (grant no. 52108447, 52468073), the Gansu Province Science Foundation for Youths (grant no. 23JRRA804), and the Fund for Excellent Young Scholars of LUT (grant no. 04-062210).

References

[1] Smith R J, Willford M R. The damped outrigger concept for tall buildings[J]. The Structural Design of Tall and Special Buildings, 2007, 16(4): 501-517. https://doi.org/10.1002/tal.413

[2] Smith M.C. Synthesis of mechanical networks the inerter. IEEE Transactions on automatic control, 2002, 47(10): 1648-1662. https://doi.org/10.1109/TAC.2002.803532

Structural Health Monitoring: 10APWSHM   Materials Research Forum LLC
Materials Research Proceedings 50 (2025) 163-171   https://doi.org/10.21741/9781644903513-20

[3] Ikago Kohju, Saito Kenji, Inoue Norio. Seismic control of single-degree-of-freedom structure using tuned viscous mass damper. Earthquake Engineering & Structural Dynamics, 2012, 41(3): 453-474. https://doi.org/10.1002/eqe.1138

[4] Li Dawei, Ikago Kohju, Yin Ao. Structural dynamic vibration absorber using a tuned inerter eddy current damper. Mechanical Systems and Signal Processing, 2023, 186. https://doi.org/10.1016/j.ymssp.2022.109915

[5] Asai Takehiko, Watanabe Yuta. Outrigger tuned inertial mass electromagnetic transducers for high-rise buildings subject to long period earthquakes. Engineering Structures, 2017, 153: 404-410. https://doi.org/10.1016/j.engstruct.2017.10.040

[6] Liu Liangkun, Tan Ping, Ma Haitao, Yan Weiming, Zhou Fulin. A novel energy dissipation outrigger system with rotational inertia damper. Advances in Structural Engineering, 2018: 1-14. https://doi.org/10.1177/1369433218758475

[7] Wang Meng, Liu Chao, Zhao Mi, Sun Fei-Fei, Nagarajaiah Satish, Du Xiu-Li. Damping dissipation analysis of damped outrigger tall buildings with inerter and negative stiffness considering soil-structure-interaction. Journal of Building Engineering, 2024, 88: 109225. https://doi.org/10.1016/j.jobe.2024.109225

[8] Xie Liyu, Yang Zijian, Xue Songtao, Gong Ling, Tang Hesheng. Topology optimization analysis of a frame-core tube structure using a cable-bracing-self-balanced inerter system. Journal of Building Engineering, 2024, 95.110210. https://doi.org/10.1016/j.jobe.2024.110210

Structural Health Monitoring: 10APWSHM   Materials Research Forum LLC
Materials Research Proceedings 50 (2025) 172-179   https://doi.org/10.21741/9781644903513-21

Using physics-informed graph neural networks for modal identification of a population of structures

Xudong Jian[1,a], Wei Liu[1,2,b], Kiran Bacsa[1,3,c], Eleni Chatzi[1,3,d*]

[1]Future Resilient Systems, Singapore-ETH Centre, 138602, Singapore

[2]Department of Industrial Systems Engineering and Management, National University of Singapore, 117576, Singapore

[3]Department of Civil, Environmental and Geomatic Engineering, ETH Zurich, Zurich, 8049, Switzerland

[a]xudong.jian@sec.ethz.ch, [b]weiliu@u.nus.edu, [c]kiran.bacsa@sec.ethz.ch, [d]chatzi@ibk.baug.ethz.ch

Keywords: Population-Based Structural Health Monitoring, Modal Identification, Physics Informed Deep Learning, Graph Neural Network, Transformer

Abstract. Modal identification is crucial for almost every downstream task related to structural health monitoring (SHM), in the sense that it contains information that is vital across the levels of the SHM hierarchy, including damage identification and prediction of future performance. This study proposes a deep learning model that is built on the Transformer module and the GraphSAGE module for modal identification across a population of structures. The model processes structural dynamic measurements and topology of structural systems to output the decomposed modal responses and corresponding mode shapes of monitored structures. Based on the model input (structural dynamic measurements) and output (modal responses and mode shapes), we exploit the modal decomposition theory and independence of the structural modes, to train the model in a physics-informed manner. We perform numerical simulation to verify the proposed model. Results show that the proposed model can decompose dynamic response for structural configurations from both the training set and the unseen testing set, demonstrating its accuracy and generalization ability in terms of modal decomposition. The decomposed modal responses can further be used to identify natural frequencies and damping ratios.

Introduction

Civil engineering structures like bridges are vital to economic development, public safety, and sustainability. However, exposure to harsh environments, improper interventions, and ageing inevitably lead to their degradation. Structural Health Monitoring (SHM) has emerged as a crucial tool for assessing these structures, with prevailing data-driven SHM approaches benefiting significantly from advances in machine learning. Nevertheless, data-driven SHM remains limited by the scarcity of comprehensive monitoring data, especially for damage-state scenarios, as current methods often focus on individual structures.

The recent advancement of sensing technologies, including mobile and contactless methods, has made it feasible to monitor populations of structures rather than isolated cases. This shift introduces Population-Based Structural Health Monitoring (PBSHM), pioneered by Worden et al.[1], which enables information sharing across structure populations. Unlike traditional SHM, PBSHM leverages data from multiple similar structures to minimize environmental and operational variability and improve data availability. For instance, by monitoring a group of bridges, PBSHM can utilize consistent conditions across structures to establish baselines and detect common damage patterns. This approach enhances data-driven SHM by expanding the pool of damage-state data, addressing data limitations, and improving damage detection and prognosis for diverse structural types.

While various data-driven methods have been explored for PBSHM, key challenges persist, especially with heterogeneous populations in which population members are similar but possess different geometry and topology. This study addresses two main issues: 1) *Lack of interpretability*. Interpretable models are essential for understanding uncertainties and identifying transferable features within PBSHM. However, most current methods rely on black-box models that obscure internal processes and limit insight. 2) *Performance limitations with topologically heterogeneous populations*: Structural variations pose challenges for traditional models reliant on fixed-dimension inputs, which are ill-suited for structures with varying topologies.

To address these interpretability and topological challenges, we propose a combined approach utilizing Graph Neural Networks (GNNs) and Physics-Informed Neural Networks (PGNNs). GNNs are well-suited to handle graph-structured data, enabling flexible input dimensions across diverse topologies [2], while PGNNs enhance interpretability by embedding physics-based information, aligning predictions with established physical principles [3]. Together, GNNs and PGNNs offer a robust, interpretable framework for capturing complex structural properties across heterogeneous populations, improving both accuracy and generalizability within PBSHM.

This paper specifically focuses on combining PGNNs and GNNs for a fundamental SHM task, namely Operation Modal Analysis (OMA), aiming to identify modal properties of a structure based solely on measured vibration response data. In this regard, there have been several studies. For example, Liu et al. [4] pioneered using neural networks for modal identification by proposing a self-coding neural network to identify modal responses and mode shapes from the vibration data of structures. They designed a physics-informed loss function that integrates the modal decomposition theory and the properties of modal responses (such as independence and non-Gaussianity) to train the neural network. Following this idea, Shu et al. [5] proposed a new loss function to constrain the neural network during model training. Bao et al. [6] proposed a new neural network that takes both vibration data and its time-frequency representation as the model input. Accordingly, they designed a physics-informed loss function that is different but similar to previous studies to train the model. Though the above studies have been verified by numerical and real-world data, the trained neural network models do not show generalization ability to other structure, meaning it would be time consuming to apply these models because civil engineering structures usually differ from each other, and different models must be trained for different structures. To solve this problem, Jian et al. [7] introduced GNNs to conduct automatic OMA. After the GNN-based model is trained, it can identify modal parameters of a population of structures in which each structure is similar but different due to the flexibility of GNNs. This method, however, requires knowing the ground truth of modal parameters during the training stage, which limits its application because the ground truth of structural modal parameters is unknown in practice. Besides, this model can only identify absolute mode shapes, whereas signs of mode shapes are important when reconstructing structural responses.

Based on the studies mentioned above, this work proposes a GNN-based deep learning model that can identify modal responses and mode shapes for different structures in heterogeneous populations. To overcome the limitation of requiring ground truth data, we propose a physics-informed loss function that enables training without known modal parameters. The proposed approach is validated through numerical experiments, demonstrating its effectiveness for real-world PBSHM applications.

Methodology

Problem Formulation. Modal identification is crucial in SHM, aiming to determine a structure's natural frequencies, damping ratios, and mode shapes by analyzing dynamic response measurements—such as accelerations—obtained from a limited and often sparse set of Degrees of

Freedom (DOFs). Operational Modal Analysis (OMA) methods achieve this identification using output-only data, under the assumption of unmeasured, broadband, and random excitations. Accordingly, this study models excitation sources as Gaussian white noise in numerical simulations.

To define the problem in this study, we denote the available acceleration measurements as $\mathbf{X}(t) \in \mathbb{R}^{N \times P}$, where t represents the time components, N denotes the amount of monitored DOFs, and P denotes the amount of available time samples per signal. In this study, we aim to learn a function $func(\cdot)$ that can identify modal responses and mode shapes of detectable vibration modes from $\mathbf{X}(t)$. Therefore, the identification process can be expressed as:

$$\left[\widehat{\mathbf{Q}}(t), \widehat{\mathbf{\Phi}}\right] = func(\mathbf{X}(t)) \tag{1}$$

where $\widehat{\mathbf{Q}}(t) \in \mathbb{R}^{M \times P}$ and $\widehat{\mathbf{\Phi}} \in \mathbb{R}^{N \times M}$ denotes the identified modal responses in the time domain and mode shapes, and M is the number of vibration modes to be identified. It is noteworthy that, in this paper, we do not further process $\widehat{\mathbf{Q}}(t)$ to identify modal frequencies and damping ratios, since $\widehat{\mathbf{Q}}(t)$ contains the dynamic response of single DOFs systems, and identifying frequencies and damping ratios of single systems is quite simple.

Model Architecture. To get the modal identification function $func(\cdot)$ shown in Eq. (1), we design a deep learning model to learn the mapping between $\mathbf{X}(t)$ and $\left[\widehat{\mathbf{Q}}(t), \widehat{\mathbf{\Phi}}\right]$. Figure 1 visualizes the architecture of the proposed model.

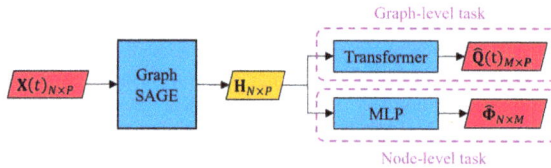

Figure 1 Architecture of the proposed model, in which in which the model input and output are marked in red, hidden features are marked in yellow, and deep learning blocks are marked in blue.

As shown in Figure 1, there are three deep learning blocks (marked in blue) used in the proposed model. Details about them are given below:

● GraphSAGE: The GraphSAGE model, proposed by Hamilton et al.[8], is a type of GNNs, and it serves the key building block in the proposed architecture. Research has shown that GraphSAGE can outperform other types of GNNs in the context of PBSHM. Therefore, this study chooses GraphSAGE to process acceleration signals $\mathbf{X}(t)$ on structural nodes, and thereby extract the spatial information $\mathbf{H}$ of those signals. As illustrated in Figure 2, for each structure in a population, an attributed graph representation can be created by designating joint locations of the structure as nodes of the GraphSAGE and structural elements connecting those nodes (e.g. beams, truss bars) to correspond to the GraphSAGE edges. It is worth mentioning that, unlike common neural networks, where the number of neurons is fixed, GraphSAGE can process graphs with varying numbers of nodes and edges, allowing it to generalize among a population of structures that possess different geometry and topology. Therefore, the number of nodes, which is N, can vary for different structures, providing GraphSAGE a great advantage in generalization to unseen structures over other deep learning model.

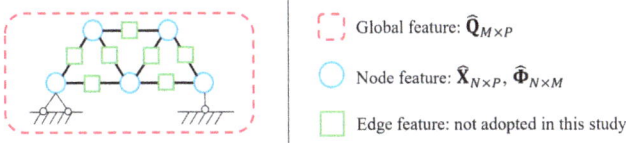

Global feature: $\widehat{\mathbf{Q}}_{M \times P}$

Node feature: $\widehat{\mathbf{X}}_{N \times P}$, $\widehat{\mathbf{\Phi}}_{N \times M}$

Edge feature: not adopted in this study

Figure 2 An example of the graph dataset used in this study. A truss structure can be naturally represented by an attributed graph, where node acceleration and mode shapes are node features, and modal responses are global features.

- Transformer: Transformer is a well-known and powerful deep learning model that can extract temporal information from data. In this study, we use a transformer to convert the output of GraphSAGE, which is **H**, into modal responses $\widehat{\mathbf{Q}}(t)$ in the time domain. By doing so, the dimension of **H** is reduced from N to M. Since $\widehat{\mathbf{Q}}(t)$ is a graph-level (also known as global) feature, this process is named as graph-level task in Figure 1.
- MLP: At last, a multi-layer perceptron (MLP) is employed to transform the hidden feature **H** to mode shapes $\widehat{\mathbf{\Phi}}$ on each node of the graph.

Training Strategy. Inspired by previous studies, this study adopts the physics-informed strategy to train the proposed deep learning model. We design a loss function as follows:

$$\mathcal{L} = \text{MSE}\big(\widehat{\mathbf{\Phi}} \times \widehat{\mathbf{Q}}(t), \mathbf{X}(t)\big) + \text{MSE}\Big(\text{R}\big(\widehat{\mathbf{Q}}(t)\big), \mathbf{I}\Big) + \text{MSE}\Big(\text{R}\big(\big|\text{FFT}\big(\widehat{\mathbf{Q}}(t)\big)\big|\big), \mathbf{I}\Big) \tag{2}$$

where MSE($\cdot$) denotes the mean squared error function, R($\cdot$) denotes the correlation coefficient matrix, |FFT($\cdot$)| denotes the amplitude spectrum of fast Fourier transform (FFT), and **I** denotes the identity matrix.

Eq. (2) consists of three terms, each informed by a distinct piece of domain knowledge from the field of structural dynamics. Details about the utilized domain knowledge are as follows:

- Term 1: Based on the modal decomposition theory, the first term in the loss function minimizes the difference between observed acceleration $\mathbf{X}(t)$ and the acceleration reconstructed with identified modal responses and mode shapes $\widehat{\mathbf{\Phi}} \times \widehat{\mathbf{Q}}(t)$.
- Term 2 & 3: Civil engineering structures are linear and classically damped in general, which means their modal responses are independent from each other in both the time and frequency domains. Mathematically, independence in the time domain implies the correlation coefficient matrix of $\widehat{\mathbf{Q}}(t)$ should be an identity matrix, and independence in the time domain means the correlation coefficient matrix of $\widehat{\mathbf{Q}}(t)$'s FFT amplitude spectrum should also be an identity matrix. Leveraging this domain knowledge, term 2 and 3 in Eq. (2) minimize the discrepancy between the correlation coefficient matrix of $\widehat{\mathbf{Q}}(t)$, $\big|\text{FFT}(\widehat{\mathbf{Q}}(t))\big|$ and the identify matrix **I**, thus maximizing the independence of modal responses.

Using Eq. (2) as the loss function to train the model eliminates the need for ground truth modal parameters, significantly enhancing the practicality of the proposed modal identification approach.

Numerical Experiment

This section introduces the numerical experiment conducted to preliminarily verify the proposed model and physics-informed training strategy. The experiment is designed to answer the following research questions:

- Question 1: Can the proposed model identify modal responses and mode shapes of structures in the training set, without any ground truth of modal responses and mode shapes.
- Question 2: After the model is trained, can it directly output modal responses and mode shapes of structures which are unseen during the training phase?

Structural Health Monitoring: 10APWSHM   Materials Research Forum LLC
Materials Research Proceedings 50 (2025) 172-179   https://doi.org/10.21741/9781644903513-21

Dataset Description. In this work, we simulate a population of 50 trusses, arranged within a trapezoidal boundary to approximate the geometry of simply-supported beam structures. Each truss is generated by randomly meshing the trapezoidal area using Delaunay triangulation. The first 32 trusses are used for model training, the subsequent 8 trusses for model validation, and the final 10 trusses are reserved for testing. The geometric boundary and examples of generated trusses are shown in Figure 3.

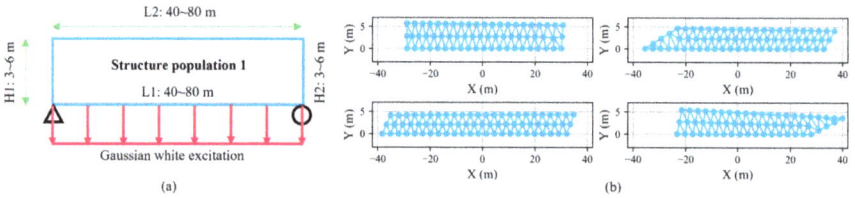

Figure 3 Visualization of the simulated dataset: (a) Geometric configuration meant to approximate a simply-supported truss population; (b) Some representative truss samples from the dataset (only nodes and elements are displayed).

Based on the generated geometric configurations, finite element models are created using truss elements with a constant density of 8015 kg/m^3 and an area of 0.5 m^2. To introduce variability in material properties, the Young's modulus for each truss element is randomly assigned between 100 GPa and 300 GPa. Figure 3 (a) illustrates the boundary conditions (simply-supported) and external excitation (Gaussian white noise) applied along the bottom boundary. Linear time-history analyses using the Newmark-β method are performed to generate in-plane vertical node acceleration time series of 10 seconds, sampled at a 0.005-second interval (200 Hz). Additionally, an eigenvalue analysis is conducted on all 50 models to produce reference modal properties for testing. Since it is standard to focus on the primary modes, a 20 Hz low-pass filter is applied to the generated acceleration signals, ensuring that only modes with frequencies below 20 Hz are identified in this study. For the simulated dataset, most trusses have less than 5 modes below 20 Hz. As a result, the number of modes to be identified is set as 7 in this experiment, which means $M = 7$ for $\widehat{\mathbf{Q}}(t)$ and $\widehat{\mathbf{\Phi}}$.

Implementation Details. The model is implemented in an environment using PyTorch 2.1.2, CUDA 12.1, and DGL 2.0.0. We use the Adam optimizer for training, with a default configuration of a 0.0005 learning rate, first momentum decay of 0.9, and second momentum decay of 0.999. Training is accelerated with an NVIDIA GeForce RTX 3060, using a batch size of 64 over 5,000 epochs. Additional implementation details will be available in our GitHub repository, which will be made public following publication [10].

Results. We first train the model with acceleration data from 32 trusses, while data from 8 trusses are used as the validation set to monitor the training process. Loss curves of the training process are shown in Figure 4.

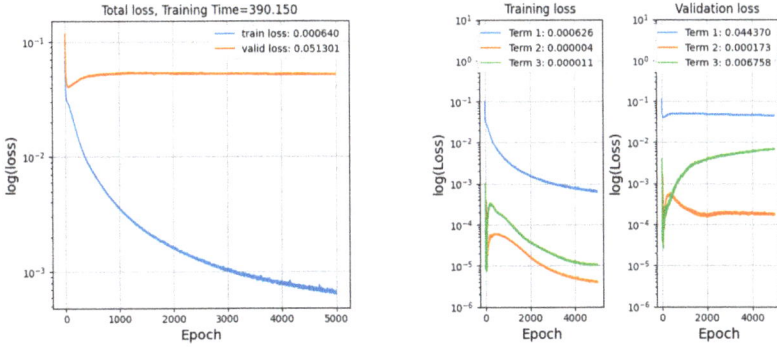

Figure 4 Loss curves of the training process: (a) Total training and validation loss; (b) Different loss terms in the training and validation loss.

From the loss curves above, we can conclude:

● In Figure 4 (a), the total training loss continues to decrease while the validation loss plateaus, indicating that additional training may be unnecessary. Furthermore, the training loss is considerably lower than the validation loss, suggesting that the model possesses a degree of generalization capability, though limited.

● In Figure 4 (b), the elevated validation loss is likely due to the increase term 3, which enforces the independence of modal responses in the frequency domain.

The trained model is then applied to acceleration data from both the training and testing sets. Figure 5 presents two typical examples of the identification results: one from the training set and one from the testing set. Due to space limitations, identification results for the entire dataset are not included in this paper.

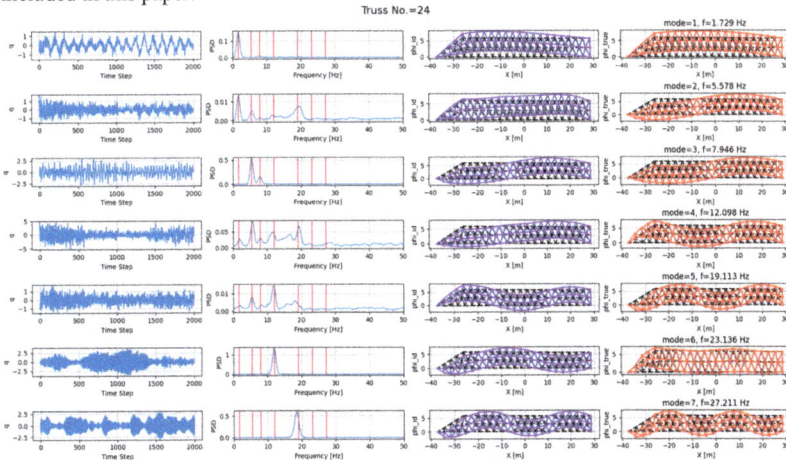

Structural Health Monitoring: 10APWSHM
Materials Research Proceedings 50 (2025) 172-179

Materials Research Forum LLC
https://doi.org/10.21741/9781644903513-21

Truss No.=47

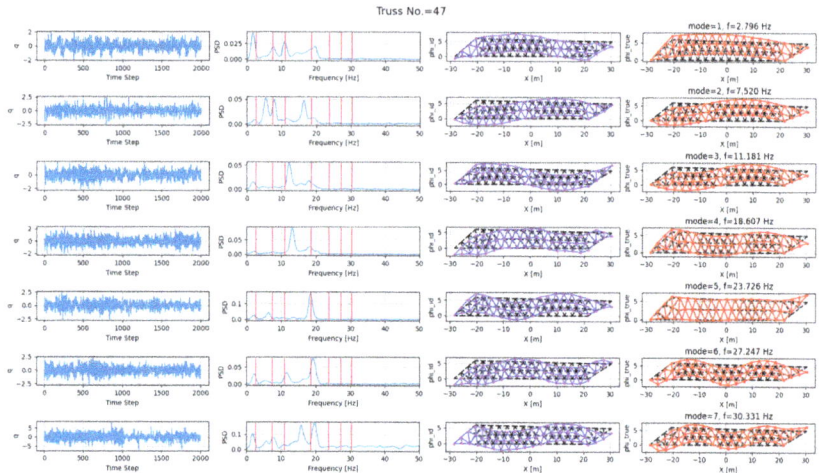

Figure 5 Modal decomposition results. Truss No. 24 and 47 are from the training set and the testing set, respectively. The four columns in the figure show identified modal responses, power spectral density (PSD) of modal responses, identified mode shapes, and ground truth of mode shapes, respectively. In addition, red vertical lines in the second column represent the first 7 true modal frequencies of the truss.

In Figure 5, modal frequencies can handily be identified by reading the frequency of the peak in the PSD of modal responses. Mode shapes are directly output by the trained model. To quantitatively show the performance of the trained model, Table 1 provides identification results for modal frequencies and mode shapes of the two trusses in Figure 5. The modal assurance criterion (MAC) of identified mode shapes is calculated to evaluate the identification accuracy. The closer MAC is to 1, the better the mode shape is identified.

Table 1 Identification results for modal frequencies and mode shapes

		Mode 1	Mode 2	Mode 3	Mode 4	Mode 5
Truss No. 24 (training)	True frequency (Hz)	1.729	5.577	7.945	12.098	19.112
	Identified frequency (Hz)	1.757	5.468	Not identified	11.914	18.554
	Modal assurance criterion	0.981	0.990		0.999	0.998
Truss No. 47 (testing)	True frequency (Hz)	2.795	7.520	11.181	18.607	23.725
	Identified frequency (Hz)	1.757	5.468	12.304	18.554	Not detectable due to the 20 Hz loss-pass filtering
	Modal assurance criterion	0.612	0.715	0.916	0.684	

Observing results shown in Figure 5 and Table 1, we can find that:
● The trained model effectively decomposes the dynamic responses of trusses in the training set, yielding clearly separated modal responses and accurately identified mode shapes—an encouraging outcome, especially since no ground truth modal responses or mode shapes were used during training.

- The model also demonstrates the ability to decompose dynamic responses of unseen trusses in the testing set, albeit with reduced performance compared to the training set. Nonetheless, this is still promising, as it suggests the model exhibits a degree of generalization across new structures within the population, and that is a capability which existing studies have yet to achieve.

Summary

This study introduces a deep learning model for modal identification across a population of structures, incorporating both Transformer and GraphSAGE modules. The model utilizes structural dynamic measurements and topology to yield decomposed modal responses and corresponding mode shapes. By leveraging modal decomposition theory and modal independence, the model is trained in a physics-informed manner, using structural measurements as inputs and modal characteristics as outputs. Numerical simulations confirm that the model can decompose dynamic responses for both training and unseen test structures, demonstrating generalization across a population. Future work will focus on enhancing the model's generalization capability, addressing incomplete measurements from limited DOFs, and validating the approach with real-world data.

References

[1] Worden, K., Bull, L. A., Gardner, P., Gosliga, J., Rogers, T. J., Cross, E. J., ... & Dervilis, N. (2020). A brief introduction to recent developments in population-based structural health monitoring. Frontiers in Built Environment, 6, 146. https://doi.org/10.3389/fbuil.2020.00146

[2] Zhao, M., Taal, C., Baggerohr, S., & Fink, O. (2024). Graph neural networks for virtual sensing in complex systems: Addressing heterogeneous temporal dynamics. arXiv preprint arXiv:2407.18691. https://doi.org/10.2139/ssrn.4941745

[3] Haywood-Alexander, M., Liu, W., Bacsa, K., Lai, Z., & Chatzi, E. (2023). Discussing the Spectra of Physics-Enhanced Machine Learning via a Survey on Structural Mechanics Applications. arXiv preprint arXiv:2310.20425. https://doi.org/10.1017/dce.2024.33

[4] Liu, D., Tang, Z., Bao, Y., & Li, H. (2021). Machine-learning-based methods for output-only structural modal identification. Structural Control and Health Monitoring, 28(12), e2843. https://doi.org/10.1002/stc.2843

[5] Shu, J., Zhang, C., Gao, Y., & Niu, Y. (2023). A multi-task learning-based automatic blind identification procedure for operational modal analysis. Mechanical Systems and Signal Processing, 187, 109959. https://doi.org/10.1016/j.ymssp.2022.109959

[6] Bao, Y., Liu, D., & Li, H. (2024). A mechanics-informed neural network method for structural modal identification. Mechanical Systems and Signal Processing, 216, 111458. https://doi.org/10.1016/j.ymssp.2024.111458

[7] Jian, X., Xia, Y., Duthé, G., Bacsa, K., Liu, W., & Chatzi, E. (2024). Using Graph Neural Networks and Frequency Domain Data for Automated Operational Modal Analysis of Populations of Structures. arXiv preprint arXiv:2407.06492.

[8] Hamilton, W., Ying, Z., & Leskovec, J. (2017). Inductive representation learning on large graphs. Advances in neural information processing systems, 30.

[9] Hembert, P., Ghnatios, C., Cotton, J., & Chinesta, F. (2024). Assessing Sensor Integrity for Nuclear Waste Monitoring Using Graph Neural Networks. Sensors, 24(5), 1580. https://doi.org/10.3390/s24051580

[10] Jian, X. (2024). Github repository [https://github.com/JxdEngineer].

Structural Health Monitoring: 10APWSHM
Materials Research Proceedings 50 (2025) 180-188

Materials Research Forum LLC
https://doi.org/10.21741/9781644903513-22

LFFNet: Layered feature fusion network enhanced passive infrared structural analysis in large-scale geomembrane covers

Yue Ma[1,a] *, Wenhao Huang[1,b], Benjamin Steven Vien[2,c], Thomas Kuen[3,d] and Wing Kong Chiu[2,e]

[1]College of Mechatronics and Control Engineering, Shenzhen University, Shenzhen, Guangdong 518000, China

[2]Department of Mechanical & Aerospace Engineering, Monash University, Clayton, VIC 3008, Australia

[3]Melbourne Water Corporation, 990 La Trobe Street, Docklands, VIC 3008, Australia

[a]yue.ma@szu.edu.cn, [b] huangwenhao2023@email.szu.edu.cn, [c] ben.vien@monash.edu, [d]Thomas.Kuen@melbournewater.com.au, [e]wing.kong.chiu@monash.edu

Keywords: Thermal Image, Image Segmentation, Structural Health Monitoring, Deep Learning, Sewage Treatment

Abstract. Anaerobic lagoons at sewage treatment plants are covered with multiple sheets of high-density polyethylene (HDPE) geomembranes to prevent the emission of odorous gases and to harness biogas as a renewable energy source. Over time, raw sewage can accumulate and solidify into a mass, which can deform the covers and potentially affect their structural integrity. Currently, traditional passive thermography is limited by insufficient thermal excitations and has difficulty identifying features and substances beneath the covers. This study proposes a novel segmentation model, Layered Feature Fusion Network (LFFNet), to identify structural anomalies on large-scale geomembranes in sewage treatment plants by integrating infrared imaging and deep learning techniques. Notably, the model performs exceptionally well in addressing class imbalance issues, reaching a maximum Mean Intersection over Union (mIoU) of 94.68% and a maximum Mean Pixel Accuracy (mPA) of 97.49%.

Introduction

The operational efficacy of sewage treatment plants in Melbourne Water Corporation, Victoria, Australia is contingent upon the integrity of the high-density polyethylene (HDPE) geomembranes that cover anaerobic lagoons [1]. These large-scale floating covers serve dual purposes: preventing the emission of odorous gases and capturing biogas for use as a renewable energy source. However, the accumulation of scum, a mass formed by the solidification of raw sewage, beneath these covers poses a significant threat to their structural integrity. The wrinkles, erosions and elevation caused by scum and biogas can lead to compromised covers, which in turn can disrupt the efficiency of the treatment process and pose environmental and safety concerns [2]. Despite the critical nature of this issue, existing methods for monitoring the accumulation of scum beneath geomembranes are rudimentary and often ineffective.

Traditional outdoor passive infrared thermography has often fallen short in its effectiveness for detecting structural anomalies, primarily due to the inherent limitations of the imaging technique itself. The method relies on the natural thermal emissions of objects without the benefit of active thermal excitation, which typically results in images with a narrow dynamic range and subdued contrast [3]. This is particularly problematic in outdoor environments where ambient conditions, such as solar radiation and wind, can introduce additional thermal noise, further complicating the differentiation between small structural defects and background thermal radiation [4]. As a result, the static nature of traditional thermography fails to capture the temporal evolution of thermal

Structural Health Monitoring: 10APWSHM
Materials Research Proceedings 50 (2025) 180-188

Materials Research Forum LLC
https://doi.org/10.21741/9781644903513-22

signatures that are indicative of structural issues. These challenges necessitate the development of advanced thermal imaging techniques that can overcome the constraints of passive thermography and provide a more accurate and reliable means of structural health monitoring.

The aim of this paper is to develop a novel segmentation model that integrates infrared imaging with deep learning techniques to identify different features in HDPE membrane covers, enhancing the structural health monitoring results from outdoor large-scale infrastructures that are inspected without sufficient external heat excitations.

Methodology

a. Data preparation.

As shown in Fig. 1, a long-term thermal imaging system was set up at the sewage treatment plant of Melbourne Water, where a thermal camera was securely mounted on a tripod at an 85-degree observation angle to monitor temperature changes within the target area. The tripod was positioned at a fixed vertical height of 3.1 meters above the horizontal surface of the covers, and set at a horizontal distance of 2.25 meters from the edge of the covers. The thermal camera was set to monitor four regions on the covers in this paper, and the sampling rate was set as 10 minutes/read to allow long-term monitoring under natural thermal cycle of the outside environment. For each region, Approximately 1,000 single-channel raw images with a size of 640×480 were collected. The pixel values in these raw images are represented as 16-bit integers, corresponding to temperature values at each pixel location. In this study, although the observed scene remains consistent within the same area, the temperature of each segment changes variably over time due to fluctuations in thermal radiation intensity. The proposed method primarily relies on segmenting structural features based on temperature gradient variations. For training and validation, we selected 15 images representing diverse temperature conditions, with 12 images allocated to the training set and 3 to the validation set, resulting in a total of 48 images for training purposes.

To enhance the diversity of the dataset, image augmentation techniques were applied during data loading. Specifically, random flipping and cropping were used, where random cropping extracts a 160×120 pixels region, maintaining the original image's aspect ratio and creating approximately 173,641 unique combinations. The cropped images were then resized to 512×512 pixels through bicubic interpolation to align with the model's input requirements. This augmentation process was repeated in each training epoch, producing a fresh set of 48 augmented images per round. Over the 500 training epochs, this yielded a total of 24,000 augmented images, effectively enriching the training dataset for robust model learning.

Fig. 1. Illustration of the thermography setup next to the floating covers

b. Feature Detection Method

To ensure the maximum functionality and structural reliability of the HDPE geomembrane covers, this study proposes a feature detection method for its structural health monitoring. The approach is divided into three steps: data preprocessing, model training, and visualization of prediction results. The raw images captured by the thermal camera of the floating cover are 16-bit single-channel images. To enhance the visualization of temperature values, the single-channel image is mapped using a color table, resulting in a more pronounced color gradient that reflects temperature variations, as illustrated by the thermal image reconstruction in Fig. 2.

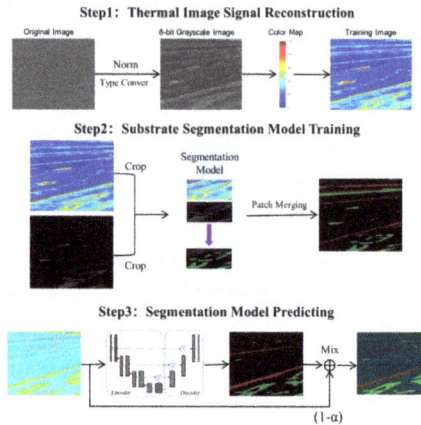

Fig. 2. Flowchart of substrate detection method beneath the floating cover structure

The segmentation results generated by the model, based on infrared images, are initially represented as single-channel grayscale images. These images use the values "0", "1", and "2" to encode the predicted pixel categories, each with three regions. The substrate detection task is defined as a pixel classification problem, where pixels are labeled as "0" for background, "1" for flotation, and "2" for dirt and wrinkles.

Although the categories are distinguished by different pixel values in the image output by the segmentation model, the grayscale output visually appears to be an image that is predominantly black because the visual difference between the pixel values is limited. In order to improve the interpretability of the segmentation output, grayscale images are converted to RGB format, that is, the pixel values "0", "1" and "2" representing different categories of areas in the image are converted to "black", "red" and "green" one-to-one. This image, which was converted to RGB format, is displayed as a suitable model output, with different colors representing different areas: black for the background, red for flotation, and green for dirt and wrinkles.

To illustrate and demonstrate the effect of segmentation, the RGB segmentation output is then overlaid onto the original image, which has been processed through a color table mapping. This overlay allows for the precise integration of segmentation results within the context of the thermal scene, making the segmented regions visually distinct and clearly identifiable. This approach effectively combines the thermal and segmentation information, providing a comprehensive and interpretable representation of the detected regions.

$$\text{Output} = (1 - \alpha) \times \text{Image1} + \alpha \times \text{Image2} \qquad (2)$$

In this process, α is the blending factor, controlling the degree of integration between the segmentation output and the original thermal image.

c. Segmentation Model.

Semantic segmentation aims to group image pixels into different categories based on their semantic meaning and classify pixels corresponding to underlying objects. In natural image segmentation tasks, classifiers identify and categorize natural objects within the image, taking natural images as input and outputting labeled predictions. These classifiers are typically not trained from scratch; instead, they utilize the fine-tuning of feature extractors or other components from widely used pre-trained neural networks as the backbone. Common backbone networks include VGG[5], ResNet[6], MobileNets[7], DenseNet[8], and others. These backbones are meticulously designed feature extractors that demonstrate outstanding performance in image classification tasks and can be effectively transferred to segmentation tasks. Leveraging these backbone networks, semantic segmentation models significantly enhance performance in pixel-level classification tasks. Unet is a representative encoder-decoder structure widely applied in fields such as medical imaging, remote sensing, and industrial defect detection[9]. Its distinctive architecture enables it to excel in a variety of image processing tasks. The encoder component of Unet (the contracting path) extracts image features through a series of convolutional and pooling operations, gradually reducing the spatial resolution of the image. Convolutional operations capture local information within the image, while pooling operations downsample the image, reducing its dimensions and mapping the features to a higher-dimensional space. As the encoder progresses through its layers, it effectively captures high-level semantic features at smaller spatial resolutions. Based on these advantages in the field of segmentation, LFFNet chooses to improve on the basis of Unet.

In this study, LFFNet is applied to the semantic segmentation of infrared thermal images, using VGG as the backbone network to replace the encoder component. The decoder retains the conventional Unet structure, utilizing upsampling and skip connections to restore spatial resolution while preserving high-resolution features. When the data is loaded, image enhancement is applied and the input size is set to 512×512 (the best input size obtained after comparing multiple sizes) to fit the input of the VGG encoder, and random image enhancement is performed in each training round. The model is trained using a combination of Cross-entropy loss[10], Dice loss[11], and Focal loss functions[12]. Cross-entropy loss serves as the foundational loss function, commonly used in multi-class semantic segmentation tasks. To enhance the model's performance further, the Dice loss function is incorporated to minimize discrepancies between the predicted and ground truth segmentation.

To optimize the model's performance and computational efficiency, further enhancements were made based on Unet's architecture, following insights from the literature[13, 14], as shown in Fig. 3. In the standard Unet design, skip connections are utilized to recover details lost during the decoding process. However, the feature maps obtained from the encoder often contain redundant information that does not contribute meaningfully to the segmentation task. To address this, an A-PPA (Attention-Parallelized Patch-Attention) module is introduced within the skip connections. This module comprises two key components: an Attention Gate and a Parallelized Patch-Attention mechanism. Attention Gate integrates the soft attention mechanism into the skip connection of the Unet, suppresses the irrelevant regions in each hierarchical scale while highlighting the salient features of specific local regions, and uses the information from the decoder to guide the model to focus on the relevant regions in the encoder feature map, thereby reducing the impact of redundant information.

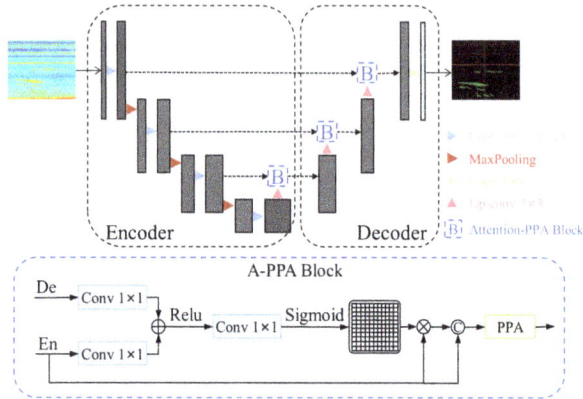

Fig. 3. Substrate detection and segmentation model structure based on infrared thermal images

d. Ablation Analysis.

The LFFNet model uses the VGG network as the backbone for its encoder and fine-tunes it based on pre-trained weights from ImageNet. Other hyperparameters include a learning rate of 0.0001 and a batch size of 8. Models were saved and evaluated at every 10 epochs. The best-performing model was updated based on evaluation metrics, ensuring selection of the optimal model for accurate predictions and visual interpretation of the test dataset. To compare the impact of different components of the infrared thermal imaging semantic segmentation algorithm proposed in this paper on segmentation performance, the ablation analysis on the dataset was conducted.

Fig. 4 shows the results of using LFFNet combined with different loss functions for training. Fig. 4a illustrates the training process achieved by modifying only the encoder component while keeping the loss function constant. Figure 4b presents the training outcomes with various combinations of loss functions, following the replacement of VGG19 with the encoder structure and the addition of our A-PPA module. As shown in the figure, the introduction of the attention mechanism increases the model's computational complexity, affecting its convergence rate. Consequently, the training rounds were set to 500 in the final configuration to ensure full convergence and obtain the optimal model. The results indicate that the loss function combining cross-entropy and focal loss converged the fastest. However, in terms of mIoU and F1 scores, the combination of cross-entropy, Dice, and focal loss performed best. By comparing the curves in Fig. 4b, it can be observed that while the introduction of focal loss had limited improvement on the final model performance, it did accelerate the model's convergence during training. Additionally, comparing the curves shows that although adding the Dice loss slightly delayed the model's convergence, it significantly improved the segmentation performance based on the evaluation metrics.

Structural Health Monitoring: 10APWSHM   Materials Research Forum LLC
Materials Research Proceedings 50 (2025) 180-188   https://doi.org/10.21741/9781644903513-22

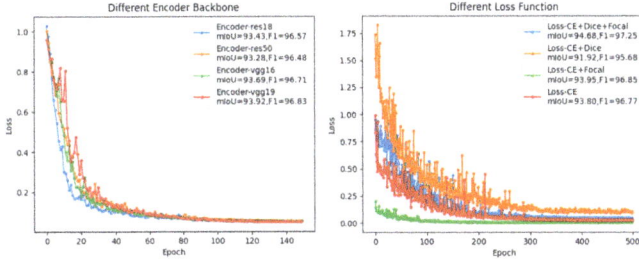

Fig. 4. Validation dataset loss variation curves. (a) Comparison of different backbone networks combined with the cross-entropy function. (b) Comparison of different loss functions

Table 1 presents a performance comparison of several common segmentation models on the same dataset, quantified through relevant accuracy metrics including Pixel Accuracy (PA), Mean Pixel Accuracy (MPA), Mean Intersection over Union (mIoU), and the F1 score. The results clearly show that the Unet models outperform others in segmentation accuracy. In contrast, the performance of FPN, PSPNet, and the DeepLab series is comparatively weaker. Specifically, both FPN and PSPNet are noticeably less effective than the Unet models across all metrics. The DeepLab series shows slightly better results, particularly with DeepLabV3+, which approaches the average performance level of the Unet models and even surpasses some specific metrics. Among the Unet variants, UNet++, an improved version of the original Unet, also shows strong performance, achieving a PA of 98.60, MPA of 97.02, and mIoU of 93.77, indicating its effectiveness in classification accuracy and intersection-over-union measures.

Table 1 Accuracy Metrics for Different Models

Model	Accuracy Metrics			
	PA	MPA	MIoU	F1
UNet+V16	98.59	96.89	93.69	96.71
Unet+V19	98.64	96.98	93.92	96.83
Unet +Attention	98.80	96.55	94.53	97.16
Unet +A-PPA(Ours)	**98.81**	**97.49**	**94.68**	**97.25**
UNet++[15]	98.60	97.02	93.77	96.76
FPN[16]	98.34	96.43	92.64	96.13
PSPNet[17]	97.42	94.46	88.93	94.05
DeepLabV3[18]	98.19	96.62	92.09	95.84
DeepLabV3+[19]	98.55	96.87	93.52	96.62

The LFFNet model proposed in this study builds upon the VGG backbone encoder by incorporating the A-PPA module. By introducing a soft attention mechanism in the skip connections, the model is guided to focus on task-relevant regions, while the PPA method effectively mitigates information loss during the downsampling process. This design enables the LFFNet model to perform well in detecting and segmenting different substrates in infrared thermal images. The ablation experiments presented in Table 1 compare Unet models using VGG networks of different depths as the encoder, models incorporating the Attention Gate (soft attention mechanism), and models integrating the proposed A-PPA module. The results indicate that when the A-PPA module is not used, VGG19 as the encoder outperforms VGG16.

Structural Health Monitoring: 10APWSHM
Materials Research Proceedings 50 (2025) 180-188

Materials Research Forum LLC
https://doi.org/10.21741/9781644903513-22

Results and Discussion

Beneath large-scale HDPE geomembrane covers, there are typically three types of substrates: scum, biogas, and sewage, all of which may adversely affect the structure of the geomembrane. However, under the cover of the membrane, these substrates are generally indistinguishable in RGB images. Based on the principle that different substrates exhibit varying rates of temperature change under the same thermal radiation, thermal imaging is used to capture data. By analyzing the temperature gradient changes in the infrared thermal images, the model can detect and segment the distribution areas of the three substrates.

Fig. 5 provides a more detailed comparison of the structural health monitoring results based on the segmentation model. The first row shows visible light images from real-world scenarios, while the second row presents the region segmentation results generated by the model, highlighting areas of structural anomalies. In a segmented image, the black area represents the surrounding background, which is normal and takes up the largest portion of the image. The green and red areas indicate structural anomalies: the green areas indicate fouling accumulation or plastic film folds, and the red areas refer to the pipe-like flotation structures formed due to the formation of waterways. These anomalous regions are often difficult to identify in visible light images because they are largely covered by the geomembrane. However, with infrared imaging technology, the differences in the rate of temperature change between various mediums allow these structural anomalies to be clearly detected. To further verify the model's effectiveness, portions of the anomalous areas in the visible light images are marked with dashed boxes in the same colors as the segmentation map, demonstrating the feasibility of the proposed LFFNet model for structural health monitoring in this scenario.

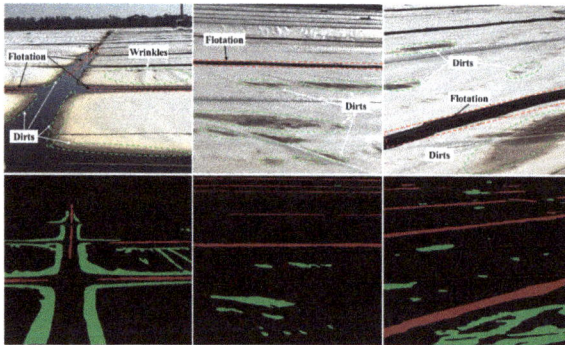

Fig. 5 Comparison of structural health detection results based on LFFNet.

Conclusions

This study presents an approach to addressing the challenges of structural health monitoring of large-scale HDPE geomembrane covers in sewage treatment plants. By utilizing thermal imaging in conjunction with a U-Net-based deep learning model, enhanced with VGG as its backbone and augmented with advanced mechanisms such as soft attention and Parallelized Patch-Attention, the research demonstrates significant improvements in detecting and segmenting structural anomalies beneath the covers. The proposed LFFNet model effectively identifies and distinguishes between scum, sewage, and biogas regions, achieving high segmentation accuracy with metrics.

Acknowledgement

This research was funded by the Australian Research Council Linkage Grant (ARC) LP170100108, the in-kind contributions from Melbourne Water Corporation are also gratefully acknowledged.

Structural Health Monitoring: 10APWSHM
Materials Research Proceedings 50 (2025) 180-188

Materials Research Forum LLC
https://doi.org/10.21741/9781644903513-22

References

[1] B. S. Vien, T. Kuen, L. R. F. Rose, and W. K. Chiu, "Image Segmentation and Filtering of Anaerobic Lagoon Floating Cover in Digital Elevation Model and Orthomosaics Using Unsupervised k-Means Clustering for Scum Association Analysis," Remote Sensing, vol. 15, no. 22, p. 5357, 2023. https://doi.org/10.3390/rs15225357

[2] B. S. Vien, L. Wong, T. Kuen, F. Courtney, J. Kodikara, and W. K. Chiu, "Strain monitoring strategy of deformed membrane cover using unmanned aerial vehicle-assisted 3D photogrammetry," Remote Sensing, vol. 12, no. 17, p. 2738, 2020. https://doi.org/10.3390/rs12172738

[3] Y. Ma, L. Wong, B. S. Vien, T. Kuen, J. Kodikara, and W. K. Chiu, "Quasi-Active Thermal Imaging of Large Floating Covers Using Ambient Solar Energy," Remote Sensing, vol. 12, no. 20, p. 19, 2020. https://doi.org/10.3390/rs12203455

[4] S. Chaudhuri, "EvalTherm-Detectability of internal defects in wind turbine rotor blades using passive infrared thermography," 2023.

[5] K. Simonyan and A. Zisserman, "Very deep convolutional networks for large-scale image recognition," arXiv preprint arXiv, 2014.

[6] K. He, X. Zhang, S. Ren, and J. Sun, "Deep residual learning for image recognition," presented at the Proceedings of the IEEE conference on computer vision and pattern recognition, 2016. https://doi.org/10.1109/CVPR.2016.90

[7] A. G. Howard et al., "MobileNets: efficient convolutional neural networks for mobile vision applications (2017)," arXiv preprint arXiv:.04861, vol. 126, 2017.

[8] G. Huang, Z. Liu, L. V. D. Maaten, and K. Q. Weinberger, "Densely Connected Convolutional Networks," presented at the 2017 IEEE Conference on Computer Vision and Pattern Recognition (CVPR), 21-26 July, 2017. https://doi.org/10.1109/CVPR.2017.243

[9] O. Ronneberger, P. Fischer, and T. Brox, "U-Net: Convolutional Networks for Biomedical Image Segmentation," Cham, 2015. https://doi.org/10.1007/978-3-319-24574-4_28

[10] A. Mao, M. Mohri, and Y. Zhong, "Cross-entropy loss functions: theoretical analysis and applications," presented at the Proceedings of the 40th International Conference on Machine Learning, Honolulu, Hawaii, USA, 2023.

[11] X. Li, X. Sun, Y. Meng, J. Liang, F. Wu, and J. Li, "Dice loss for data-imbalanced NLP tasks," arXiv preprint arXiv:.02855, 2019. https://doi.org/10.18653/v1/2020.acl-main.45

[12] T. Lin, "Focal Loss for Dense Object Detection," arXiv preprint arXiv:.02002, 2017. https://doi.org/10.1109/ICCV.2017.324

[13] O. Oktay et al., "Attention u-net: Learning where to look for the pancreas," arXiv preprint arXiv:.03999, 2018.

[14] S. Xu et al., "HCF-Net: Hierarchical Context Fusion Network for Infrared Small Object Detection," arXiv preprint arXiv:.10778, 2024. https://doi.org/10.1109/ICME57554.2024.10687431

[15] Z. Zhou, M. M. Rahman Siddiquee, N. Tajbakhsh, and J. Liang, "Unet++: A nested u-net architecture for medical image segmentation," presented at the Deep Learning in Medical Image Analysis and Multimodal Learning for Clinical Decision Support: 4th International Workshop, DLMIA 2018, and 8th International Workshop, ML-CDS 2018, Held in Conjunction with MICCAI 2018, Granada, Spain, September 20, 2018, Proceedings 4, 2018.

[16] T.-Y. Lin, P. Dollár, R. Girshick, K. He, B. Hariharan, and S. Belongie, "Feature pyramid networks for object detection," presented at the Proceedings of the IEEE conference on computer vision and pattern recognition, 2017. https://doi.org/10.1109/CVPR.2017.106

[17] H. Zhao, J. Shi, X. Qi, X. Wang, and J. Jia, "Pyramid scene parsing network," presented at the Proceedings of the IEEE conference on computer vision and pattern recognition, 2017. https://doi.org/10.1109/CVPR.2017.660

[18] L.-C. Chen, "Rethinking atrous convolution for semantic image segmentation," arXiv preprint arXiv:.05587, 2017.

[19] L.-C. Chen, Y. Zhu, G. Papandreou, F. Schroff, and H. Adam, "Encoder-decoder with atrous separable convolution for semantic image segmentation," presented at the Proceedings of the European conference on computer vision (ECCV), 2018. https://doi.org/10.1007/978-3-030-01234-2_49

Structural Health Monitoring: 10APWSHM
Materials Research Proceedings 50 (2025) 189-200

Materials Research Forum LLC
https://doi.org/10.21741/9781644903513-23

Wind turbine structure health monitoring through zero-shot learning with supervised variational autoencoders

Kiran Bacsa[1,3,a] *, Gregory Duthé[3,b], Wei Liu[1,2,c], Xudong Jian[1,d] and Eleni Chatzi[1,3,e]

[1]Future Resilient Systems, Singapore-ETH Centre, 138602, Singapore

[2]Department of Industrial Systems Engineering and Management, National University of Singapore, 117576, Singapore

[3]Department of Civil, Environmental and Geomatic Engineering, ETH Zurich, Zurich, 8049, Switzerland

[a]kiran.bacsa@sec.ethz.ch, [b]duthe@ibk.baug.ethz.ch, [c]weiliu@u.nus.edu, [d]xudong.jian@sec.ethz.ch, [e]chatzi@ibk.baug.ethz.ch

Keywords: Deep Learning, Zero-Shot Learning, Structure Health Monitoring, Wind Turbines

Abstract. Real-world Structural Health Monitoring (SHM) applications can be classified to Zero-Shot Learning (ZSL) tasks, since structural data hardly ever reveal or label the full extent of damage that may be incurred to the system. This aligns with the definition of ZSL tasks, where part of the classes (or even all but one class) of the problem are hidden during the training time. Thus, the model must be able to generalize to classes that it has not been exposed to during the training. In SHM, this translates to a model of the studied structure generalizing the states/damages that are not present within the training data. This type of generalization is crucial to design resilient digital twins of infrastructure. Variational inference models are a prominent set of models for engineering features for ZSL tasks. This is because they learn latent features for the model with regularization for independence. Thus new clusters corresponding to unseen data would be more clearly identifiable in such a latent space. Specifically, Conditional Variational Autoencoders (CVAE) are a popular method used in ZSL tasks, as they allow the conditioning of the latent variables on external information pertaining to the data samples. Recent works have shown that Supervised VAEs (SVAE) can learn features that generalize just as well as CVAEs, yet they do not require the labels as an input during deployment. The SVAE conditions the latent space with an auxiliary tasks, in this case the GZSL task. Both the latent features and the GSZL task are learned jointly. Thus the labels are only needed during the training phase. We validate our method on a synthetic dataset consisting of simulated measurements on a wind turbine subject to stiffness degradation. In this case, the different classes correspond to different levels of erosion of the blades. Part of these damage levels are removed from the training set. We show that, when the unseen classes can be expressed by an interpolation of the seen classes, that the SVAE is able to learn global features. Moreover, our study demonstrates this generalization by tackling the damage detection task in a ZSL setting.

Introduction

Wind turbines are essential in advancing to a more sustainable production of electricity. To ensure their dependable and efficient operation, the application of efficient Structural Health Monitoring (SHM) strategies is essential. However, the complex nature of wind turbine dynamics presents challenges for traditional modeling techniques. A particular concern in maintaining wind turbines is the effect of aerodynamic erosion, which significantly affects turbulence but is difficult to detect [1, 2]. This gradual process requires thorough inspections, often resulting in high costs when

Structural Health Monitoring: 10APWSHM   Materials Research Forum LLC
Materials Research Proceedings 50 (2025) 189-200   https://doi.org/10.21741/9781644903513-23

utilizing inspection teams. The nonlinear nature of aerodynamic erosion further complicates the design of efficient monitoring methods.

The development of a wireless, non-intrusive, MEMS-based pressure and acoustic measurement system for large-scale operating wind turbine blades allow for the automated collection of large-scale time-series wind turbine datasets for the monitoring of erosion [3]. Analyzing and making sense of this data requires advanced data-driven approaches. While current methods have proven effective, they often involve extensive preprocessing before the data can be used in diagnostic models. Traditional preprocessing methods struggle to scale, leading to increased interest in neural network architectures like Recurrent Neural Networks (RNNs), Long Short-Term Memory (LSTM) networks [3] and Transformers [4]. These models are highly effective at capturing temporal relationships and nonlinear time-dependent patterns, making them particularly suitable for SHM. The RNNs' sequential modeling and the LSTMs' ability to handle vanishing gradient problems in long sequences are especially advantageous for dealing with the intricate dynamics of wind turbines.

Structural Health Monitoring methods rely on data gathered from a wide array of heterogeneous sensors, many of which often provide redundant information. To address this, it is beneficial to first learn a compressed latent representation of the data as a preprocessing step before engaging in specific learning tasks. These sensors can however be prone to noise, which needs to be filtered out. In this scenario, a Variational Auto-encoder (VAE), optimized through Evidence Lower Bound Optimization (ELBO) [5] can be used to capture the noise within the model's uncertainty. The VAE framework encodes the input data into a latent statistical variable via an encoder neural network, and then reconstructs the data using a decoder neural network. This approach effectively learns a compressed representation that reduces redundancies, while the modeled uncertainty within the framework helps account for noise. Bacsa et al. [6] have shown that the pairing of the VAE with a supervised learning auxiliary task such as damage detection can improve both the feature learning and the supervised task jointly.

Zero-Shot Learning (ZSL) is the machine learning task of processing and classifying instances of categories that have not been included during the training process. ZSL is an important task to incorporate when integrating machine learning methods in SHM, as an infrastructure may be subjected to loads/constraints that shift its condition from the regular regime to which the infrastructure is subjected. Variants of the VAE have been applied to ZSL problems [7] such as the Conditional VAE (CVAE) [8] with the CVAE-SZL [9], GDAN [10] and CADA-VAE [11] methods. The CVAE is a modification of the VAE where the input can be conditioned on an exogenous variable to modify the latent space. Bacsa et al. [6] has shown that the SVAE can learn better features than the CVAE while at the same time tackling auxiliary classes. In this work, our contribution is to study the ZSL potential of the SVAE given that the CVAE has been used before in this field.

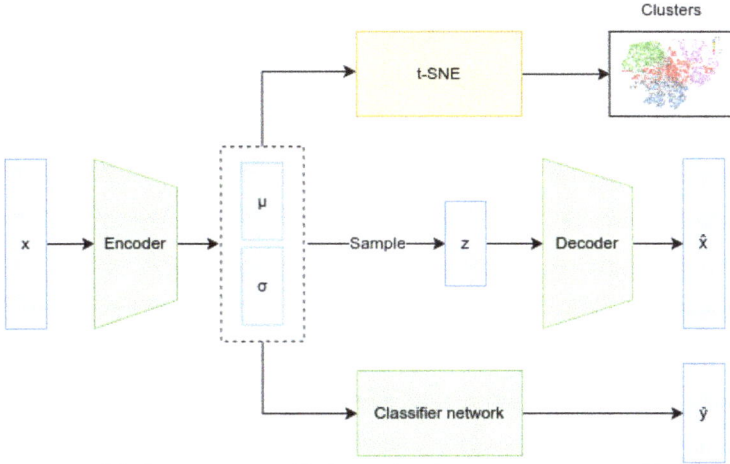

Fig 1. SVAE model: the parameters of the latent distribution μ,σ are extracted from the sensor data using the encoder. The latent variable z is sampled from these parameters and mapped back to the input data using the decoder. Jointly, distribution parameters are used as inputs for the classification task. Once trained, we extract clusters from the latent distribution using t-SNE.

Data

We utilize a numerical aero-servo-elastic model to produce synthetic measurements that represent the progressive delamination of a wind turbine blade, followed by structural damage at the base of the wind turbine tower. The ongoing damage to both the blade and tower is simulated as a reduction in equivalent cross-sectional stiffness. The simulations are designed to cover a period of 10 months in the operational life of the wind turbine. During the initial 6 months, only the blade experiences delamination deterioration, after which significant structural damage is inflicted on the tower base. Both components continue to deteriorate over the subsequent 4 months of operation.

Additionally, it is widely recognized that wind inflow over a wind turbine's rotor exhibits stochastic behavior in the short term (over 10-minute intervals) due to turbulence and changes in the long term due to seasonal effects. To enhance the realism of our case study, we incorporate both short- and long-term variability into the sampled turbulent wind inflow conditions. The aero-servo-elastic simulations were performed using OpenFAST, an open-source software package. This platform provides a wealth of dynamic and operational response signals, which we categorized as follows:

-   Operational signals: wind speed, electrical power output, rotor speed, and blade pitch.
-   Vibration signals: nine 2-channel accelerometer sensors distributed along the blade span and six 2-channel accelerometer sensors along the height of the tower. The blade sensors measure in both flapwise and edgewise directions, while the tower sensors capture data in the fore-aft and side-to-side directions.
-   Strain signals: nine strain sensors along the blade span and six strain sensors along the tower height.

Structural Health Monitoring: 10APWSHM   Materials Research Forum LLC
Materials Research Proceedings 50 (2025) 189-200   https://doi.org/10.21741/9781644903513-23

We established the following classes of damage severity:

- Healthy (both the blade and tower are undamaged)
- 6% reduction in blade stiffness (tower remains healthy)
- 14% reduction in blade stiffness (tower remains healthy)
- 24% reduction in blade stiffness and 20% reduction in tower stiffness
- 45% reduction in blade stiffness and 65% reduction in tower stiffness

Further references on the dataset can be found in Aballah et al [11].

Methodology
We reuse the model from Bacsa et al. [6]. This model is known as a Supervised VAE. The VAE component learns features using the ELBO [5] but optimizing the reconstruction loss, which is simply the mean square loss. The MSE is given in Eq. 1, where x is the input data, $\hat{x}$ is the reconstruction, and N is the number of elements in x:

$$MSE(x,\hat{x}) = \frac{1}{N}||x - \hat{x}||_2^2. \tag{1}$$

In addition to the reconstruction, we add an auxiliary classification task, hence the term Supervised VAE, coined by Ji et al. [12]. While this task can be perceived to offer an exogenous bias to make our features more robust, we here consider this auxiliary task to be the primary task of interest, with the reconstruction offering a regularization of the input features. The classification is done with a RNN followed by a 2-layer Multi-layer Perceptron (MLP) neural network trained with a cross-entropy loss. The cross-entropy loss is given in Eq. 2 for the labels y and the predictions $\hat{y}$ for n classes.

$$CE(y,\hat{y}) = -\sum_i^n y_i log(\hat{y}_i) \tag{2}$$

Training on entire dataset
Our initial dataset comprises 482 simulations. Each sample consists of 60,000 time-steps from 173 channels sampled at a frequency of 100 Hz. The dataset is segmented into 10-second samples, and randomly partitioned into training, validation, and test sets with an 80/10/10 split. Ideally, samples from the same simulation are kept separate across sets; however, due to certain labels appearing only within specific simulations, these samples are duplicated across sets for comprehensive model evaluation. To prevent any potential data leakage during segmentation, non-overlapping windows are enforced when dividing simulations into samples. To enhance robustness and prevent overfitting, a 20% dropout is implemented on each LSTM layer, as well as on the encoder and decoder layers. Training utilizes the Adam optimizer [13], a variant of stochastic gradient descent (SGD) with momentum, with a batch size of 256. The initial learning rate is set to 0.001 and exponentially decayed to 0.00001 over 300 epochs to facilitate convergence. Furthermore, L2-norm weight decay with a coefficient of 0.00001 is employed to regularize the network.

We employ 3 measurement configurations described in [6], known as S4, S5 and S6. In the case of S4, the sensors that are considered are the following: top tower acceleration, shaft rotation speed, root blade strain, middle tower acceleration, blade tip acceleration. This measurement scenario reflects a typical monitoring configuration for a real turbine. In contrast, S5 and S6 are theoretical configurations, where the former includes all blade sensors and the latter all tower sensors (more details on these sensors can be found in [11]). The confusion matrices on the test set are given on Fig. 2, 3, and 4.

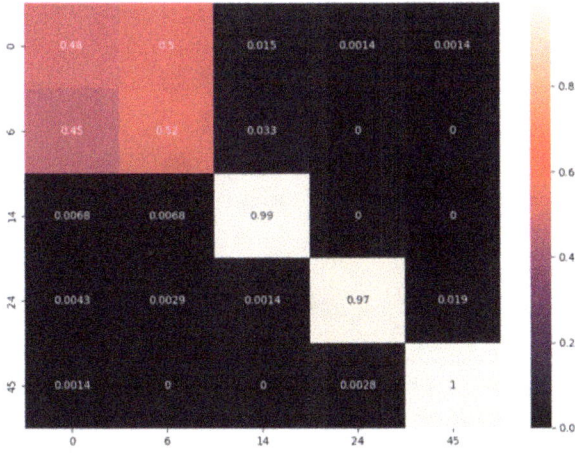

Fig 2. Confusion matrix for S4.

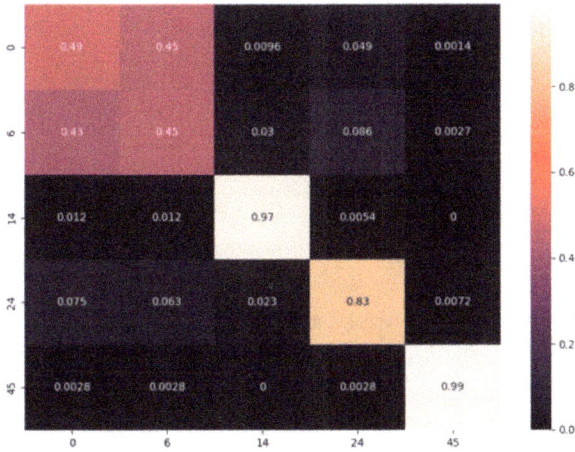

Fig 3. Confusion matrix for S5.

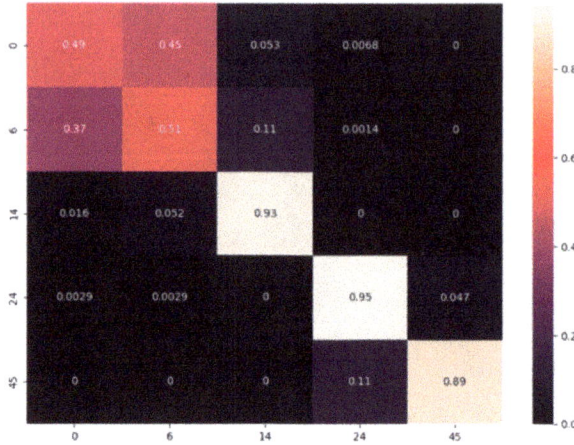

Fig 4. Confusion matrix for S6.

We can notice that in all 3 measurement configurations, the classes 0% and 6% cannot be distinguished as they are confused close to a 50/50 ratio. Therefore for the rest of our work, we will consider these two classes to be identical. If a sample of one class were to be confused with the other class, we will consider it as a successful classification.

Zero-shot learning
We now retrain our models with datasets where some of the classes are hidden using the same measurement scenarios. These define 3 different zero-shot learning configurations:
- partial obfuscation, classes 6% and 24% are unseen
- interpolation, classes 6%, 14% and 24% are unseen
- extrapolation, classes 0% and 45% are unseen

Since the unseen classes are not present during training, the classifier trained with a cross-entropy loss cannot correctly classify these samples during the testing phase. However, we would like to visualize the associated latent features of such classes to see if these can be separated from the seen classes. To visualize our high-dimensional feature space, we use the t-SNE [14] data visualization to project the sample features into a 2-dimensional space. The results can be seen on Fig. 5-13. We will consider the features apt for ZSL if the unseen classes have little overlap with the seen classes. We make an exception for the class 6% which presents too many similarities with class 0%. In this case, we will consider it a success if the classes do indeed overlap.

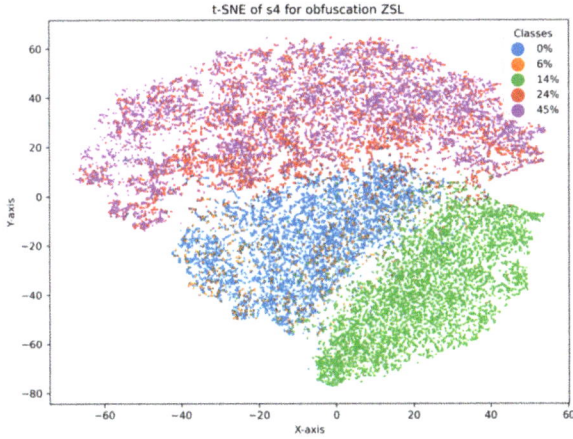

Fig. 5 The t-SNE features for the obfuscations (6% and 24% unseen) for S4. Class 6% successfully overlaps with class 0%. On the other hand, class 24% has a significant overlap with 45%. These features are not suitable for ZSL.

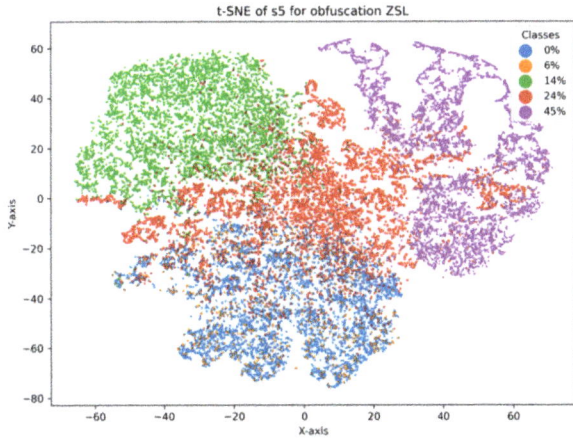

Fig. 6 The t-SNE features for the obfuscations (6% and 24% unseen) for S5. Class 6% successfully overlaps with class 0%. Furthermore, we can see that class 24% has formed its own cluster between the seen classes. These features are suitable for ZSL.

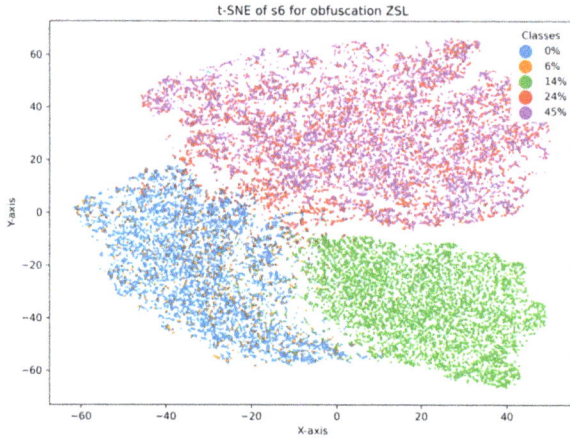

Fig. 7 The t-SNE features for the obfuscations (6% and 24% unseen) for S4. Class 6% successfully overlaps with class 0%. On the other hand, class 24% has a significant overlap with 45%. These features are not suitable for ZSL.

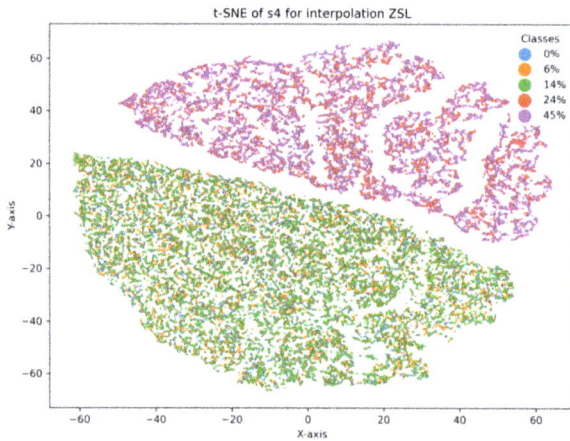

Fig. 8 The t-SNE features for the interpolations (6%, 14% and 24% unseen) for S4. None of the unseen damage classes can be distinguished from the seen classes. These features are unsuitable for ZSL.

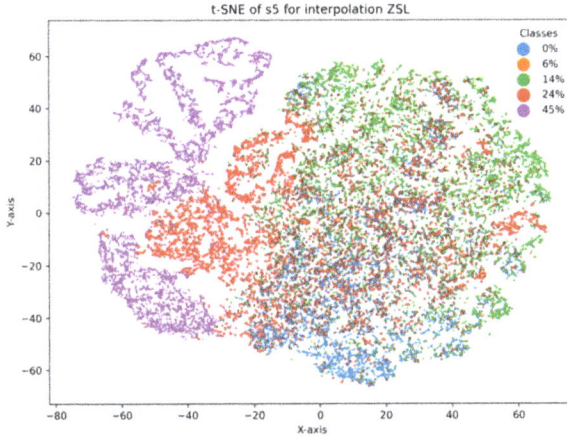

Fig. 9 The t-SNE features for the interpolations (6%, 14% and 24% unseen) for S5. Class 6% successfully overlaps with class 0%. We can see that both class 14% and 24% do appear in distinct regions. However, both these classes have a significant overlap. These features are only partially suitable for ZSL.

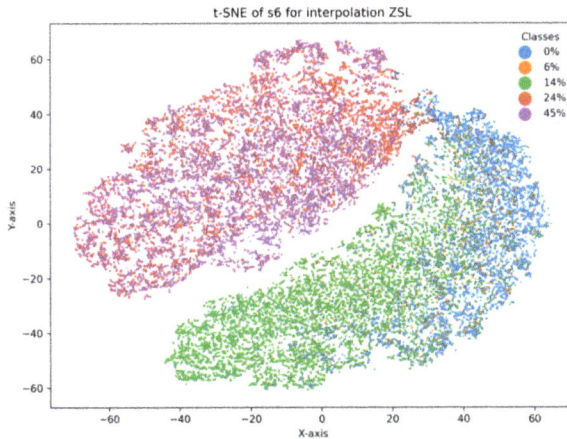

Fig. 10 The t-SNE features for the interpolations (6%, 14% and 24% unseen) for S6. Class 6% successfully overlaps with class 0%. Class 14% has successfully been separated from the other classes as its own cluster. On the other hand, class 24% cannot be differentiated from class 45%. These features are only partially suitable for ZSL.

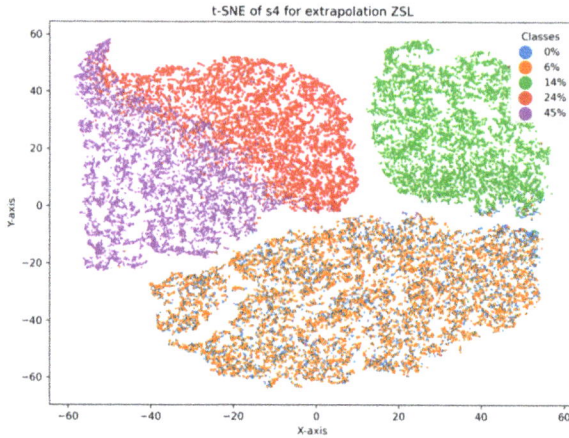

Fig. 11 The t-SNE features for the extrapolations (0% and 45% unseen) for S4. Class 0% successfully overlaps with class 6%. Furthermore, class 45% has been successfully separated into its own cluster and can be seen as a shift from class 24%. These features are suitable for ZSL.

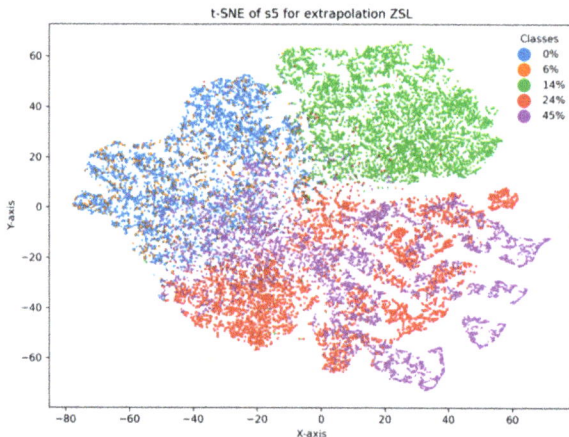

Fig. 12 The t-SNE features for the extrapolations (0% and 45% unseen) for S5. Class 0% successfully overlaps with class 6%. Moreover, class 45% forms a distinctive cluster, albeit a little overlap with class 24%. These features are suitable for ZSL.

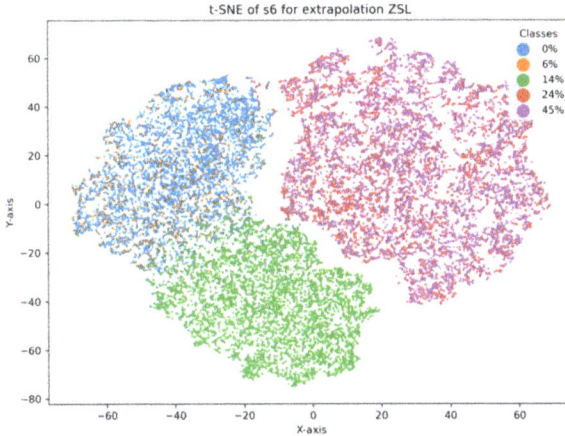

Fig. 13 The t-SNE features for the extrapolations (0% and 45% unseen) for S6. Class 0% successfully overlaps with class 6%. However, class 45% is indistinguishable from class 24%. These features are not suitable for ZSL.

Unsurprisingly, dense measurements collected from the blades directly yield the best features for unseen damage classes, while monitoring the tower itself does not. This is crucial information to select the sensor configuration, as training with S6 yields a high accuracy on seen classes and thus would usually be considered as a good model for the SHM of wind turbines. However, the failure to generalize for unseen damages shows that such a configuration is not robust to unexpected scenarios.

Summary

We have presented a study of the application of the SVAE in ZSL configurations for the SHM of wind turbines. Once we have trained our SVAE on a specific configuration, the t-SNE has successfully been used to help us determine whether novel classes appear within our latent features or not. We show that the measurement configuration is crucial to detect different types of damage that do not appear in the training dataset. Typically, high amounts of damage to the wind turbine cannot be extrapolated without measuring the blades of the turbine directly. The discovery is important to consider when deploying monitoring systems on wind turbines, as different parts of the turbine are more expensive to monitor than others (typically, monitoring the blade itself is more expensive than monitoring the tower). Thus, the monitoring team will have to take into account that monitoring the blade directly provides an additional robustness to the diagnosis model in the form of an extrapolation to unseen classes, which could justify its additional monitoring cost.

Acknowledgments

The research was conducted as part of the Future Resilient Systems (FRS) program at the Singapore-ETH Centre, which was established collaboratively between ETH Zurich and the National Research Foundation Singapore. This research is supported by the National Research Foundation, Prime Minister's Office, Singapore under its Campus for Research Excellence and Technological Enterprise (CREATE) programme.

Structural Health Monitoring: 10APWSHM   Materials Research Forum LLC
Materials Research Proceedings 50 (2025) 189-200   https://doi.org/10.21741/9781644903513-23

References

[1] Mishnaevsky L., Hasager C., Bak C. et al. (2021), Leading edge erosion of wind turbine blades: Understanding, prevention and protection, Renewable Energies, 169, pp. 953-969 https://doi.org/10.1016/j.renene.2021.01.044

[2] Keegan M. H. (2014), Wind Turbine Blade Leading Edge Erosion: An Investigation of Rain Droplet and Hailstone Impact Induced Damage Mechanisms

[3] Barber, S., Deparday, J., Marykovskiy, Y., Chatzi, E., Abdallah, I. and Duthé, G., Magno, M., Polonelli, T., Fischer, R., Mueller, H. (2022), Development of a wireless, non-intrusive, MEMS-based pressure and acoustic measurement system for large-scale operating wind turbine blades, Wind Energy Science, 7(4), pp 1383 - 1398 https://doi.org/10.5194/wes-7-1383-2022

[4] Simpson T., Dervilis N. and Chatzi E. (2021), Machine Learning Approach to Model Order Reduction of Nonlinear Systems via Autoencoder and LSTM Networks, Journal of Engineering Mechanics, 147, 04021061 https://doi.org/10.1061/(ASCE)EM.1943-7889.0001971

[5] Li Z., Xu P., Xing J. et al. (2022), SDFormer: A Novel Transformer Neural Network for Structural Damage Identification by Segmenting the Strain Field Map, Sensors (Basel, Switzerland), 22(6), 2358 https://doi.org/10.3390/s22062358

[6] Kingma D. P. and Welling M. (2013), Auto-Encoding Variational Bayes

[7] Bacsa K., Liu W., Abdallah I., Chatzi E., Structural Dynamics Feature Learning Using a Supervised Variational Autoencoder, Journal of Engineering Mechanics 151 (2), 04024106 https://doi.org/10.1061/JENMDT.EMENG-7635

[8] Pourpanah F., Abdar M., Luo Y., Zhou X., Wang R., Lim C. P., Wang X. Z., Wu Q. M. J. (2022), A review of generalized zero-shot learning methods, IEEE transactions on pattern analysis and machine intelligence https://doi.org/10.1109/TPAMI.2022.3191696

[9] Sohn K., Lee H., and Yan X. (2015), Learning structured output representation using deep conditional generative models, in Proc. Adv. Neural Informat. Process. Syst., pp. 3483-3491

[10] Mishra A., Krishna R. S., Mittal A, and Murthy H. A. (2018), A generative model for zero shot learning using conditional variational autoencoders, in Proc. IEEE Conf. Comput. Vis. Pattern Recognit. Workshops, pp. 2188-2196 https://doi.org/10.1109/CVPRW.2018.00294

[11] Huang H., Wang C., Yu P. S., and Wang C. D. (2019), Generative dual adversarial network for generalized zero-shot learning, in Proc. IEEE Conf. Comput. Vis. Pattern Recognit., pp. 801-810 https://doi.org/10.1109/CVPR.2019.00089

[12] Abdallah I., Natarajan A. and Sørensen J. (2015), Impact of uncertainty in airfoil characteristics on wind turbine extreme loads, Renew. Energy, 75, pp. 283-300 https://doi.org/10.1016/j.renene.2014.10.009

[13] Ji T., Vuppala S. T., Chowdhary G. et al. (2020), Multi-Modal Anomaly Detection for Unstructured and Uncertain Environments, in Conference on Robot Learning (CoRL)

[14] Kingma D. P. and Ba J. (2015), Adam: A Method for Stochastic Optimization, ICLR (Poster)

[15] Van der Maaten L. and Hinton H. (2008), Visualizing Data using t-SNE, JMLR 9(86), pp. 2579–2605

Structural Health Monitoring: 10APWSHM
Materials Research Proceedings 50 (2025) 201-211

Materials Research Forum LLC
https://doi.org/10.21741/9781644903513-24

Experimental investigation of mode veering phenomena in a bolted connection

Dashty Samal Rashid[1,a*], Francesco Giorgio Serchi[1], Naoki Hosoya[2], David Garcia Cava[1,b]

[1]University of Edinburgh, United Kingdom

[2]Shibaura Institute of Technology

[a]D.S.Rashid@ed.ac.uk, [b]david.garcia@ed.ac.uk

Keywords: Veering Phenomena, Bolted Connections, Modal Analysis, Structural Dynamics, Engineering Mechanics

Abstract. Mode veering phenomena, sometimes accompanied by mode localization, refers to the sudden and abrupt changes in the trajectories of eigenvalues in a system, often caused by small variations in system parameters or boundary conditions. In this paper, mode veering phenomena is studied with the focus on the dynamics of bolt connection geometry. A combination of experimental and analytical modeling is used to identify and validate the conditions under which veering occurs in bolted connections. The parameter varied here is the length ratio of the shank and protruding end of a bolt, which serves to characterize the bolt's geometry. The experimental results demonstrate that the observed behavior corresponds well with the analytical model, indicating the presence of localized modes. In particular, the mode shape variation showed consistency with the model's predictions. The implications of these findings with respect to the agreement between the model and observed experimental results are discussed.

1. Introduction

Bolted connections are fundamental components in many structural systems, commonly used in a wide range of engineering applications from civil infrastructure to aerospace and automotive industries [1]. Due to their versatility and ease of maintenance, they serve as critical joints in structures subjected to different types of loading. However, bolted connections are not without limitations, as their dynamic behaviour can significantly affect the overall performance of a system, particularly under cyclic loading or vibrations. Structural vibrations, if left unchecked, can lead to fatigue damage, loosening, and even catastrophic failure of the entire structure. Consequently, understanding the dynamic characteristics of bolted connections is essential for ensuring the long-term reliability of these systems.

Vibration-induced loosening has been identified as a major challenge for bolted connections. When bolts vibrate at or near their natural frequencies, resonance can amplify dynamic forces, leading to progressive loosening through mechanisms such as cyclic slipping at the thread interface. Studies have shown that small geometric variations, changes in loading conditions, and boundary constraints, can alter a bolted system's natural frequencies and stiffness [2, 3]. Nevertheless, most of this research primarily focuses on how these factors influence the global joint dynamics.

Mode veering is a well-documented phenomenon in structural dynamics [4-9], observed when small changes in system parameters result in sharp, sudden changes in natural frequencies. This behaviour is typically seen in system with coupled dynamic modes, where interaction between closely spaced modes cause their frequencies to shift abruptly as parameters like stiffness, geometry, or boundary conditions are varied. The literature shows several works on experimental evidence of this phenomena. Mode veering has been extensively studied, with significant attention

Structural Health Monitoring: 10APWSHM
Materials Research Proceedings 50 (2025) 201-211

Materials Research Forum LLC
https://doi.org/10.21741/9781644903513-24

given to it. Preliminary investigations were conducted on beams by Petyt and Fleisxer [4], and on curved plates by Nair and Durvasula [5]. The phenomena observed was described as modes approaching each other but, instead of crossing over, they veer away. The other observation is that the two corresponding eigenvectors interchange. A detailed explanation of this phenomenon was later provided by Leissa [6], who coined the term "veering".

Mode veering is frequently associated with mode localization, where the energy or displacement in certain modes becomes concentrated in specific areas of the structure. Pierre [7] introduced a perturbation technique, demonstrating that the occurrence of strong mode localization and eigenvalue loci veering are both manifestations of the same phenomena, typically caused by small structural irregularities in systems with closely spaced eigenvalues. Further advancements in theoretical and experimental works can be attributed to Balmes [8] and du Bois et al [9]. While mode veering has been extensively studied in simpler systems like beams and plates, its occurrence in more complex structures, such as bolted connections have been limited. However, recent studies have started to explore these phenomena in the context of bolted connections. For instance, Wang et al. [10] examined veering in multi-plate bolted structures, observing localized frequency veering due to non-uniform contact pressure between the bolted plates. Similarly, Du et al. [11] analyzed both global and local veering in rotating cylindrical shells connected by multiple bolts. The study found that as rotational speed increased, mode interactions across the entire shell caused global veering. However, in both studies, the veering is attributed to the interaction of the entire multi-bolted structure, rather than to the behavior of individual bolts.

This paper aims at investigating mode veering in bolted connections, focusing on how bolt geometry influences its dynamic behaviour. Unlike prior research that mainly considered the entire bolted structure, this study isolates the effects specific to the bolt itself. The primary objective is to experimentally demonstrate how variations in the shank and protruding end lengths impact natural frequencies and mode shapes. This research could potentially contribute to passive control strategies in the design of bolted connections, allowing engineers to influence the system's dynamic behaviour by adjusting the bolt's geometric parameters. A theoretical model from our earlier work [12] will be employed to validate and benchmark the experimental observation of the veering phenomena, providing a comparison between theoretical predictions and experimental data.

This paper is structured as follows: Section 2 provides a theoretical background on the bolt model. Section 3 describes the experimental setup used in this investigation, detailing the methodology and measurement techniques. Section 4 presents the experimental results and a comparison with model predictions, followed by a discussion of their implications. Finally, conclusions and potential directions for future work are presented in Section 5.

2. Theory

In this section, a model based on Euler-Bernoulli beam theory is presented to model the transverse vibrations of a bolt (see Fig. 1). This model is identical to one of the models from our earlier work in [12]. The continuous beam with its shank and protruding end is modeled as two segments. The first segment is restrained by a translational K_t and rotational K_r stiffness and under axial load p to represent the applied tension. The effects of mass M_i and rotary inertia J_i of the bolt head and but are also considered. Both segments are assumed to have uniform density ρ, second moment of area I, elastic modulus E and cross-sectional area A. The beam's transverse displacements can be described by $U(X, t)$, where $X \in (0, l)$ is the axial coordinate and t is the time. ξ is defined as the shank length over the entire bolt length; however, from here on, ξ will be referred to as shank/protruding end length ratio. To simplify the calculations, we can introduce the following nondimensional quantities:

$$x = \frac{X}{l}, \quad \xi = \frac{l_1}{l}, \quad u(x,\tau) = \frac{U(X,t)}{l}, \quad \tau = \omega_0 t, \quad \widetilde{M}_{1,2} = \frac{M_{1,2}}{\rho Al},$$

$$\tilde{J}_{1,2} = \frac{J_{1,2}}{\rho Al}, \quad \omega_0 = \sqrt{\frac{EI}{\rho Al^4}}, \quad p = \frac{Nl^2}{EI}, \quad k_t = \frac{K_t l^3}{EI}, \quad k_r = \frac{K_r l}{EI} \tag{1}$$

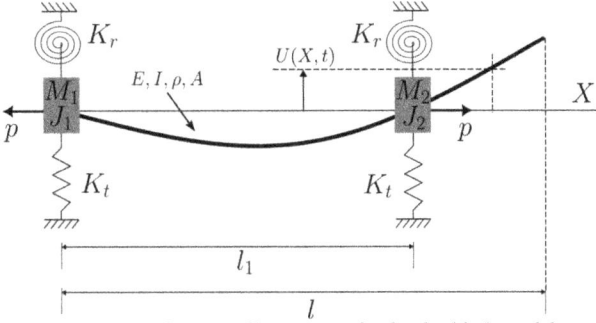

Figure 1: *Schematic illustration of individual bolt model*

The partial differential equations for a prestressed and non-prestressed Euler-Bernoulli beam can be given by:

$$EI\frac{\partial^4 u(x,\tau)}{\partial x^4} - p\frac{\partial^2 u(x,\tau)}{\partial x^2} + m\frac{\partial^2 u(x,\tau)}{\partial^2 t} = 0 \tag{2}$$

$$EI\frac{\partial^4 u(x,\tau)}{\partial x^4} + m\frac{\partial^2 u(x,\tau)}{\partial^2 t} = 0 \tag{3}$$

Where m is the mass per unit length of the beam. Eq. 2 and Eq. 3 refer to the first and second segment of the beam respectively. The system can be solved by the superposition of eigen-solutions:

$$u(x,\tau) = \phi(x)e^{i\omega\tau} \tag{4}$$

To simplify the analysis, the beam is analyzed over the spatial and temporal domains, where the spatial solution for the two segments is expressed as:

$$\phi(x) = \begin{cases} \phi_1(x), & 0 \le x \le l_1 \\ \phi_2(x), & l_1 \le x \le l \end{cases} \tag{5}$$

The general solutions for the sub-functions in Eq. 5 can then be given by:

$$\phi_1(x) = A_1\cos(\alpha_1 x) + A_2\sin(\alpha_1 x) + A_3\cosh(\alpha_2 x) + A_4\sinh(\alpha_2 x) \tag{6}$$

$$\phi_2(x) = B_1\cos(\beta x) + B_2\sin(\beta x) + B_3\cosh(\beta x) + B_4\sinh(\beta x) \tag{7}$$

Structural Health Monitoring: 10APWSHM   Materials Research Forum LLC
Materials Research Proceedings 50 (2025) 201-211   https://doi.org/10.21741/9781644903513-24

The beam vibration eigenvalues are contained in α_1, α_2, and β. They can be obtained by the following expressions:

$$\alpha_1 = \sqrt{\sqrt{\left(\frac{p}{2}\right)^2 + \omega^2} - \frac{p}{2}}, \quad \alpha_2 = \sqrt{\sqrt{\left(\frac{p}{2}\right)^2 + \omega^2} + \frac{p}{2}}, \quad \beta = \sqrt{\omega} \tag{8}$$

After inserting the general solutions from Eq. 6 and Eq. 7 into the boundary conditions, the boundary conditions are transformed into the following form:

$$\frac{\partial^2 \phi_1}{\partial x^2}(0) = k_r \frac{\partial \phi_1}{\partial x}(0) - \tilde{J}_1 \omega^2 \frac{\partial \phi_1}{\partial x}(0) \tag{9}$$

$$\frac{\partial^3 \phi_1}{\partial x^3}(0) = p \frac{\partial \phi_1}{\partial x}(0) - k_t \phi_1(0) + \widetilde{M}_1 \omega^2 \phi_1(0) \tag{10}$$

$$\phi_1(l_1) = \phi_2(l_1) \tag{11}$$

$$\frac{\partial \phi_1}{\partial x}(l_1) = \frac{\partial \phi_2}{\partial x}(l_1) \tag{12}$$

$$\frac{\partial^2 \phi_1}{\partial x^2}(l_1) + k_r \frac{\partial \phi_1}{\partial x}(l_1) - \tilde{J}_2 \omega^2 \frac{\partial \phi_1}{\partial x}(l_1) = \frac{\partial^2 \phi_2}{\partial x^2}(l_1) \tag{13}$$

$$\frac{\partial^3 \phi_1}{\partial x^3}(l_1) - p \frac{\partial \phi_1}{\partial x}(l_1) - k_t \phi_1(l_1) + \widetilde{M}_2 \omega^2 \phi_1(l_1) = \frac{\partial^3 \phi_2}{\partial x^3}(l_1) \tag{14}$$

A set of 8 simultaneous equations from the boundary conditions can be obtained with the unknown integration constants:

$$\mathbf{A}\mathbf{c} = 0 \tag{15}$$

Where $\mathbf{A}$ is an 8x8 matrix and $\mathbf{c}$ is a vector containing the unknown integration constants. One can solve Eq. 11 by the nontrivial solution of $\mathbf{c}$ by setting the determinant of $\mathbf{A}$ equal to zero:

$$|\mathbf{A}| = 0 \tag{16}$$

The natural frequencies ω_n can be determined by solving for the roots of the characteristic frequency function obtained from Eq. 16 using any appropriate numerical method. For a given dimensionless natural frequency $\check{}_n$, the corresponding integration constants can be obtained by solving Eq. (15). Once these constants are determined, they can be substituted back into the general solutions from Eq. (6) and Eq. (7). This process will yield the corresponding mode shape of the beam. To investigate veering behaviour in a bolted connection, we employ the dynamic bolt model, treating it as a function whose outputs (ω_n and $\phi(x)$) depend on key parameters. The primary inputs for this function are the shank/protruding end length ratio ξ, translational and rotational stiffness k_t and k_r and the applied tension p.

Structural Health Monitoring: 10APWSHM
Materials Research Proceedings 50 (2025) 201-211

Materials Research Forum LLC
https://doi.org/10.21741/9781644903513-24

$$f(\xi, k_t, k_r, p) = \{\omega_n, \phi(x)\} \tag{17}$$

By systematically solving the function above for a range of ξ values ranging from 0 to 1, a relationship between the model outputs and the shank/protruding end length ratio can be obtained. The stiffness parameters k_t and k_r typically needs to be fitted based on measurement data. In this study, k_t and k_r was fitted only to measurements from the highest ξ case and was kept constant for all the other ξ cases.

3. Experimental set up and testing procedure

3.1 Description of the bolt and subject body

The test bolt of this study is an M12 bolt encased in an aluminum plate. The bolt has a nominal diameter of 12 mm, a property class of 8.8, and is made of high-tensile steel with a bright zinc-plated finish. The nominal length of the bolt is 250 mm. Table 1 summarizes the bolt specifications, while Table 2 provides material properties and key parameters, including density, elastic modulus, and the mass and rotary inertias for both the bolt head and nut. The subject body was specifically designed to be modular, allowing adjustments to the shank/protruding end length ratio ξ by adding or removing small plates. ξ is now allowed to vary due to the modularity of the subject study. The clearance for the bolt was 1 mm. Fig. 2 provides a schematic representation of the modular subject body, illustrating how different shank/protruding end length ratios can be achieved. The diagram also includes information on the dimensions of the subject body.

Table 1: *Specifications of the M12 test bolt*

Bolt Type	Nominal diameter d [mm]	Property class	Finish	Steel type	Nominal Length [mm]
M12	12	8.8	Bright Zinc Plated	High tensile steel	250

Table 2: *Bolt material specifications and parameters*

Density Ò [kg/m³]	Elastic Modulus E [N/m2]	$\widetilde{M_1}$ [-]	$\widetilde{J_1}$ [-]	$\widetilde{M_2}$ [-]	$\widetilde{J_2}$ [-]
7850	200×10^9	6.923×10^{-2}	4.147×10^{-5}	7.797×10^{-2}	6.748×10^{-5}

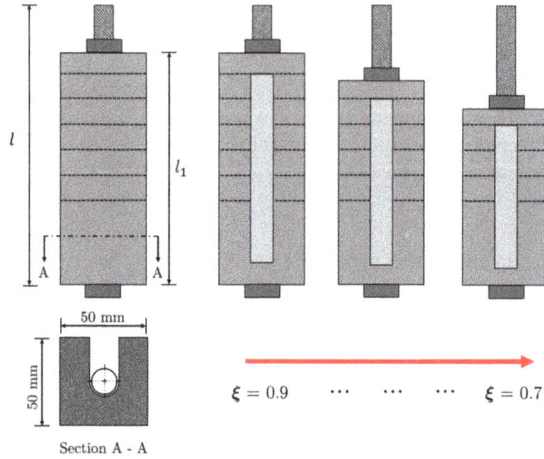

Figure 2: Graphical representation of subject body with varying ξ.

3.2 Experimental setup

The experimental setup, illustrated in Fig. 3, involves modal testing of an M12 bolt encased in the modular subject body. To measure the velocity response of the bolt, a Polytec OFV 303 laser Doppler vibrometer (LDV) was used. The LDV device was manually adjusted to multiple locations along the bolt's length, ensuring a comprehensive capture of vibrational displacements at various points for each configuration. The bolt was excited by a Brüel & Kjær impulse hammer (Type 8206-002) to provide a consistent input force. To ensure consistent preload conditions across all tests, the bolt was torqued to approximately 50 Nm using a calibrated Norbar torque wrench. For excitation, an impact was applied to the tip of the protruding end, with measurements taken at eight distinct points along the bolt for each configuration. The sampling frequency for the modal testing was set to 17066 Hz which was sufficient to capture the first four modes.

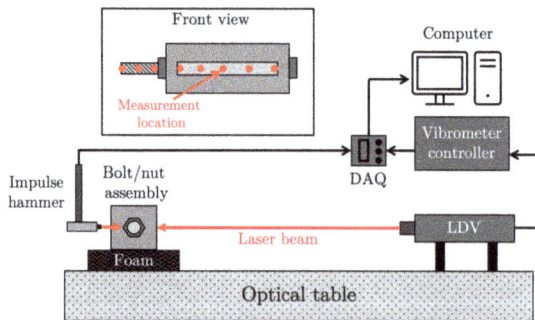

Figure 3: Schematic diagram of the experimental setup

Structural Health Monitoring: 10APWSHM

Materials Research Forum LLC

Materials Research Proceedings 50 (2025) 201-211

https://doi.org/10.21741/9781644903513-24

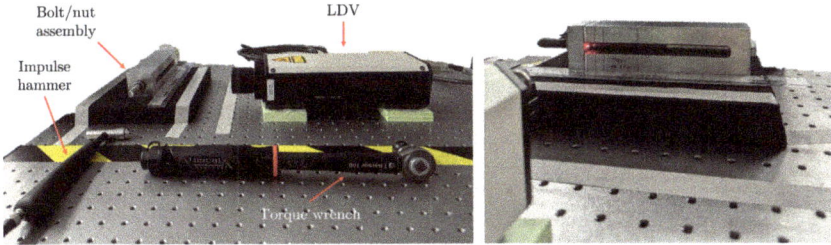

Figure 4: Images of the experimental setup during testing

3.3 Data processing

For each configuration of ξ, an impact was applied to the protruding end of the bolt, and measurements were taken at eight distinct points along its length. At each measurement location, five frequency response functions (FRFs) were computed and averaged to improve data reliability. These averaged FRFs at each point served as the basis for the modal parameter extraction. To extract mode shapes, the averaged FRFs from each measurement point were processed using MATLAB's `modelfit` function, applying the Least Squares Rational Fraction (LSRF) method. This approach enabled precise fitting and identification of mode shapes across the eight points along the bolt. For determining natural frequencies, the mean of the eight averaged FRFs (one from each measurement point) was calculated for each configuration. This averaging process provided a reliable estimate of the natural frequencies for each configuration of ξ. Fig. 5 shows example averaged FRFs for different ξ configurations measured at protruding end of the bolt.

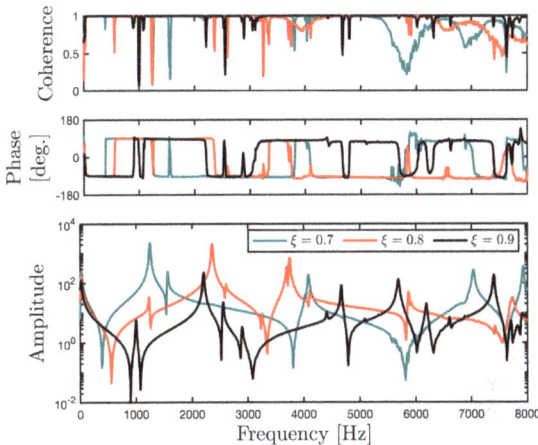

Figure 5: Frequency response functions for different ξ configurations.

4. Results and discussion

This section presents the dynamic behaviour of the M12 bolt under varying shank/protruding end length ratios ξ has been investigated experimentally using modal analysis techniques. Experimental natural frequencies and corresponding mode shapes obtained from modal testing were compared with theoretical predictions, with a focus on the veering regions where significant shifts in frequency and mode localization are observed.

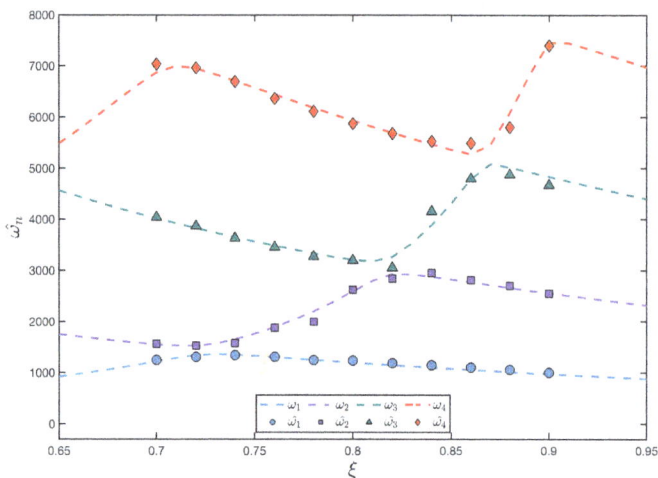

Figure 6: *Comparison of experimentally obtained and theoretically computed natural frequency for varying* $\boldsymbol{\xi}$ *(* $\boldsymbol{k_t}$ *= 15754,* $\boldsymbol{k_r}$ *= 46,* $\boldsymbol{p}$ *= 7.68).*

The experimental setup enabled systematic variation of ξ by adjusting the detachable plates, creating multiple configurations with ξ values ranging from 0.9 to 0.7. Fig. 6 shows the experimental and theoretical natural frequencies across these configurations. Each vibration mode exhibits a distinct trend as ξ decreases. Notably, certain frequency pairs demonstrate a sharp change in behaviour within specific ξ ranges, highlighting regions of veering. Considering the first four modes, three primary veering regions were identified, occurring when ξ is approximately 0.87, 0.82, and 0.72. The veering effect is further evidenced by mode localization, where the spatial displacement profile of the modes become concentrated in specific regions of the bolt. Fig. 7 illustrates the normalized mode shapes for configurations before and after the veering regions, showing that the mode shapes associated with the protruding end seems to be travelling from higher modes to lower modes exchanging with adjacent modes during the veering regions. Furthermore, this localization is visually apparent in the mode shapes, where the displacement amplitude sharply increases at the protruding ends and the displacements at the shank decrease. A comparison of the experimental and theoretical natural frequencies reveals good agreement across most configurations. Although minor discrepancies are present, the small deviations between experimental and theoretical results may be attributed to slight variations in bolt tension, deviations in subject body representation of ξ, and slight offset of measurement points of the LDV.

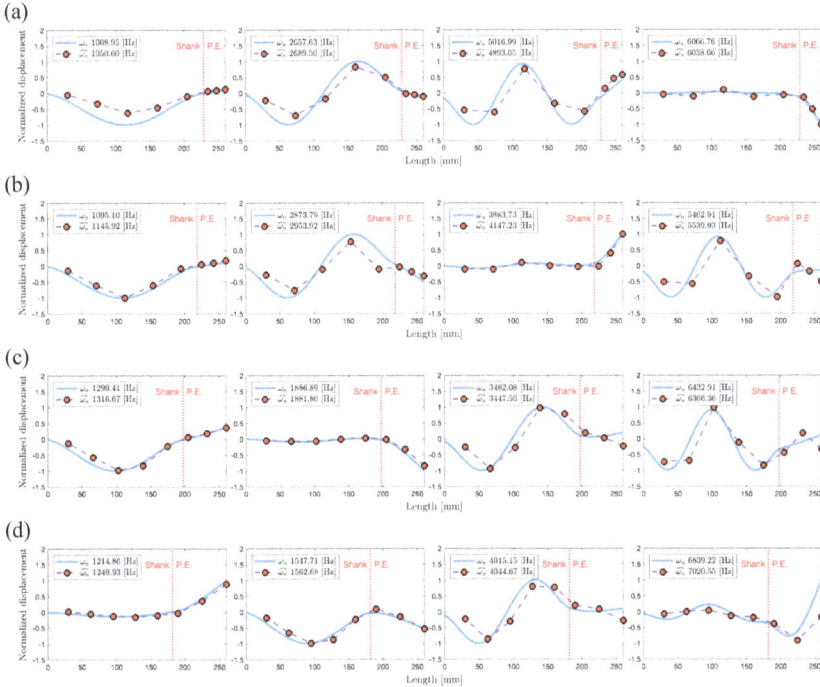

Figure 7: *Comparison of the mode shapes between theoretical ω_n and experimental $\widehat{\omega_n}$; (a)*
$\xi = 0.88$; *(b)* $\xi = 0.84$; *(c)* $\xi = 0.76$; *(d)* $\xi = 0.7$.

The findings of this study suggests that the shank/protruding end length ratio ξ has influence on the vibrational behaviour of the M12 bolt. Across different geometric configurations, distinct shifts in natural frequencies were observed as ξ was varied, confirming that the bolt geometry significantly affects the dynamics characteristics. Specifically at the highlighted geometric regions, frequency veering was noted, where pairs of natural frequencies approached and diverged away from each other. The mode shapes associated with these veering regions showed localization effects. As ξ was being varied, mode shapes were altered, leading to specific modes exhibiting an increase of displacement in the protruding end. Every time this localized mode was passing through the veering regions, there was a mode shape transition where the mode shape interchanged with its adjacent mode. The comparison between experimental results and theoretical predictions demonstrated satisfactory agreement in the natural frequencies and mode shapes, with minor discrepancies attributed to slight variations in bolt tension, measurement alignment, and non-optimized stiffness values. The experimental findings validate the theoretical model's capacity to capture essential veering behaviour and mode localization effects due to shank/protruding end length ratio. These observations suggest that modifying bolt geometry can be an effective means of influencing and controlling mode shapes and frequencies. The mode localization achieved through specific configurations could be beneficial in applications where isolating or reducing the propagation of vibrations is crucial, enhancing stability in dynamically loaded bolted connections.

5. Conclusions

This study demonstrates the influence of the shank/protruding end length ratio on the dynamic behaviour of an M12 bolt with experimental evidence of mode veering phenomena. Significant shifts in natural frequencies and mode localization were observed. Considering the first four modes, three primary veering regions were identified, occurring when ξ approximately was around 0.87, 0.82, and 0.72. The experimental analysis shows that as the bolt transitions through veering zones, the mode shapes associated with the protruding end travel from the fourth mode to the first mode indicating an interchange with adjacent modes. The findings demonstrate that geometry by means of ξ could potentially allow the isolation of specific vibrational modes. Further research could build on this work by conducting detailed sensitivity analyses that can contribute to a robust framework for optimizing bolt design in critical applications, ensuring the reliability of bolts under dynamic conditions.

Acknowledgements

This work was supported by the UK's Engineering and Physical Research Council via the Wind Marine Energy Systems Centre for Doctoral Training, grant number EP/S023801/1. We also extend our gratitude to the Japan Society for the Promotion of Science for their support under the Fostering Joint International Research (B) (Grant No. JP22KK0053).

References

[1] Bickford, John H. Introduction to the design and behavior of bolted joints: non-gasketed joints. CRC press, 2007. https://doi.org/10.1201/9780849381874

[2] Zadoks, R. I., and X. Yu. "An investigation of the self-loosening behavior of bolts under transverse vibration." Journal of sound and vibration 208.2 (1997): 189-209. https://doi.org/10.1006/jsvi.1997.1173

[3] Hosoya, Naoki, et al. "Axial force measurement of the bolt/nut assemblies based on the bending mode shape frequency of the protruding thread part using ultrasonic modal analysis." Measurement 162 (2020): 107914. https://doi.org/10.1016/j.measurement.2020.107914

[4] Petyt, M., and C. C. Fleischer. "Free vibration of a curved beam." Journal of Sound and Vibration 18.1 (1971): 17-30. https://doi.org/10.1016/0022-460X(71)90627-4

[5] Nair, P. S., and S. Durvasula. "On quasi-degeneracies in plate vibration problems." International Journal of Mechanical Sciences 15.12 (1973): 975-986. https://doi.org/10.1016/0020-7403(73)90107-0

[6] Leissa, Arthur W. "On a curve veering aberration." Zeitschrift für angewandte Mathematik und Physik ZAMP 25 (1974): 99-111. https://doi.org/10.1007/BF01602113

[7] Pierre, Christophe. "Mode localization and eigenvalue loci veering phenomena in disordered structures." Journal of Sound and Vibration 126.3 (1988): 485-502. https://doi.org/10.1016/0022-460X(88)90226-X

[8] Balmes, E. "High modal density density, curve veering, localization: a different perspective on the structural response." Journal of Sound and Vibration 161.2 (1993): 358-363. https://doi.org/10.1006/jsvi.1993.1078

[9] Du Bois, Jonathan Luke, Sondipon Adhikari, and Nick AJ Lieven. "Eigenvalue curve veering in stressed structures: An experimental study." Journal of Sound and Vibration 322.4-5 (2009): 1117-1124. https://doi.org/10.1016/j.jsv.2008.12.014

Structural Health Monitoring: 10APWSHM   Materials Research Forum LLC
Materials Research Proceedings 50 (2025) 201-211   https://doi.org/10.21741/9781644903513-24

[10] Wang, Yan Qing, et al. "Theoretical and experimental studies on vibration characteristics of bolted joint multi-plate structures." International Journal of Mechanical Sciences 252 (2023): 108348. https://doi.org/10.1016/j.ijmecsci.2023.108348

[11] Du, Dongxu, et al. "Vibration characteristics analysis for rotating bolted joined cylindrical shells considering the discontinuous variable-stiffness connection." Thin-Walled Structures 177 (2022): 109422. https://doi.org/10.1016/j.tws.2022.109422

[12] Rashid, Dashty Samal, et al. "Coupled Dynamics of Shank and Protruding End for Bolt Loosening." Available at SSRN 4778676.

Structural Health Monitoring: 10APWSHM
Materials Research Proceedings 50 (2025) 212-218

Materials Research Forum LLC
https://doi.org/10.21741/9781644903513-25

Analysis of multiple reflected ultrasonic waves generated during a drilling process

Jonathan Liebeton[1,a] * and Dirk Söffker[1,b]

[1]University of Duisburg-Essen, Duisburg, Germany

[a]jonathan.liebeton@uni-due.de, [b]soeffker@uni-due.de

Keywords: Ultrasonic Waves, Wave Propagation, Reflection, Signal Processing

Abstract. The propagation behavior of ultrasonic waves plays a crucial role in Structural Health Monitoring and non-destructive testing applications. Understanding the reflections and interactions of these waves within a material is essential for accurately diagnosing the integrity of structures, especially when using time-domain and non-stationary features. This study focuses on the analysis of multiple reflected ultrasonic waves generated during thread forming. The measured ultrasonic waves are multiple times reflected at pre-drilled bore holes on the path between their origin and the sensor position. The extracted features derived from signal processing are used to train regression models to predict the propagation distance between bore hole and sensor position. The predictions of the regression models are evaluated and used to investigate the propagation paths based on the prediction accuracy. The findings reveal a relationship between the propagation distance and the derived features from the signal processing of the Acoustic Emission data indicating a change of ultrasonic wave characteristics due to reflections at material boundaries. The results of this study contribute to the development of more accurate and robust Structural Health Monitoring and non-destructive testing techniques. By considering the behavior of reflected ultrasonic waves, the detection and localization of defects can be enhanced.

Introduction

The use of ultrasonic waves in the context of Structural Health Monitoring is a proven and widely used method to inspect structures during operation. Ultrasonic waves have the ability to travel over large distances, which makes them a suitable tool to monitor entire structures. The ultrasonic waves testing methods are divided into an active and passive approach, namely Guided Waves and Acoustic Emission, respectively. The Acoustic Emission approach relies on the passive generation of ultrasonic waves, when the mechanical threshold of a material is exceeded under load and the potential energy is partly released as ultrasonic waves. The detection of the passively generated ultrasonic waves is the indication of an existing damage and by analyzing the detected Acoustic Emission signal damage characteristics are obtained. The physical process of generating ultrasonic wave is described by the Felicity effect. For the Guided Waves approach transducers are operating in pitch-catch configuration to measure actively generated ultrasonic waves. The Guided Waves method depends on the reflection and interaction of the generated ultrasonic wave at and with material boundaries and material defects. The comparison of the received wave with a baseline measurement in the damage-free state indicates the presence, location, and severity of damages of the monitored structure.

Wave propagation is a crucial aspect of both methods, because the diagnostic performance is directly affected due to material properties [1]. Changes of the material properties caused by environmental, loading, and boundary conditions affect the propagation behavior [2,3,4]. In materials exhibiting bending stresses ultrasonic waves experience an increased attenuation [5]. Reflections occur at sudden changes of acoustic impedance. The reflection and transmission of the incident wave depend on the ratio of the acoustic impedances of the adjacent materials [6]. Reflections at material boundaries also impact the propagation behavior, because of interference

Structural Health Monitoring: 10APWSHM
Materials Research Proceedings 50 (2025) 212-218

Materials Research Forum LLC
https://doi.org/10.21741/9781644903513-25

between the directly propagating wave and its reflection [7]. Reflections at material boundaries depend on the surface properties, because the surface profile and roughness impact the attenuation and causes mode conversion and scattering [8,9]. The objective of this contribution is to predict the distance between sensor and source position by analyzing the frequency spectrum of multiple reflected waves.

The paper is divided into five sections. In section 2, the theoretical background and methodology of wave propagation in solids, the used signal processing and machine learning method are explained. The experimental setup and the measured data are detailed in section 3. In section 4 the results are presented and discussed. Section 5 contains a summary and conclusions of the presented results and an outlook on future research projects.

Theoretical Background and Methodology

Ultrasonic waves propagating in homogeneous and isotropic solids are called Bulk waves. Waves causing particles to oscillate in propagation direction are referred to as longitudinal waves. Shear waves are characterized by particles oscillating perpendicular to the direction of propagation. The longitudinal wave speed is described by

$$c_L = \sqrt{\frac{1-v}{(1+v)(1-2v)}\frac{E}{\rho}} \qquad (1)$$

and the speed of shear waves by

$$c_S = \sqrt{\frac{1}{2(1+v)}\frac{E}{\rho}}, \qquad (2)$$

where v describes the Poisson ratio, E the modulus, and ρ the material density [10]. At the material surface the waves are called Rayleigh waves and their approximated wave speed is described by

$$c_R = c_S\left(\frac{0.87+1.12v}{1+v}\right), \qquad (3)$$

where c_S is the wave speed of shear waves, described in Eq. 3, and v the Poisson ratio [10].

Signal processing is an elemental aspect in ultrasonic wave-based Structural Health Monitoring, because relevant information about the state of the monitored structure can be extracted from the raw measurement signals. Fast Fourier Transform is used to transform signal from time domain into frequency domain. The Fourier transform of a signal x in time domain consisting of n samples is defined by

$$X_m = \sum_{k=0}^{n-1} x_m e^{-i2\pi km/n} \quad \text{with } m = 0, \dots, n-1. \qquad (4)$$

The Short-Time Fourier Transform enables a combined signal analysis in time and frequency domain. The Fourier Transform of a signal x is defined by

$$X(m,\omega) = \sum_{n=-\infty}^{\infty} x[n]w[n-m]e^{-j\omega n}, \qquad (5)$$

where w is a window function, which is shifted across the signal. The Fourier Transform is performed on each windowed signal section resulting in a homogenous time-frequency resolution. The Short-Time Fourier Transform is used to analyze signal energy of a specific frequency range and features are extracted of the Fourier transformed measurement signals.

A neural network regression model relates Acoustic Emission measurement features to the distance between their origin and the sensor position. The model has five layers; an input layer, a fully connected layer, a ReLu layer, another fully connected layer, and an output layer. The size of the input layer matches the number of features and the first fully connected layer has ten outputs. The outputs of the second fully connected layer match the output variables of the model. Bayesian optimization determines appropriate weights and biases in the fully connected layers.

Experimental Setup and Data

The data set used is generated to investigate the tapping torque test according to ASTM D5619 using water-mixed metalworking fluids [11,12]. In [13], the data set is used to discriminate metalworking fluids using convolutional neural networks and transfer learning. A C45E (1.1191) platform with predrilled holes is used in the experiment. A threading tool is used to produce M6 threads. The tool has an active length of 8 mm, a cutting lead of 2-3 mm, a thread pitch of 1 mm, and is operated at 1061 rpm. The pre-drilled holes are 5.6H7 in diameter and 28 mm deep. The platform (Fig. 1) consists of 368 holes arranged in 23 columns and 16 rows. In the experiment, ten different metalworking fluids are tested with a naphthenic-based mineral oil as the reference fluid. The series of measurements for each metalworking fluid are made column wise, in even-numbered columns between two and twenty. The corresponding holes are marked in yellow in figure 1. The odd numbered columns used for the reference fluid are marked in white in figure 1.

A tribometer is used for the tapping process. A piezoelectric transducer is attached to the front of the platform to measure the acoustic emission generated during the tapping torque tests. The ultrasonic waves are converted by the transducer from micro-mechanical displacements on the platform surface into an electrical signal. The signal is amplified by a preamplifier and sampled at 4 MHz by an analog-to-digital converter. The entire thread-forming process is measured, including the run-in and run-out phases. The two high energy phases are detectable in the signal energy as shown in figure 2. In the regression analysis only the run-in phase is considered, because this includes the thread forming process and therefore the plastic deformation of the material. The last two measurement series are corrupted, and therefore not included in the analysis, and marked with a blue color in figure 1. In the majority of the measurements the direct propagation path between the Acoustic Emission origin and the transducer is blocked by the predrilled holes, therefore the measurements are assumed to be super positioned reflections.

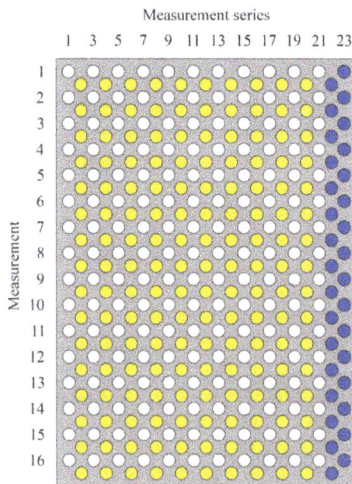

Fig. 1: Bore hole positions

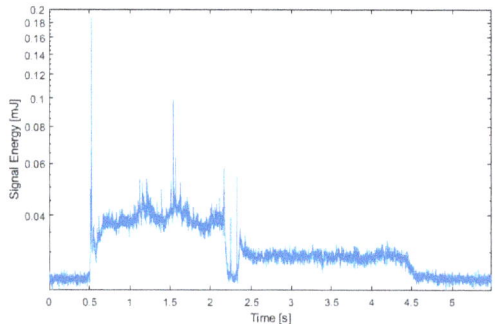

Fig. 2: Signal energy during run-in and run-out phase during thread forming

Structural Health Monitoring: 10APWSHM   Materials Research Forum LLC
Materials Research Proceedings 50 (2025) 212-218   https://doi.org/10.21741/9781644903513-25

The data pre-processing and feature extraction consists of a series of applied filters. First, the run-in phase is determined by analyzing the signal energy. Short-Time Fourier Transform is applied to the raw measurement signal. The resulting coefficients in the range of 1 to 500 kHz are used to calculate the signal energy. Based on the calculated signal energy, the phase of the tapping torque test is detected. The raw measurement does not contain separable wave packages because of the continuous thread forming process and the superposition of wave reflections. The detected section is divided into segments, each containing the samples of two full tool revolutions. For each measurement, 21 segments are obtained for forward tool motion.

The Fast Fourier Transform is used to transform each segment into frequency domain. Only the coefficients related to frequencies between 1 and 200 kHz are considered because lower frequency ranges do not contain acoustic emission and higher frequency ranges are dominated by noise. To reduce the resulting number of coefficients from 22507, the considered frequency spectrum of 1 and 200 kHz is divided into 71 frequency bands. The number 71 is the smallest prime factor of the number of coefficients and therefore allows for an even distribution of coefficients across the frequency ranges. For each frequency range, the frequency-weighted average of the coefficients is calculated and used as features in the regression analysis.

The resulting data set consists of eleven measurement series of 16 measurements each. From each measurement, 21 segments of the run-in phase are derived and each segment is described by 71 features, which are used during the training and test of the regression models to predict the distance between Acoustic Emission source and sensor position.

The frequency spectrum of each measurement is analyzed to detect the effect of propagation distance and number of reflections across measurement series. Therefore, each frequency spectrum is normalized. The normalization enables the assessment of relative amplitudes per frequency within a measurement. The average frequency spectrum across an entire measurement series is subtracted to analyze the relative changes between measurements in a measurement series. This allows the extraction of changes in the frequency spectrum depending on the distance between the source and transducer.

Table 1: Regression Results

Model of measurement series	1	3	5	7	9	11	13	15	17	19	21
RMSE	1.302	1.670	1.602	2.005	3.557	2.668	1.961	3.854	2.884	3.322	1.354
Relative error	4.544	5.729	4.733	7.148	10.968	10.489	7.587	11.462	8.951	8.990	5.430

Results

The boundary conditions are changing between the measurement series, because of the formed threads in the pre-drilled holes. The altered surface profiles change the reflected waves, therefore a regression model is trained for each measurement series. The low diversity of the data set presents an additional challenge that only neural networks can overcome compared to support vector machine regression, Gaussian process regression, and regression trees shown in preliminary analysis. The sub-datasets for each model consist of 336 samples. The training process includes 273 samples and the remaining 63 samples are used for testing. To avoid a manipulation of the results all 21 samples of a single measurement are used during either training or testing. The optimization included the variation of combinations of training and testing data.

Structural Health Monitoring: 10APWSHM
Materials Research Proceedings 50 (2025) 212-218

Materials Research Forum LLC
https://doi.org/10.21741/9781644903513-25

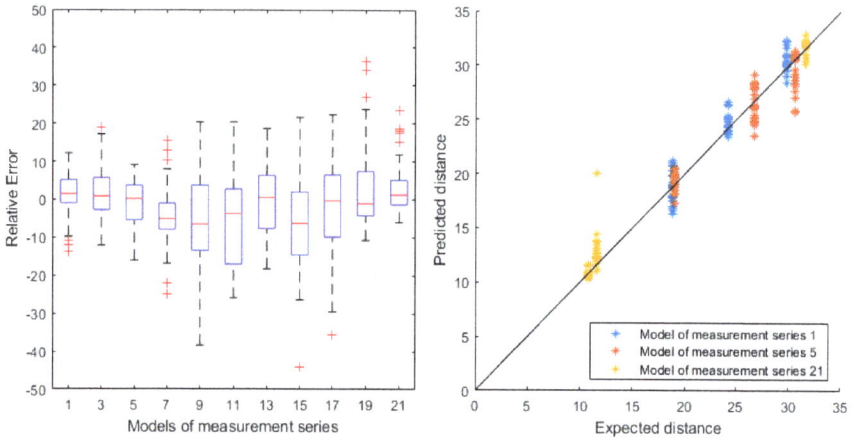

Fig. 4: Statistic analysis of regression results (left) and predictions of selected models (right)

The test results for each model are listed in table 1. The relative error is used to compare the test results of the models, because each model is tested with data of the related measurement series. The best result is obtained for measurement series 1. The predictions of the models related to measurement series 3, 5, and 21 are most accurate. The largest relative error is obtained for the model trained on measurement series 9, 11, and 15. With the exception of model 13, all below-average performing models are located close the transducer. Their propagation paths are characterized by large angle between thread and transducer position. In figure 3, the distribution of test results of each model and the predicted distances of the models of measurement series 1, 5, and 21 are shown. The average deviation between the predicted and actual distance tends to increase in negative direction with decreasing distance to the transducer for measurement series 1

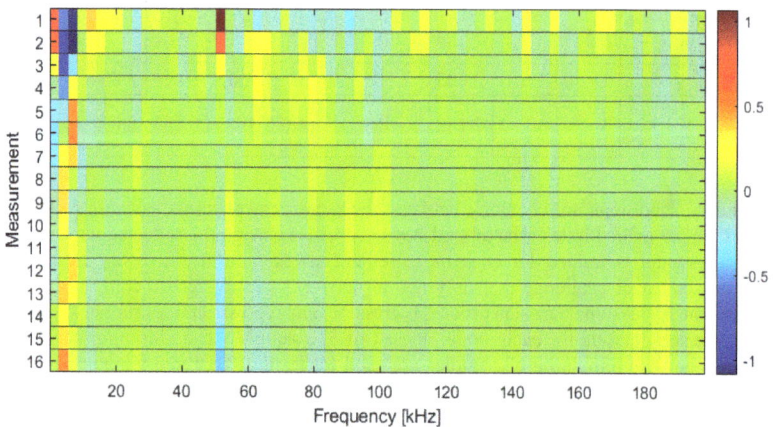

Fig. 3: Normalized frequency spectrum of measurement series

to 9 and 21 to 15. Measurement series 11 and 13 are closest located to the transducer and show an increasing average deviation in comparison to the adjacent measurement series 9 and 15.

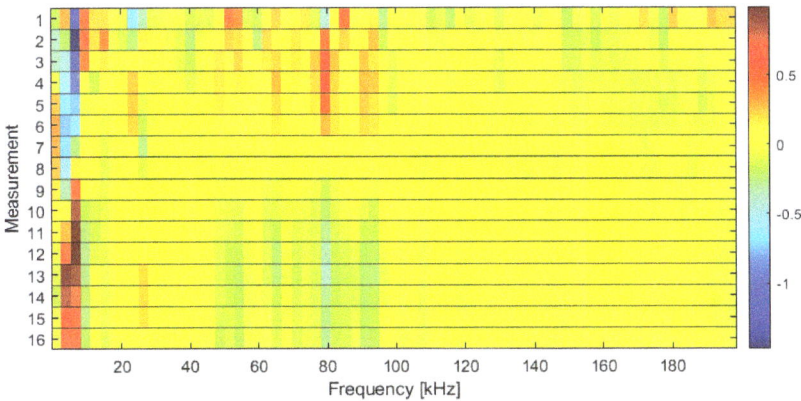

Fig. 5: Normalized frequency spectrum of measurement series 1

The results of the frequency spectrum analysis are visualized in figure 4 and 5. The color shade for a measurement indicates a lower relative amplitude in blue and a higher relative amplitude in red. The same applies to the relative amplitudes across measurements at a specific frequency. Two effects across all measurement series can be detected. Measurements 1 to 7 show a lower amplitude at low frequencies, 3 to 6 kHz, in comparison to the amplitude of 50 to 55 kHz. The amplitudes of measurements 2 to 6 at 79 kHz is also relatively elevated in comparison to the low frequency range. Measurement series 9 to 16 show the opposite effect, amplitudes at low frequencies are larger than at 50 to 55 kHz and 79 kHz. The amplitude of lower frequencies is in comparison to amplitudes of higher frequencies increasing with decreasing distance between Acoustic Emission source and sensor position and decreasing number of reflections. This indicates a stronger attenuation of lower frequencies for increasing number of reflections and increasing propagation distance.

Summary, Conclusions, and Outlook

In this contribution highly reflected and super positioned Acoustic Emission signals measured during thread forming are analyzed to investigate the effect of reflections at material boundaries. Based on features extracted from the frequency spectrum, models for each measurement series are trained to predict the distance between the thread and transducer position. The results demonstrate the trained neural network is able to overcome the low diversity of the dataset and predict the distance with an average accuracy of up to 4.5 %.

The analysis of the frequency spectrum shows the amplitudes at low frequencies are relatively small in comparison to the amplitudes at higher frequencies, when the Acoustic Emission are generated at a larger distance to distance to the transducer. In the same measurement series, the frequency spectrum of Acoustic Emission originated at a position with smaller distance to the transducer shows relatively high amplitudes at low frequencies. The comparison of the frequency spectra indicates a higher attenuation of the low frequency content with increasing propagation distance and increasing number of reflections.

The frequency spectrum of ultrasonic waves is altered during propagation by the propagation path, distance, and reflections and also contains implicit information of these contributing factors. This states that the individual changes during transmission and specific reflections are unique and allow a specific distance-related qualification.

In future work, the reflections are to be further analyzed by designing experiments to isolate the effects described. The frequency-dependent attenuation due to reflections can be quantified by comparing the frequency spectra of incident waves and their reflections.

References

[1] M. Anaya1, D. A. Tibaduiza, M. A. Torres-Arredondo, F. Pozo, M. Ruiz, L. E. Mujica1, J. Rodellar and C.-P. Fritzen, Data-driven methodology to detect and classify structural changes under temperature variations, Smart materials and structures 23.4 (2014) 045006. https://doi.org/10.1088/0964-1726/23/4/045006

[2] M. Rennoch, J. Moll, M. Koerdt, A. S. Herrmann, T. Wandowski, and W. M. Ostachowicz, How Temperature Variations Affect the Propagation Behaviour of Guided Ultrasonic Waves in Different Materials, Advances in Condition Monitoring and Structural Health Monitoring (2021) 695-703. https://doi.org/10.1007/978-981-15-9199-0_67

[3] S. F. Wirtz, S. Bach, and D. Söffker, Experimental Results of Acoustic Emission Attenuation Due to Wave Propagation in Composites, Annual Conference of the PHM Society 11.1 (2019) 821 https://doi.org/10.36001/phmconf.2019.v11i1.821

[4] J. Liebeton and D. Söffker, Experimental Analysis of the Reflection Behavior of Ultrasonic Waves at Material Boundaries, Latam-SHM (2023). https://doi.org/10.23967/latam.2023.036

[5] M. O. Si-Chaib, H. Djelouah, and T. Boutkedjirt, Propagation of ultrasonic waves in materials under bending forces, NDT&E international 38.4 (2005) 283-289. https://doi.org/10.1016/j.ndteint.2004.09.008

[6] D. Lootens, M. Schumacher, M. Liard, S. Z. Jones, D. P. Bentz, S. Ricci, and V. Meacci, Continuous strength measurements of cement pastes and concretes by the ultrasonic wave reflection method, Construction and Building Materials 242 (2020) 117902. https://doi.org/10.1016/j.conbuildmat.2019.117902

[7] G. Kim, C.-W. In, J.-Y. Kim, K. E. Kurtis, and L. J. Jacobs, Air-coupled detection of nonlinear Rayleigh surface waves in concrete-Application to microcracking detection, NDT & E International 67 (2014) 64-70. https://doi.org/10.1016/j.ndteint.2014.07.004

[8] P. B Nagy and L. Adler, Surface roughness induced attenuation of reflected and transmitted ultrasonic waves, Journal of the Acoustical Society of America 82.1 (1987) 193-197. https://doi.org/10.1121/1.395545

[9] S.C. Olisa, M. A. Khan, and A. Starr, Review of Current Guided Wave Ultrasonic Testing (GWUT) Limitations and Future Directions, Sensors 21 (2021) 811 https://doi.org/10.3390/s21030811

[10] V. Giurgiutiu, Structural health monitoring: with piezoelectric wafer active sensors, Elsevier, 2007. https://doi.org/10.1016/B978-012088760-6.50008-8

[11] A. L. Demmerling and D. Söffker, Improved examination and test procedure of tapping torque tests according to ASTM D5619 using coated forming taps and water-mixed metalworking fluids, Tribology International 145 (2020) 106151. https://doi.org/10.1016/j.triboint.2019.106151

[12] ASTM Committee D5619, Standard test method for comparing metal removal fluids using the tapping torque test machine (2011).

[13] X. Wei, A. L. Demmerling, and D. Söffker, Metalworking Fluid Classification Based on Acoustic Emission Signals and Convolutional Neural Network, European Conference of the PHM Society 6.1 (2021), 471-476. https://doi.org/10.36001/phme.2021.v6i1.2798

Structural Health Monitoring: 10APWSHM
Materials Research Proceedings 50 (2025) 219-232

Materials Research Forum LLC
https://doi.org/10.21741/9781644903513-26

A tuneable piezoelectric vibration energy harvester for helicopter-gearbox sensing

Jess D. Flicker[1,a], David J. Munk[1,b], Matthew J. Shipper[1,c] and Scott D. Moss[2,1,*]

[1]Defence Science and Technology Group, Platforms Division, Victoria, Australia

[2]Monash University, Dept. Of Mechanical Aerospace Engineering, Victoria, Australia

[a]jess@flickers.com.au, [b]david.munk@defence.gov.au,
[c]matthew.schipper@defence.gov.au, [d]scott.moss@monash.edu

© 2024 Commonwealth of Australia

Keywords: Piezoelectric Energy Harvesting, Condition Based Monitoring, Finite Element Analysis

Abstract. Piezoelectric vibration energy harvesters are critical for self-powered sensing on aircraft, however tuning a harvester to the desired source frequency is challenging. This is because aerospace ap-plications usually require energy harvesters to be small, lightweight and highly durable, making mechanical tuning methods impractical. Frequency tuning via shunting a piezoelectric transducer with variable electrical loads, as explored in this work, is practicable since the tuning mechanism is low mass and has no moving parts. This paper presents the modelling and experimental characterisation of a cantilevered bimorph configuration comprising a steel beam with two relaxor ferroelectric single crystal transducers for harvesting and tuning. A one-dimensional model for the tuneable harvester was developed and two finite element analyses (FEA) were conducted. The effect of temperature up to 90°C is discussed, being the upper gearbox temperature expected operationally. The first FEA model only couples the vibration to the direct production of charge in the transducers, while the second model includes both the direct and converse piezoelectric effects. A prototype harvester was tested on an electrodynamic shaker at 1g peak acceleration at frequencies around 1900 Hz which was the helicopter-gearbox meshing-frequency of interest. The predictions of the one-dimensional model and both finite element models are found to agree favourably with experimental measurements of the resonant frequency. The approach successfully tuned the resonant frequency of the prototype piezoelectric harvester by up to ±1.7%, matching the harvester's resonant frequency to 1900 Hz gearbox frequency of interest.

1. Introduction

Piezoelectric vibration energy harvesters (PVEHs) are a type of vibration energy harvester (VEH) that are useful for powering wireless sensors for structural health monitoring (SHM) on board vehicles or in industrial settings [1-3]. Modelling is frequently employed as an intermediate step in the development of VEHs and different models exist for the various types of harvesting applications and boundary conditions [4]. Even relatively simple analytical models are useful, as they allow different designs that vary in working principle or topology to be quickly assessed, or serve as the basis for optimisation [5]. This paper focuses on a non-uniform piezoelectric bimorph harvester with frequency matching capability. Frequency matching is achieved by modulating the beam stiffness via the characteristic load-dependent electro-mechanical elasticity of the piezoelectric transducers. The base-excited cantilever beam PVEH is extremely prevalent in the literature and, in particular, the modelling formulation of the uniform piezoelectric bimorph topology has been thoroughly explored [6-8]. As a precursor to complex distributed-parameter models or finite element analysis (FEA), some prior studies [9, 10] establish a sin-gle-degree-of-

Structural Health Monitoring: 10APWSHM   Materials Research Forum LLC
Materials Research Proceedings 50 (2025) 219-232   https://doi.org/10.21741/9781644903513-26

freedom (SDOF) model which enables a clear understanding of the underly-ing physics with less mathematical complexity than methods such as modal analysis. Yet, analytical models alone are challenging for complex harvesting configurations where geometry, material positioning and electrode patterns are irregular. In these situations, numerical simulations are needed [11]. Hence, studies often present the results of analytical calculations along with FEA [10, 12 - 14].

This study will compare the resonant frequency and power output predictions of the SDOF model introduced in [15] with partially-coupled FEA (PCFEA) and fully-coupled FEA (FCFEA) models of a high-frequency PVEH operating at temperatures between room temperature and 90∘C. Since temperature can strongly affect a PVEH's resonant frequency and power output [16], predicting the effects of temperature on a PVEH design is crucial. Results from the SDOF model and FEA are validated against measured frequency response and power data from a prototype PVEH tested on an electrodynamic shaker for a range of temperatures under controlled laboratory conditions.

A defining characteristic of high-Q resonance-based harvesters is the proclivity for the power output to decrease significantly if the source (or host) frequency fluctuates from the harvester's fixed resonant frequency [17]. In controlled laboratory conditions, maintaining resonant performance is achieved relatively easily. However, in an operational setting the situation is quite different. On board an aircraft, a PVEH will be exposed to source vibration with a known frequency range and due to operational conditions, the harvester will also experience temperature changes, which alter the frequency at which it naturally resonates [16]. Recently, a paper described a PVEH for use on a helicopter main rotor trans-mission casing and estimated that a reasonably insulated harvester could be expected to be exposed to a temperature range of 50-60∘C [18], when the gearbox internal oil temperature range is 80-100∘C [19]. This necessitates a PVEH that can frequency match with the source vibration.

Load shunting is a way to tune the PVEH without mechanical actuators which can add unwanted mass and are difficult to implement in an autonomous tuning scheme for high-frequency harvesting [20]. However, the power produced is also dependent on the electrical load and, in general, the load for optimal power production is a function of the harvester's natural resonant frequency and the frequency of the source vibration [9]. There is the opportunity to leverage the piezoelectric transducer for both harvesting energy and tuning the vibrating mechanism if separate transducers are used for each purpose. The PVEH examined herein relies principally on two piezoelectric transducers that are shunted using variable resistors to allow optimal power production. One transducer is used for harvesting power and the other for tuning the beam by exploiting the well-known piezoelectric short-circuit and open-circuit stiffness.

The harvester is a '3-2' mode cantilever device designed to harvest from low-kilohertz, high-acceleration vibrations such as those commonly found on aircraft [21]. A photograph of a prototype device is shown in Figure 1. The basis of the PVEH is a bimorph cantilever with co-located disk-shaped piezoelectric transducers near the clamped (root) end of the beam. The piezoelectric transducers are diameter 6.35 mm × thickness 0.5 mm 'High-Performance Single Crystal' (HPSC200-145) [011] poled Mn-Pb(Mg1/3Nb2/3)O3-Pb(ZrxTi1−x)O3 (Mn-PMN-PZT) disks (Ceracomp Co. Ltd.) with nominal properties listed in reference [22]. The transducers have parallel flat faces, which are (011) planes normal to the poling direction, also called the '3'-direction. Normal to this direction are the '1'- and '2'-directions. The transducer faces have sputtered gold electrodes. The metal beam is 0.51 mm thick cut with electric discharge machining (EDM) from spring steel (AISI 1095). The surface of the beam was prepared by dry-sanding an area roughly equal to the transducer size along the beam's longitudinal direction and cleaning with acetone. A silver loaded epoxy (Chemtronics CW2400) was used to bond the transducers to the prepared surface. Flexural deformation of the beam due to vibration results in a shear strain along the longitudinal axis increasing in magnitude towards the clamped end, so the transducers were positioned as close as practicable (1 mm) to the root with the [100] ('2'-direction) being the most mechanically compliant axis nominally aligned with the beam's length. The inward transducer faces, conductive epoxy and beam collectively form an electrically grounded node, with the outward transducer faces producing an electric potential during vibration due to charge generation on the electrodes per the direct piezoelectric effect [23]. Section 2 will discuss and compare the approaches taken to model the harvester, following which experimental characterisation of the harvester's resonant frequency as a function of temperature and the shunt resistances is presented in Section 3.

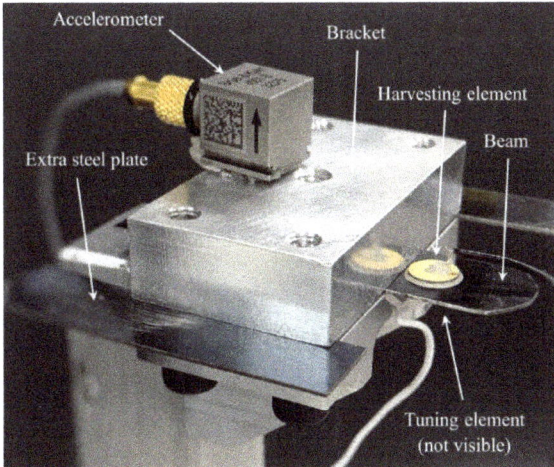

Figure 1. Hamrvester mounted in bracket.

2. Modelling

This work focuses on modelling the tuneable harvester's resonant frequency at temperatures up to 90°C. Modelling PVEH behaviour at high temperature requires detailed knowledge of the effects of temperature on each of the piezoelectric matrix coefficients of Mn-PMN-PZT. As a starting point, the SDOF model uses measurements of harvester volt-age output and independent piezoelectric properties at temperatures up to 90°C to scale the lumped stiffness, damping ratio, piezoelectric properties and force-voltage transfer factor α (introduced in Section 2.1). The details of these measurements are not covered in this paper. The PCFEA model attempts to bridge the gap between the SDOF model and a future temperature-inclusive FCFEA model by estimating the temperature-dependence of the piezoelectric stiffness coefficients and α using measured data. In this paper, the FCFEA approach is restricted to the room temperature case. The following sections will discuss the three modelling approaches in further depth.

2.1 Analytical SDOF model

The analytical SDOF model of a tuneable harvester introduced in reference [13] is used in the current work with the same parameters. The modelled bimorph cantilever has effective stiffness and mass k_e and m_e respectively and the fundamental resonant frequency is given by $\omega_n = \sqrt{k_e/m_e}$. At room temperature, the corresponding resonant frequency is 1900 Hz. Separate electrical loads, R_L and R_T, are connected in parallel across the harvesting and tuning transducers, respectively, which each have capacitance C_p. The source vibration is harmonic with acceleration amplitude a(t) and gives rise to tip displacement u(t), velocity $\dot{u}$(t) and acceleration $\ddot{u}$(t). Voltages

Structural Health Monitoring: 10APWSHM   Materials Research Forum LLC
Materials Research Proceedings 50 (2025) 219-232   https://doi.org/10.21741/9781644903513-26

$v_H(t)$ and $v_T(t)$ are developed across the electrodes of the harvesting and tuning transducers, respectively. The converse piezoelectric effect results in the electrically-induced force f_e on the mechanical oscillator, which is coupled to the voltages via the force-voltage transfer factor α defined in reference [24], and found to have a value of approximately 10^{-3} from earlier analysis in reference [15]. Viscous damping is implemented via the mechanical damping coefficient $b = 2\,\zeta\,\omega_n m_e$, where the damping ratio ζ is a temperature dependent parameter. It is assumed that there are no electrical losses. The equations governing the SDOF model are given in time domain by Equations (1) to (4),

$$m_e \ddot{u}(t) + b\dot{u}(t) + k_e u(t) + f_e = m_e a(t)\,, \tag{1}$$
$$R_L C_p \dot{v}_H(t) + v_H(t) - R_L \alpha \dot{u}(t) = 0\,, \tag{2}$$
$$R_T C_p \dot{v}_T(t) + v_T(t) - R_T \alpha \dot{u}(t) = 0\,, \tag{3}$$
$$f_e = \alpha\big(v_H(t) + v_T(t)\big)\,. \tag{4}$$

Solving Equations (1) to (4) allows the harvester's power output as a function of drive angular frequency ω to be readily calculated from $P_H = v_H^2 / R_L$, which yields Equation (5) after simplification and the introduction of the dimensionless frequency ratio $\Omega = \omega/\omega_n$, dimensionless load resistance $r_L = R_L C_p \omega_n$, and dimensionless tuning resistance $r_T = R_T C_p \omega_n$,

$$\left|\frac{P_H}{a^2}\right| = \cfrac{\alpha^2 k_e^2 R_L \Omega^2 (r_T^2 \Omega^2 + 1)}{\left(\begin{array}{l} \omega_n^2 k_e^2\big((4\zeta^2 - 2)\Omega^2 + \Omega^4 + 1\big)(r_L^2 \Omega^2 + 1)(r_T^2 \Omega^2 + 1) + 4\omega_n^3 \alpha^2 k_e \Omega^4 r_L^2 R_T (2\zeta - r_T \Omega^2 + r_T) \\ -2\omega_n^3 \alpha^2 k_e \Omega^2 r_L (R_L - R_T)(2\zeta r_T \Omega^2 + \Omega^2 - 1) + 2\omega_n^3 \alpha^2 k_e \Omega^2 (R_L + R_T)(2\zeta - r_T \Omega^2 + r_T) \\ + \omega_n^4 \alpha^4 \Omega^2 (4 r_L^2 R_T^2 \Omega^2 + (R_L + R_T)^2) \end{array}\right)} \cdot \tag{5}$$

As is evident by comparing the output powers depicted in Figures 2a and 2b, the SDOF model provides valuable insight into the qualitative and quantitative aspects of a harvester's performance without the high computational cost (hours to days of compute time) of three-dimensional FEA. The frequency bandwidth of the harvester is the lower and upper boundaries on ω, within which the harvester can be tuned to harvest energy optimally. In Figure 2a, this bandwidth is represented by the spread between the leftmost and rightmost peaks of the black curve, which was calculated by numerically finding the combination of R_L and R_T that maximises power output at each ω. Power output is generally maximised when the tuning piezoelectric transducer is in short-circuit ($R_T \sim 0$) or open-circuit ($R_T \sim \infty$), corresponding to the resonance and anti-resonance frequencies and associated optimal loads $R_L^{SC,opt}$ and $R_L^{OC,opt}$ respectively, as described in reference [9]. The centre peak accounts for the cases where the short-circuit optimal harvesting load R_L^{SC} is used in conjunction with the open-circuit tuning resistance R_L^{OC}, or vice-versa. The SDOF model does not account for the inherent three-dimensional nature of the stresses and strains present in the actual PVEH, which resulted in the SDOF model over-predicting the change in resonant frequency as a result of a change in R_T. By de-rating the piezoelectric charge coefficient and stiffness matrix values associated with the '2'- direction, better agreement with experimental measurements of the resonant frequency shift and FEA was achieved. However, this trade-off came at the expense of accuracy in the power output prediction, which is evident in comparing Figures 2a and 2b.

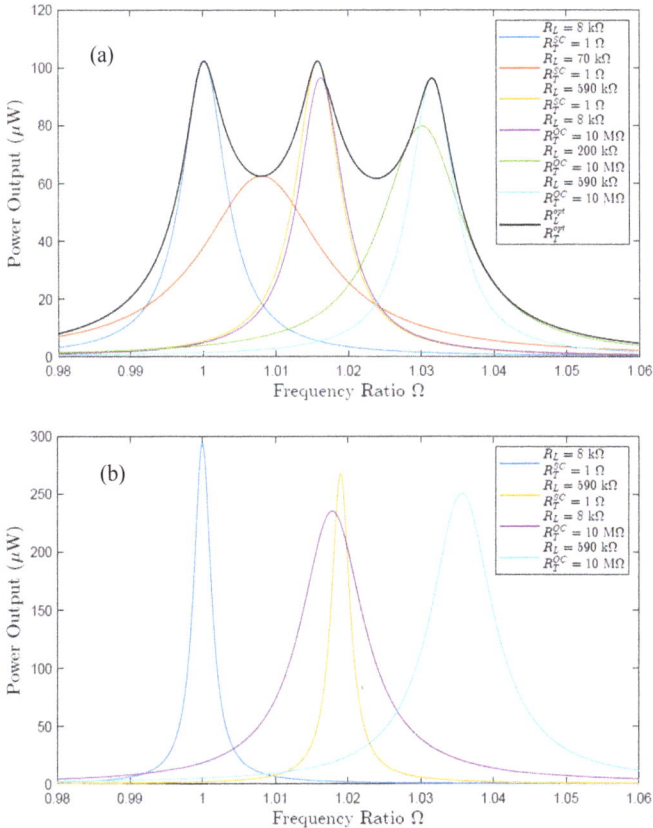

Figure 2. *Both the SDOF analytical model and FCFEA predicted similar increases in resonant frequency at 20°C: **(a)** SDOF model; **(b)** Fully-coupled FEA (Comsol Multiphysics 5.6).*

Finally, a variance-based sensitivity analysis was undertaken to understand the effects of several design parameters (excluding temperature, which is discussed later) on the resonant frequency and maximum power of a PVEH. Each parameter was independently varied by up to ±10% from a nominal value as given in Table I. From the results, shown in Figure 3, it is concluded that the frequency response is most sensitive to d_{32}, while ζ most affects the power output. An earlier study found that the highest sensitivities on the frequency response function were the thickness, density and mechanical stiffness of the piezoelectric transducers [25]. However, the order of influence of these parameters changes from the short-circuit to open-circuit condition, despite being relatively independent of drive frequency [26]. Thickness was not included in the sensitivity analysis due to the very low uncertainty in the transducer thickness.

Table I. *Nominal values of parameters assessed in the sensitivity analysis.*

Parameter	Nominal Value
HPSC charge coefficient d_{32}	-1072 pC N^{-1}
HPSC compliance s_{22}^E	182.6 GPa^{-1}
Tip mass m_e	0.1224 g
Damping ratio ζ	0.0013

2.2 Finite Element Analysis

The PCFEA and FCFEA are based on a three-dimensional structural representation of the harvester's geometry and material composition in two commercially available FEA solvers (MSC NASTRAN and Comsol Multiphysics 5.6, respectively). The epoxy bond lines are represented as material volumes and do not include the temperature dependent description of the cohesive properties [27]. The harvester is subjected to a vibration in the out-of-plane direction of frequency ω and acceleration amplitude of $1g$ at the clamped end. The modelled harvesters for the PCFEA and FCFEA are pictured in Figures 4 and 5 respectively.

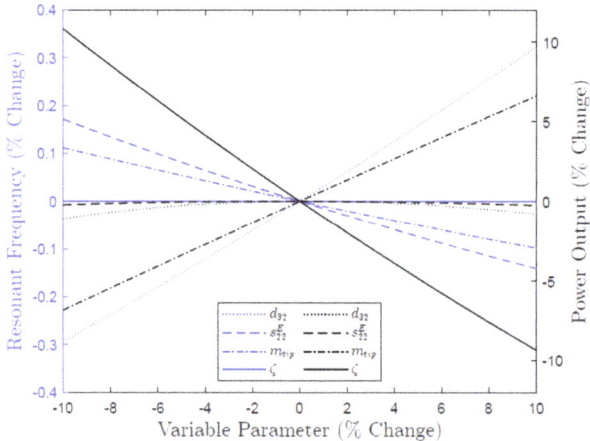

Figure 3. *Modelled harvester response to various design parameters with $R_L = 590$ $k\Omega$ and $R_T^{SC} = 1$ Ω.*

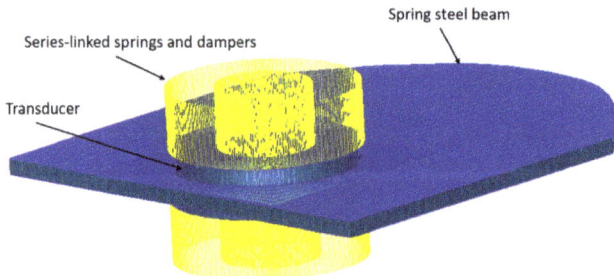

Figure 4. *Structural model of harvester in PCFEA (MSC Patran 2020).*

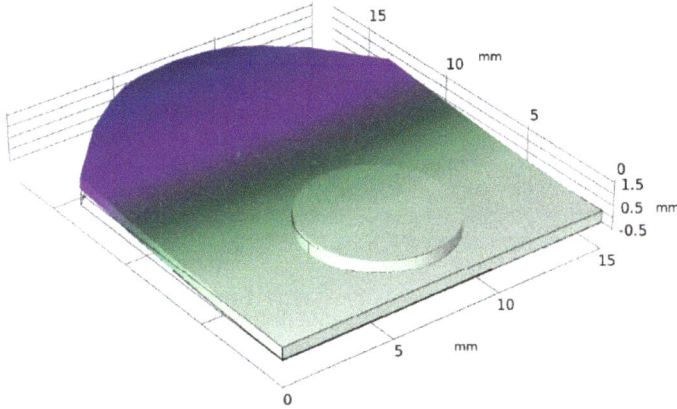

Figure 5. *Harvester exhibiting first resonant bending mode around 1900 Hz (Comsol multiphysics 5.6).*

2.2.1 Partially-coupled FEA: In the PCFEA, the transfer of energy from the structural domain to the electrical domain is implemented (MSC Patran 2020) by series-linked one-dimensional springs and dampers equivalent to a parallel capacitor and resistor used in lumped-parameter models of PVEHs [8, 9]. While this only models feedback from the electrical domain to the mechanical domain in one-dimension (the out-of-plane direction) instead of three-dimensions like other multi-physics solvers (e.g. Comsol Multiphysics 5.6), it significantly reduces the cost of computation because only one-dimension is modelled. Additionally, the PCFEA uses a single value for α determined from measurements of the piezoelectric properties for all resistance combinations. The mechanical stiffness of piezoelectric transducers changes between short-circuit and open-circuit states [28]. However, using mechanical dampers whose values vary proportionally to R_L and R_T , i.e. ignoring the feedback of the electric field on the transducer stiffness, reduces the solve time. A fully-coupled solver includes the feedback in both directions, which is why the solve time can be orders of magnitude longer. The structural response is computed at discrete excitation frequencies, corresponding to the frequency of the driving acceleration ω, by solving the damped forced vibration equation of motion defined in Equation (6),

$$[M]\ddot{x} + [C]\dot{x} + [K]x + F_e = [M]ae^{i\omega t} , \qquad (6)$$

where $[M]$, $[C]$ and $[K]$ are the mass, damping and stiffness matrices, respectively, a and F_e are the acceleration and spring-damper force vectors, respectively, and t is the time variable. To solve Equation (6), a harmonic solution is assumed with the form x = $u(\omega)$ $e^{i\omega t}$, where u is a complex displacement vector. The aforementioned harmonic solution is substituted into Equation (6) and solved for u, determining the structural response at each node in the model. Once this is known, the force in each damper is found from the velocity through the linear damping relationship, where the damping constant of each damper is an equal share of the shunt resistance in accordance with reference [13]. This is then used to determine the voltage and, hence, output power of the transducer. The structural model of the harvester is shown in Figure 4 where the series-linked springs and dampers are shown as one-dimensional linear elements and the beam, bond line and transducer are modelled as 8-node three-dimensional hexahedral elements.

Structural Health Monitoring: 10APWSHM   Materials Research Forum LLC
Materials Research Proceedings 50 (2025) 219-232   https://doi.org/10.21741/9781644903513-26

2.2.2 Fully-coupled FEA: The FCFEA (Comsol Multiphysics 5.6) accounts for the converse piezoelectric effect, a phenomenon where changes in the transducer voltage damp the oscillator due to the conversion of mechanical energy into electrical energy [29], leading to Equation (4). Upon analysing Equation (4) it is readily apparent that the importance of the converse effect relative to the direct effect is proportional to the parameter α. Including the converse effect results in a more expensive computational solve, since the process of determining the harvester's deflection and voltage relies on iteration. First, an initial estimate of the deflection at each node is ascertained by solving the purely mechanical equation of motion not accounting for piezoelectricity (Equation (6) less the F_e term). Then, the electric displacement resulting from the direct effect is determined from Equation (7),

$$S = [s^E]\, T + [d]^T\, E \,, \tag{7}$$

where S, T, and E are the strain, stress and electric field vectors, respectively, and $[s^E]$ and $[d]$ are matrices representing the elastic compliance at constant electric field and piezoelectric charge constants, respectively. The free charge Q_0 produced on the harvesting and tuning transducers' electrodes is found via the converse constitutive relation in Equation (8) and Gauss' Law in Equation (9), where D is the electric displacement vector and $[\varepsilon^T]$ is the dielectric permittivity at constant stress matrix,

$$D = [d]\, T + [\varepsilon^T]\, E \,, \tag{8}$$

$$Q_0 = \int_A D \cdot n\, dA \,. \tag{9}$$

The current I is directly proportional to the frequency of vibration as shown in Equation (10). The power P lost by the mechanical oscillator due to electrical damping, which is equal to the harvested power, is given by the familiar expression in Equation (11),

$$I = \omega Q_0 e^{i\theta} \,, \tag{10}$$

$$P = I^2 R_L \,. \tag{11}$$

This power lost is used to update the strain due to the converse effect in Equation (8) above and so on until the changes in successive estimates falls below a predetermined threshold.

3. Experimental

The harvester was tested in the laboratory using an electrodynamic shaker (TIRAvib 51120) over a temperature range of 20∘C to 90∘C, which was the maximum temperature of the laboratory oven (Thermoline). The shaker was driven by a power amplifier (TIRA BAA500) connected to a loop controller (Bruel & Kjær 7541), over a frequency range of 1000-3000 Hz, which encapsulated the first bending mode at around 1900 Hz. An accelerometer was used to provide feedback to the loop controller. The accelerometer was mounted to the assembly using a high-temperature mounting clip (Bruel & Kjær UA 1564) for experiments at high temperatures and, for ease, beeswax for experiments at room temperature. As shown in Figure 1, extra steel plate was used to prevent the bracket deforming around the beam after tightening the four bolts to 4 N.m. The PVEH-bracket assembly was mounted on a 300 mm aluminium rod host mass, which did not have any eigenmodes near the 1900 Hz beam mode. In an operational setting, the clamped end of the harvester would be rigidly attached to the host structure subjected to vibration.

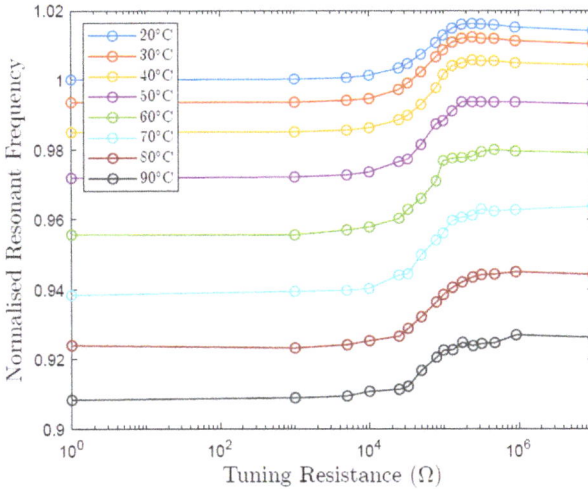

Figure 6. Measured resonant frequency response as R_T is varied with R_L = 10 MΩ up to 90°C.

Voltages from both piezoelectric transducers were recorded using an oscilloscope (Picoscope 6403) and 10 MΩ probes. Harvester load resistance and tuning resistance sweeps were performed using resistor decade boxes (Tenma 72-7270). With constant R_L and R_T, the measured resonant frequency decreased by 171 Hz as the temperature increased from 20°C to 90°C, equating to a 9% reduction. The tuning frequency bandwidth was approximately 1.7% at 20°C as shown by the vertical spread of the curves in Figure 6. The vertical overlap in the 20-40°C lines indicates that the harvester can be tuned to a constant drive frequency over this temperature range. For example, if the PVEH initially has the desired resonant frequency at 20°C with low or zero tuning resistance R_T^{SC}, then at 40°C the same resonant frequency can be achieved by adjusting the tuning resistance to the open-circuit value R_T^{OC}. Such a feature would be useful in a setting wherein the vibration source warms up over time but the frequency of the source vibration is constant. At each temperature step above 40°C the tuning bandwidth only offsets 10°C of heating.

4. Discussion

The SDOF model predicted an almost 2% increase in resonant frequency with changing the harvesting load R_L or tuning resistance R_T from short-circuit to open-circuit independently, or a nearly 4% increase when changing both loads. The PCFEA and FCFEA simulations provided similar agreement with the SDOF model and measured data, having an average difference of 0.26% for the PCFEA and 0.51% for the FCFEA when compared to the measurements, shown in Figure 7. The small force-voltage transfer factor α on the order 10^{-3} N/V for this harvester [15] suggests that the resonant response is dominated by the direct piezoelectric effect, which justifies the reasonable agreement between the PCFEA and other results in addition to encouraging the continued focus on high-d single crystals for energy harvesting [30].

The effect of temperature on the PVEH's resonant frequency [31] was investigated and a tuning bandwidth enabling frequency matching over maximum 20°C of heating (between 20-40°C) was achieved. As mentioned earlier, there was a measured 171 Hz reduction (equating to 9%) in resonant frequency from room temperature to 90°C. This temperature-induced frequency change is primarily associated with the softening of the epoxy bond layer [16], with additional frequency

Structural Health Monitoring: 10APWSHM   Materials Research Forum LLC
Materials Research Proceedings 50 (2025) 219-232   https://doi.org/10.21741/9781644903513-26

decreases due to the reduction in the stiffness of the steel beam and any load-independent bulk effects on the piezoelectric transducers. Reducing the temperature-induced frequency shift as much as possible (by using temperature-stable adhesives for example) will enable a PVEH to be 'temperature-agile' over a much larger temperature range.

The tuning bandwidth mostly depends on the authority of the variably adjustable piezoelectric stiffness over the beam, which is correlated with the transducer's ability to convert strain to charge, i.e. its [d] matrix coefficients. The α parameter in Equations (1) to (4) depends strongly on d_{32} and with the primary beam dimensions (length, width and thickness). The relevance of the converse effect is inferred by the order of magnitude of α; in the case of the examined PVEH, an α ≪ 1 implies that the induced force is negligible even for moderate transducer voltages.

Temperature was incorporated into the SDOF model using the epoxy bond line data from reference [14] by modulating the overall system stiffness parameter k_e accordingly. The results, shown in Figure 8, suggest that the frequency reduction up to 60°C is correlated with the reduction in epoxy stiffness. However, at still higher temperatures, the model does not account for the steep drop in resonant frequency observed. This is due to the epoxy stiffness data, which showed the stiffness flattening out around 60°C. The similarity of the SDOF and PCFEA predictions supports this conclusion. The damping ratio ζ was computed as a function of temperature using measured data, allowing the impact of temperature on the predicted power to be determined. As shown in Figure 9, increasing temperature lowers the frequency of the peak sets and causes them to merge and flatten. At 50°C, the splitting of the centre peak into the $R_L^{OC,opt} |R_T^{SC}$ and $R_L^{SC,opt} |R_T^{OC}$ peaks becomes noticeable. When the temperature increases further, the two parts of the centre peak then migrate outward until they overlap with the leftmost $R_L^{SC,opt} |R_T^{SC}$ peak and the rightmost $R_L^{OC,opt} |R_T^{OC}$. At this point the individual transducers have very little residual authority and therefore the two small peaks present above 70°C represent both transducers in a near short-circuit or open-circuit state.

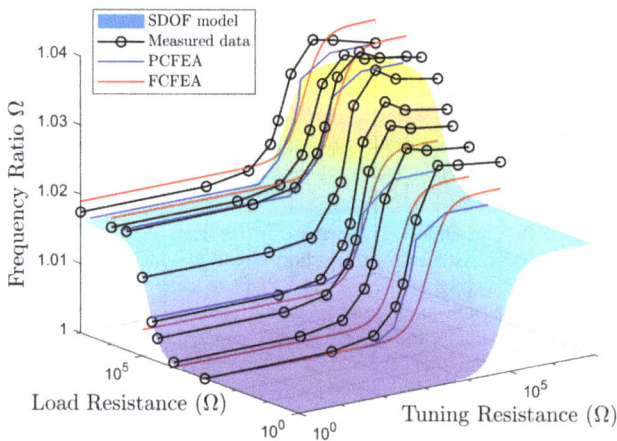

Figure 7. Measured and modelled resonant frequency 20°C.

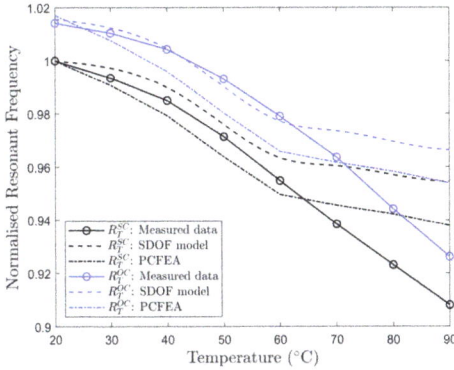

Figure 8. *Measured and modelled resonant frequency normalized to the R_T^{SC} case with $R_L = 10$ MΩ at 90°C.*

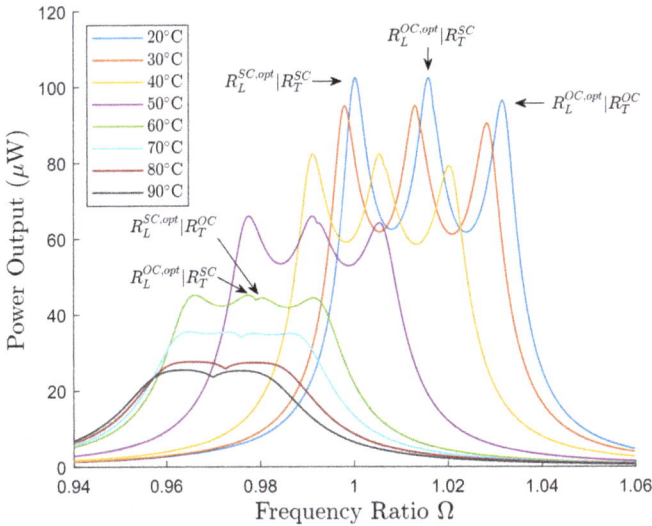

Figure 9. *Modelled power output peaks separate and flatten at higher temperatures.*

5. Conclusion

This paper compared a partially-coupled FEA approach with a fully-coupled commercial multi-physics package and applied these analysis methods to a tuneable high-frequency piezoelectric energy harvester. The partially-coupled approach (MSC Patran 2020) excelled in predictions of resonant frequency, since the requisite short-circuit and open-circuit piezoelectric stiffness could be modelled using one-dimensional dampers in the out-of-plane vibration direction. However, the unidirectional coupling of the transducer physics limited the utility of this approach, as the model did not include electrical damping caused by a changing electric field across the piezoelectric layers. Consequently, the partially-coupled model did not accurately predict the harvester output power. A separate FEA implementation (Comsol Multiphysics 5.6) that included all the necessary coupled-physics to accurately calculate power output required longer solve times (days to weeks). Both FEA approaches corroborated experimental measurements of the harvester's resonant frequency and predictions from a SDOF at room temperature. At higher temperatures, the accuracy of model predictions are mainly limited by material data. For some analyses, such as determining harvester modal frequencies, a partially-coupled approach was suitable with the advantage of greatly-reduced solve times. Other analyses, such as the prediction of harvester output power, required electromechanical models with fully-coupled physics requiring a high-performance computational solver and longer solve times.

Acknowledgments
The authors would like to thanks Mr. Daniel Bitton for his technical assistance. This work was supported by the Australian Federal Government through the Advanced Piezoelectric Materials and Applications program including DMTC Ltd., the DSTG Strategic Research Initiative in Advanced Materials and Sensors, and the DSTG Industry Experience Placement program.

References

[1] R. Caliò, U.B. Rongala, D. Camboni, M. Milazzo, C. Stefanini, G. De Petris, C.M. Oddo, Piezoelectric Energy Harvesting Solutions, Sensors. 14 (2014) 4755-4790. https://doi.org/10.3390/s140304755

[2] C. Covaci, A. Gontean, Piezoelectric Energy Harvesting Solutions: A Review, Sensors. 20 (2020) 3512. https://doi.org/10.3390/s20123512

[3] S. Priya, D.J. Inman, Energy Harvesting Technologies, 1st ed., Springer, 2009. https://doi.org/10.1007/978-0-387-76464-1

[4] S. Rafique, Piezoelectric Vibration Energy Harvesting Modeling & Experiments, 1st ed., Springer International Publishing, 2018. https://doi.org/10.1007/978-3-319-69442-9

[5] S. Roundy, P.K. Wright, J.M Rabaey, Energy Scavenging for Wireless Sensor Networks: With Special Focus on Vibrations, Springer, Boston, MA, 2003. https://doi.org/10.1007/978-1-4615-0485-6

[6] A. Erturk, D.J. Inman, On Mechanical Modeling of Cantilevered Piezoelectric Vibration Energy Harvesters, J. Intell. Mater. Syst. Struct. 19 (2008) 1311-1325. https://doi.org/10.1177/1045389X07085639

[7] A. Erturk, D.J. Inman, Issues in mathematical modeling of piezoelectric energy harvesters, Smart Mater. Struct. 17 (2008) 065016. https://doi.org/10.1088/0964-1726/17/6/065016

[8] S. Roundy, P.K. Wright, A piezoelectric vibration based generator for wireless electronics, Smart Mater. Struct. 13 (2004) 1131-1142. https://doi.org/10.1088/0964-1726/13/5/018

[9] N. E. du Toit, B.L. Wardle, S.-G. Kim, Design considerations for mems-scale piezoelectric mechanical vibration energy harvesters, Integrated Ferroelectrics. 112 (2005) 121-160. https://doi.org/10.1080/10584580590964574

[10] W. Jiang, L. Wang, L. Zhao, G. Luo, P. Yang, S. Ning, D. Lu, Q. Lin, Modeling and design of v-shaped piezoelectric vibration energy harvester with stopper for low-frequency broadband and shock excitation, Sensors and Actuators. A. Physical. (2021) 317. https://doi.org/10.1016/j.sna.2020.112458

[11] S. Ravi, A. Zilian, Time and frequency domain analysis of piezoelectric energy harvesters by monolithic finite element modeling, Int. J. Numer. Methods Eng. 112 (2017) 1828-1847. https://doi.org/10.1002/nme.5584

[12] T.M. Kamel, R. Elfrink, M. Renaud, D. Hohlfeld, M. Goedbloed, C. de Nooijer, M. Jambunathan, R. Schaijk, Modeling and characterization of memsbased piezoelectric harvesting devices, J. Micromechanics and Microengineering. 20 (2010) 105023. https://doi.org/10.1088/0960-1317/20/10/105023

[13] X. Xiong, S.O. Oyadiji,. Modal electromechanical optimization of cantilevered piezoelectric vibration energy harvesters by geometric variation, J. Intell. Mater. Syst. Struct. 25 (2014) 1177-1195. https://doi.org/10.1177/1045389X13502872

[14] O. Rubes, M. Brablc, Z. Hadas, Nonlinear vibration energy harvester: Design and oscillating stability analyses, Mechanical systems and signal processing. 125 (2019) 170-184. https://doi.org/10.1016/j.ymssp.2018.07.016

[15] J.D. Flicker, D.J. Munk, W.K. Chiu, B.S. Vien, P. Finkel, S.D. Moss Analysis and design approaches for a frequency-agile piezoelectric energy harvester, Proc. 10th Australasian Congress on Applied Mechanics (ACAM10), Melbourne, Australia (2021).

[16] J.D. Flicker, E.J.G. Ellul, S.D. Moss, D.J. Munk, D. Blunt, W. Wang, E. Lee, R. Hussein, P. Stanhope, Validation of optimised vibration energy harvesters under near operational conditions, Proc. AIAC19: 19th Australian International Aerospace Congress. Engineers Australia, Melbourne, Australia, 29 November (2021).

[17] A. Erturk, Piezoelectric Energy Harvesting, Wiley, Chichester, 2011. https://doi.org/10.1002/9781119991151

[18] T.F. Doughney, S.D. Moss, D. Blunt, W. Wang, H.J. Kissick, Relaxor ferroelectric transduction for high frequency vibration energy harvesting, Smart Mater. Struct. 28 (2019) 065011. https://doi.org/10.1088/1361-665X/ab15a5

[19] K.W. Vaughan, G.D. Baldie, Guidelines for the overhaul of a Bell 206B-1 helicopter main rotor transmission assembly, Defence Science and Technology Group (DSTO-GD-0370), 2003.

[20] S.W. Ibrahim, W.G. Ali, A review on frequency tuning methods for piezoelectric energy harvesting systems, J. Renewable and Sustainable Energy. 4 (2012) 62703. https://doi.org/10.1063/1.4766892

[21] D.J. Munk, E.J.G. Ellul, S.D. Moss, An approach for the design and validation of high frequency vibration energy harvesting devices, Smart Mater. Struct. 30 (2021) 65018. https://doi.org/10.1088/1361-665X/abfb41

[22] Information on http://www.ceracomp.com/ecatalog/CeracompBrochure.pdf (Accessed on 14 December 2022).

[23] J. Curie, P. Curie, Ph'enom'enes 'electriques des cristaux h'emi'edres 'a faces inclin'ees. J. phys. theor. appl. 1 (1882) 245-251. https://doi.org/10.1051/jphystap:018820010024500

[24] B. Richter, J. Twiefel, T. Hemsel, J. Wallaschek, Model Based Design of Piezoelectric Generators Utilizing Geometrical and Material Properties, IMECE2006 (2006) 521-530. https://doi.org/10.1115/IMECE2006-14862

[25] R.O Ruiz, V. Meruane, Uncertainties propagation and global sensitivity analysis of the frequency response function of piezoelectric energy harvesters, Smart Mater. Struct. 26 (2017) 65003. https://doi.org/10.1088/1361-665X/aa6cf3

[26] R. Aloui, W. Larbi, M. Chouchane, Global sensitivity analysis of piezoelectric energy harvesters, Composite Structures. 228 (2019), 111317. https://doi.org/10.1016/j.compstruct.2019.111317

[27] T. Carlberger, A. Biel, U. Stigh, Influence of temperature and strain rate on cohesive properties of a structural epoxy adhesive. International Journal of Fracture. 155 (2009) 155-166. https://doi.org/10.1007/s10704-009-9337-4

[28] "IEEE Standard on Piezoelectricity," ANSI/IEEE Std 176- 1978, DOI 10.1109/IEEE STD.1978.8941331 (1978) 1-58.

[29] G. Lesieutre, G. Ottman, H. Hofmann, Damping as a result of piezoelectric energy harvesting, J. Sound and Vibration. 269 (2004) 991-1001. https://doi.org/10.1016/S0022-460X(03)00210-4

[30] X. Gao, J. Wu, Y. Yu, Z. Chu, H. Shi, S. Dong, Giant piezoelectric coefficients in relaxor piezoelectric ceramic pnn-pzt for vibration energy harvesting, Adv. Funct. Mater. 28 (2018) 1706895. https://doi.org/10.1002/adfm.201706895

[31] E. Arroyo, Y. Jia, S. Du, S. Chen, A.A. Seshia, Experimental and Theoretical Study of a Piezoelectric Vibration Energy Harvester Under High Temperature, J. Microelectromechanical Systems. 26 (2017) 1216-1225. https://doi.org/10.1109/JMEMS.2017.2723626

Structural Health Monitoring: 10APWSHM
Materials Research Proceedings 50 (2025) 233-243

Materials Research Forum LLC
https://doi.org/10.21741/9781644903513-27

Single crystal linear array for modal decomposition analysis

Eliza Baddiley[1,a], Scott D. Moss[2,1,b*], Ben Vien[2,c], Pooia Lalbakhsh[1,d],
Jaslyn Gray[1,e], Nik Rajic[1,f], Cedric Rosalie[1,g], David J. Munk[1,h], Crispin Szydzik[3,i],
Arnan Mitchell[3,j], and Wing K. Chiu[2,k]

[1]Defence Science and Technology Group, Platforms Division, Victoria, Australia

[2]Monash University, Dept. Of Mechanical Aerospace Engineering, Victoria, Australia

[3]R.M.I.T University, Micro Nano Research Facility, Victoria, Australia

[a]eliza.baddiley@defence.gov.au, [b]scott.moss@monash.edu, [c]ben.vien@monash.edu,
[d]pooia.lalbakhsh@defence.gov.au, [e]jaslyn.gray@defence.gov.au, [f]nik.rajic@defence.gov.au,
[g]cedric.rosalie@defence.gov.au, [h]david.munk@defence.gov.au, [i]crispin.szydzik@rmit.edu.au,
[j]arnan.mitchell@rmit.edu.au, [k]wing.kong.chiu@monash.edu

© 2024 Commonwealth of Australia.

Keywords: Acousto Ultrasonics, Single Crystal Relaxor Ferroelectric, Modal Decomposition

Abstract. The characterisation of a 16-element acoustic emission (AE) sensor manufactured using relaxor ferroelectric single crystal (RFSC) is reported. The sensor, a *linear array for modal decomposition and analysis* (LAMDA) based on piezoelectric transduction, can be used for AE-based damage detection in aerospace, land, and sea applications. The RFSC LAMDA sensor was mounted on a 1.6 mm thick aluminium plate, the response of the sensor to AE excitations produced via pencil lead break (PLB) was examined. PLB break-energy generated by various pencil lead-types were examined experimentally. HB pencil lead 3 mm long, 0.5 mm diameter, has a measured break-energy of 6.6 µJ. Experimentally, RFSC LAMDA was able to detect the 6.6 µJ PLB excitation at a distance of 200 mm across the 1.6 mm aluminium plate. Modal decomposition of modelled and measured signals produced by the RFSC LAMDA in response to A_0 and S_0 Lamb waves, in a 1.6 mm aluminium plate, are compared and found to be in good agreement.

1. Introduction

The *linear array for modal decomposition and analysis* (LAMDA) sensor is an *acoustic emission* (AE) device that is being developed for structural sensing applications [1]. LAMDA can potentially be used for AE-based structural damage detection on land, sea, air and space platforms [2,3]. AE-based damage detection is a well-studied non-destructive testing (NDT) method that can be used for structural health monitoring [4].

Various types of acoustic waves can propagate within a thin plate [5,6,7], among which Lamb waves are scientifically significant. These are guided elastic waves characterised by distinct dispersion properties [5,8]. Lamb waves have several properties that make them effective for non-destructive inspection. They are influenced by both the material and geometrical properties of the plate they travel in, and are capable of propagating over long distances before dissipating [9]. Lamb waves typically have a wavelength on the order of magnitude of the plate thickness and can have both symmetric and antisymmetric propagation modes which are denoted by S_i and A_i (i = 0, 1, 2, …) respectively [5,10]. The key difference between the modes is that the S_i modes are compressional, with radial in-plane displacement, while A_i modes are flexural, with out-of-plane displacement [5]. In general, the A_i modes have a higher amplitude than the S_i modes for the same event [11]. Structural events such as impact or cracking can produce AEs that propagate as Lamb

Structural Health Monitoring: 10APWSHM
Materials Research Proceedings 50 (2025) 233-243

Materials Research Forum LLC
https://doi.org/10.21741/9781644903513-27

waves. The Lamb waves can be captured and analysed to determine information about the AE source, such as its location and mechanical characteristics.

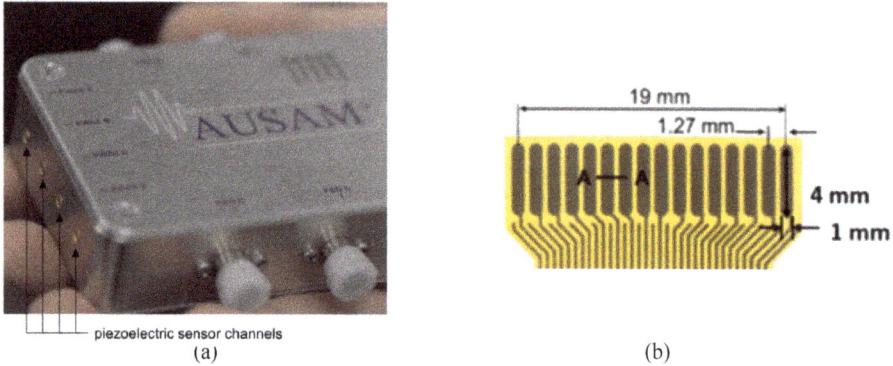

Fig. 1 (a) Image of bespoke hardware (AUSAM module [19]) used to interogate piezoelectric LAMDA sensors, and (b) a plan view schematic of RFSC LAMDA dimensions.

Analysing Lamb waves can, however, be difficult as signals often contains multiple modes, and the proportion of those modes present in the signal can be changed by mode conversion when the waves traverse a region of structural complexity [12]. The wave modes will often overlap in the recorded time history making it difficult to analyse the AE signal since the amplitude of each individual mode cannot be uniquely determined. The Lamb waves are also often dispersive, meaning that they change shape as they propagate due to different modal frequencies travelling at different velocities causing the waveform to spread out over time [12]. This, again, increases the difficulty of analysing the signal at a single point in time. The solution to these issues is to analyse the signal in both the time domain and the spatial domain. A *two-dimensional Fast Fourier Transform* (2D-FFT) takes the time histories from a series of evenly spaced points and converts this amplitude-time data into amplitude-wavenumber data at discrete frequencies [13]. This method of using temporal and spatial data means that modal decomposition can be done on multimodal and dispersive signals. This led to the development of the LAMDA sensor with its linear array of sensing elements that can be used to obtain spatial and temporal data.

A LAMDA sensor based on polyvinylidene fluoride (PVDF) transduction was previously investigated [1,15,16]. PVDF LAMDA was first reported in [1], where it was tested using ball drops on an aluminium plate. It was found that the plate modal signatures (i.e. AEs generated by a ball drop) measured by LAMDA contained enough information for the location of the ball drop impact to be calculated with reasonable accuracy [15]. The PVDF LAMDA was also tested using pencil-lead breaks (PLB) [16]. A PLB, often referred to as the Hsu-Nielson source because of their work that introduced it, is often used experimentally to produce Lamb waves in the plate that are similar to the AE's produced during mechanical plate failure [17]. Sause investigated PLB as an AE source and how varying experimental parameters, such as the free lead length and lead angle to the plate, impacted the signal [17]. It was found that while variation of these parameters didn't change the shape of the source function they did change the magnitude of the contact force and therefore the magnitude of the AE signal.

This paper reports on a LAMDA made with a relaxor ferroelectric single crystal (RFSC) material [011] $Pb(In_{1/2}Nb_{1/2})O_3$-$Pb(Mg_{1/3}Nb_{2/3})O_3$-$PbTiO_3$ (or PIN-PMN-28%PT)[18]. The RFSC material was machined into 200 µm thick x 1 mm long x 4 mm wide chiplets, with 16 of these

Structural Health Monitoring: 10APWSHM   Materials Research Forum LLC
Materials Research Proceedings 50 (2025) 233-243   https://doi.org/10.21741/9781644903513-27

chiplets arranged with a 0.27 mm pitch on a flexible polyimide carrier to form a RFSC LAMDA, as shown in Fig. 1(b). A RFSC LAMDA was then bonded to a 1.6 mm thick x 600 mm long x 600 mm wide aluminium plate for testing using a structural adhesive (M-Bond 200). The chiplets of the sensor are simultaneously interrogated by bespoke hardware shown in Fig. 1(a) [19]. The chiplets are aligned so as to take advantage of the d_{32} piezoelectric coefficient as it is the largest in-plane coefficient at -1200 pC/N, compared to d_{31} at 460 pC/N and d_{33} at 780 pC/N [18].

2. Model

This section describes the models used to characterise the behaviour of LAMDA bonded to an isotropic plate, including Lamb wave detection and the effect that the sensor has on the plate dynamics. A computationally efficient 2D plane-strain multi-physics finite element model was used and proved to be reliable when compared to theoretically determined and experimental results, as will be shown in section 4. The geometry for the model includes an aluminium plate that is 600 mm long and 1.6 mm thick matching the plate used experimentally. A LAMDA sensor was added to the top of the modelled aluminium plate. The modelled sensor consisted of an array of 16 RFSC chiplets, with array dimensions as detailed in section 1. To reduce computation time, the plate was cut in half at the centre (where the AE pulse was applied) and a symmetry constraint added, reducing the number of modelled degrees-of-freedom by half, Fig. 2(a). The total length of the array including spacing was 20.05 mm, as indicated in Fig. 2(b) the first chiplet of the modelled sensor was located 140 mm from the centre of the plate, thus the centre of the modelled sensor was ~150 mm from the centre of the plate. The modelled AE pulse was applied at the plate centre, the 140 mm distance between ensured the modelled sensor was located in the acoustic far field, minimising near-field Lamb wave interference from the location where the AE pulse was applied. In the model each of the 16 chiplets was electrically grounded at the interface between the sensor and the aluminium plate. The effect of the adhesive used to bond the sensor to the plate was ignored.

Standard material properties where used for the aluminium plate [6], and the properties for the PIN-PMN-28%PT 16 chiplets where taken from the literature [18]. As PIN-PMN-28%PT has the largest piezoelectric coefficient in the 2-direction (i.e., d_{32}), the orientation of the piezoelectric material had to be redefined so that the 2-direction is parallel to the x-axis. This was done by defining a rotated coordinate system which can then be used to orient the piezoelectric material.

2.1 Modelled Acoustic Pulse

To generate Lamb waves in the aluminium plate, a Hanning modulated function was applied at 17 points on the left-hand edge of the partial-plate shown Fig. 2(a). In order to produce specific wave modes at certain frequencies, the amplitude of the tone-burst function was scaled using the x and y displacement functions for each wave mode shape generated by commercial software to produce Lamb wave dispersion curves, DISPERSE [20] calculated for a 1.6 mm aluminium plate. For each of the 17 points on the left edge of the plate, a point load was applied. The x and y displacement functions from DISPERSE are given as a function of vertical position on the plate, so point loads were added at 17 points corresponding to the vertical positions that were included in the DISPERSE data. The x and y displacements were applied to the respective x and y components of the point load with the corresponding vertical position, and then these displacements were multiplied by the tone-burst function so that the displacements scale the amplitude of the tone-burst function.

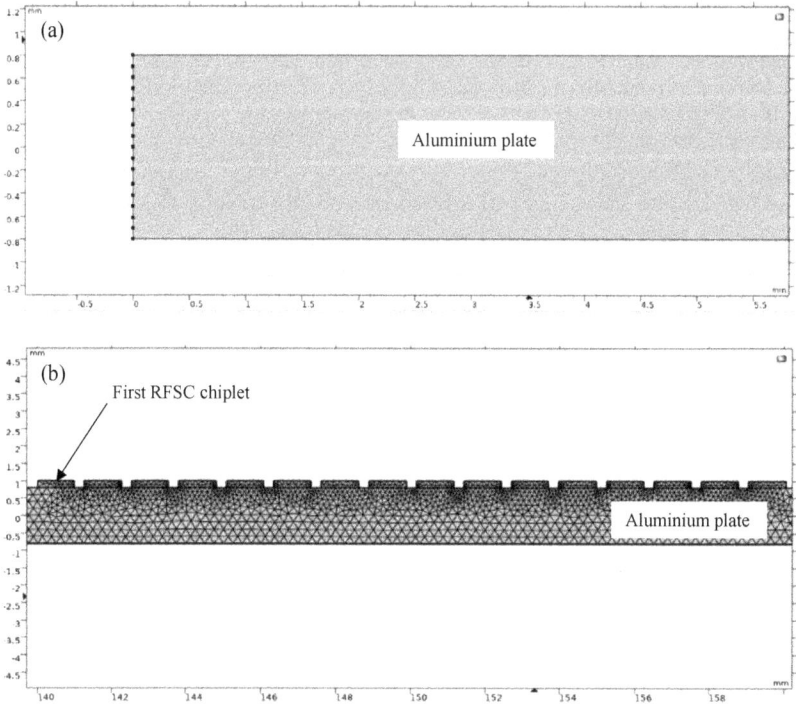

Fig. 2 *(a) Middle portion of the 2D plane-strain modelled aluminium plate, nominally 1.6 mm thick and 300 mm long, with symmetry constraint applied at centre, being the left-hand edge of plate model shown. Left-hand edge also shows points where the AE pulse was applied. (b) Modelled LAMDA sensor with first chiplet located 140 mm from the plate centre, and closest to the acoustic pulse location.*

The Hanning modulated tone-burst function is defined by the following equation,

$$F(t) = F_0 sin^2 \left\{ \frac{\pi f}{C} (t - t_0) \right\} sin\{2\pi f(t - t_0)\}, (t_0 \leq t \leq L + t_0) \qquad (1)$$

where $F(t)$ is the force of the tone-burst at a given time t, C is the number of cycles, F_0 is the amplitude and f is the centre frequency of the tone-burst. The time delay, t_0, was included to prevent instantaneous model stresses and was set to 50 ns. The number of cycles was set to 3.5. The centre frequencies chosen were dependent on the wave mode displacement shapes that could be generated by DISPERSE. These include A_0 at 302 kHz and S_0 at 597 kHz. The bandwidth of the tone-burst function is given by the following equation,

$$B = 2\frac{f}{C} . \qquad (2)$$

The effective maximum frequency of the tone-burst function, which is used to calculate the optimum solver time step in the model solver, is given by the following equation,

$$f_{max} \approx f \left(1 + \frac{2}{C} \right) . \qquad (3)$$

Structural Health Monitoring: 10APWSHM
Materials Research Proceedings 50 (2025) 233-243

Materials Research Forum LLC
https://doi.org/10.21741/9781644903513-27

2.2 Model Mesh and Solver

The mesh size, an example of which is shown in Fig. 2(b), was selected so that there are sufficient mesh elements per wavelength to recreate the wave. With a minimum wavelength of 2.5 mm, the maximum mesh size was set to 0.25 mm so that there are at least 10 mesh elements per wavelength for any of the frequencies used. A free triangular mesh was applied to the plate as it best supports modelled wave propagation [20].

A time-dependent solver was used with a solver timestep that is dependent on the effective maximum frequency of the tone-burst function being used. The timestep is given by the following equation,

$$t_s = \frac{1}{60 \times f_{max}}.$$

(4)

The maximum solve time was modified for each frequency. It was calculated using the theoretical wave mode curves from DISPERSE by taking the predicted wavenumber at the desired frequency, then converting the wavenumber to wavelength. The velocity of the wave can then be predicted by multiplying the frequency and wavelength. This velocity can be used to calculate the time it would take the wave to travel past the sensor (i.e. 0.16 mm) and these times were used as the maximum solve time for their respective frequencies.

3. Experimental

PLBs were performed according to literature standards, which specifies breaking a 3 mm long, 0.5 mm diameter, HB pencil-lead at a 45° angle to the plate [17]. A small change in this PLB angle can change the magnitude of the resulting AE [17]. To assist with the repeatability of PLB experiments the arrangement shown in Fig. 3(a) was assembled ensuring that the PLB angle remained constant at 45°. The PLB break-energy for various pencil leads was measured using the force-gauge (Mark 10 Corporation M7-50) and laser vibrometry (PolyTec PSV QTec 3D) arrangement shown in Fig. 3(b). The PLB break-energies were calculated by multiplying the measured distances and forces both averaged over 30 measurements for each pencil lead-type i.e. 4H, HB, and H.

PLB measurements, using HB pencil lead, were performed as a function of distance from the RFSC LAMDA. The range of distances across the 1.6 mm aluminium plate was 1 mm and 200 mm, in a direction normal to the first chiplet i.e. parallel with direction A-A in Fig. 1(b), as shown in Fig. 3(a). Another set of PLB measurements, also using HB pencil lead, were performed as a function of angle to the first chiplet, at a 50 mm radius from the centre of the first chiplet. The range of angles was between 0° to 30°, with 0° in the direction normal to the first chiplet, also shown in Fig. 3(a). Although the angle between the PLB and the first chiplet was changed, the pencil direction remained parallel with direction A-A as per Fig. 1(b).

Structural Health Monitoring: 10APWSHM   Materials Research Forum LLC
Materials Research Proceedings 50 (2025) 233-243   https://doi.org/10.21741/9781644903513-27

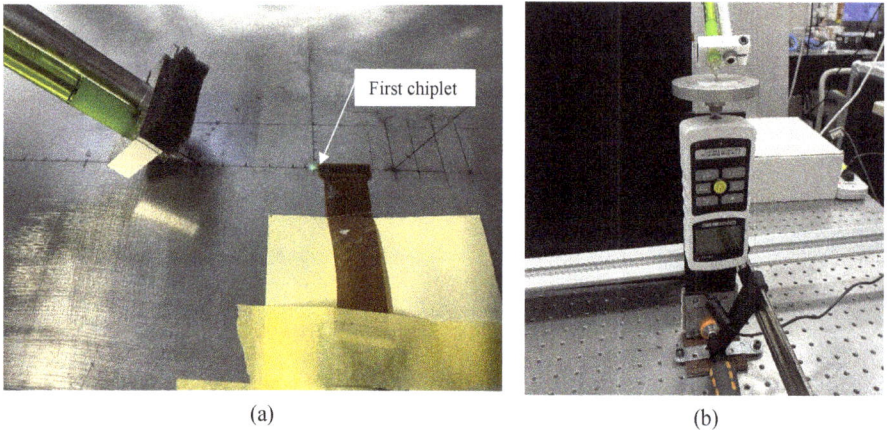

(a)   (b)

Fig. 3 (a) RFSC LAMDA experimental arrangement for measuring PLBs as a function of distance and angle, and (b) experimental force-gauge and laser vibrometer arrangement for PLB break-energy measurements.

4. Discussion and Results

Experiments characterising PLB break-energy for various pencil lead-types are presented, followed by a comparison of the modelled and measured modal decomposition of A_0 and S_0 Lamb waves generated in a 1.6 mm aluminium plate. Measurements of RFSC LAMDA signals generated by PLB as a function of distance and in-plane angle are presented and discussed.

Table 1 shows the average results of the PLB force measurements, displacement measurements, and the calculated break-energy for 4H, HB and H pencil lead-types. The 4H lead produced the largest break force, and HB the smallest. 4H is the hardest lead, with the most filler and the least graphite while HB is the softest lead tested with the most graphite and the least filler so there is a correlation between lead hardness and peak force. The 4H lead also had the highest average displacement and consequently the highest break-energy at 8.5 µJ. Although the HB lead had the lowest average peak force, it had the second highest average displacement, exhibiting the second highest break-energy at 6.6 µJ. The HB lead type was selected for AE experiments in this study due to its frequent use in the literature [17].

Pencil Lead Type	4H	HB	H
Ave. Peak Force (N)	4.7	4.1	4.3
Ave. Peak Disp. (µm)	1.8	1.6	1.4
Peak Energy (µJ)	8.5	6.6	6.0

Table 1. Force, displacement, and energy results for 4H, HB and H pencil lead.

Fig. 4 shows the modelled 2D-FFT of the signal produced when an A_0 wave mode with a centre frequency of 302 kHz and a bandwidth of approximately 173 kHz was driven in the plate, whereas Fig. 5 shows the modelled 2D-FFT of the signal produced by the model when an S_0 wave mode with a centre frequency of 597 kHz and a bandwidth of approximately 341 kHz was driven in the plate. The red region in the figures is due to the acoustic signal and the yellow lines overlayed on the plots are the theoretical predictions of the A_0, S_0 and A_1 wave modes from DISPERSE for 1.6 mm aluminium plate. The modelled results for both frequencies show good agreement with the theoretical predictions from DISPERSE but in both cases, the centre of the red regions sit slightly above the theoretical lines. This suggests that there may be a minor interaction between RFSC LAMDA and the plate wave dynamics due to the stiffness of the RFSC material.

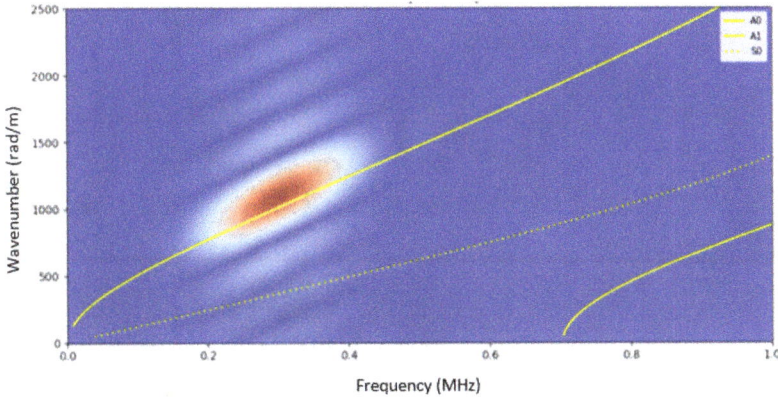

Fig. 4 Modelled modal decomposition for a plate excited by A_0 with centre frequency of 302 kHz.

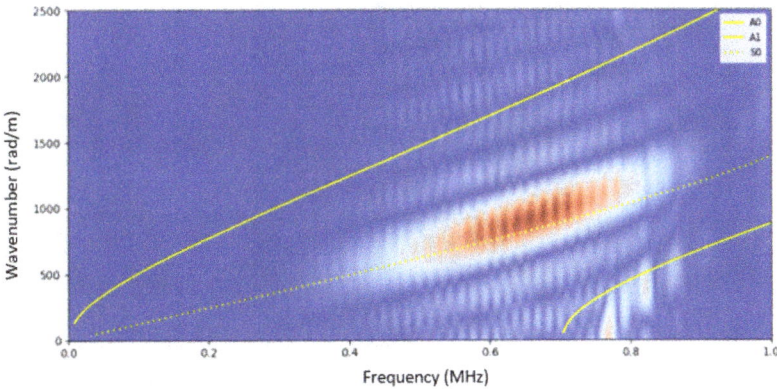

Fig. 5 Modelled modal decomposition for plate excited by S_0 with centre frequency of 597 kHz.

Fig. 6 and Fig. 7 show the measured 2D-FFTs of the signal when a PLB is performed 90 mm from the leading edge of LAMDA. The S_0 mode had a much lower amplitude than the A_0 mode, so it was challenging to filter the data and get both the A_0 and S_0 modes on the same plot, therefore they were filtered separately. For the A_0 mode depicted in Fig. 6, a bandpass filter with a lower cut-off of 300 kHz and an upper cut-off of 500 kHz was applied and for the S_0 mode depicted in Fig. 7, a bandpass filter with a lower cut-off of 600 kHz and an upper cut-off of 800 kHz was applied. The results show good agreement with both the theoretical dispersion curves and the model results shown in Fig. 5 and Fig. 6. The centre of the red regions in both plots sit slightly above the yellow prediction lines; this behaviour was also observed in the model results, and as mentioned earlier is likely attributed to the sensor participating in the plate-wave dynamics.

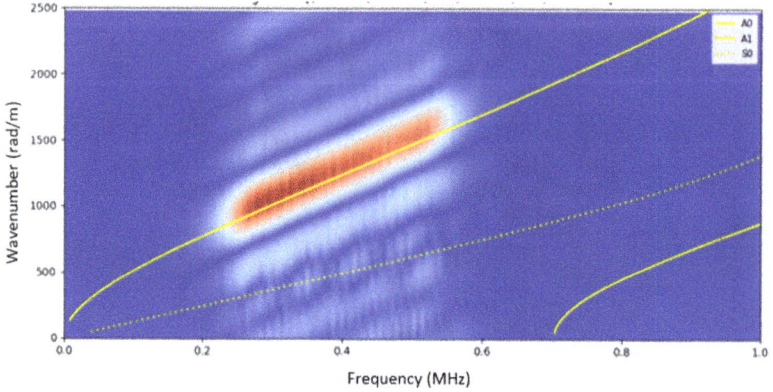

Fig. 6 Measured A_0 modal decomposition for a pencil-lead break at a distance of 90 mm.

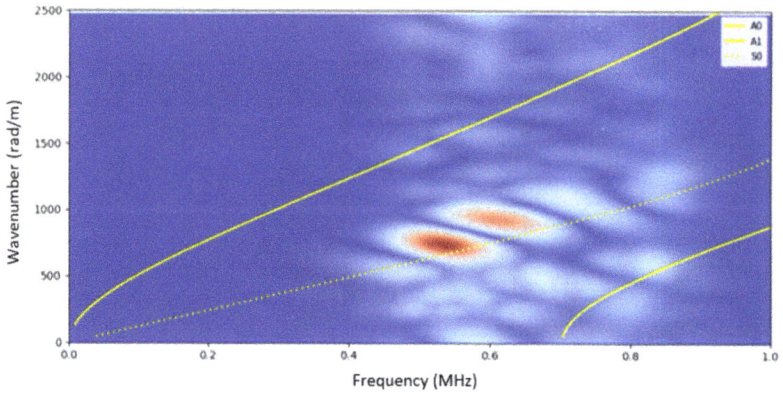

Fig. 7 Measured S_0 modal decomposition for a pencil-lead break at a distance of 90 mm.

Structural Health Monitoring: 10APWSHM
Materials Research Proceedings 50 (2025) 233-243

Materials Research Forum LLC
https://doi.org/10.21741/9781644903513-27

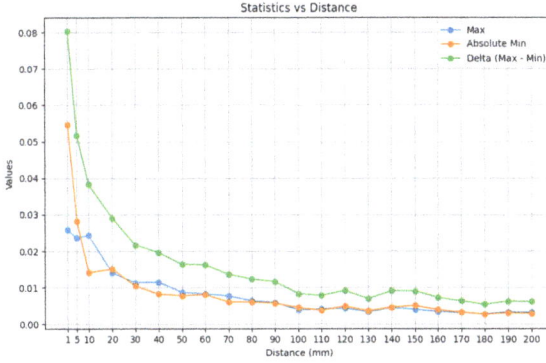

Fig. 8 Measured signal amplitude as a function of distance between PLB location and first chiplet of RFSC LAMDA.

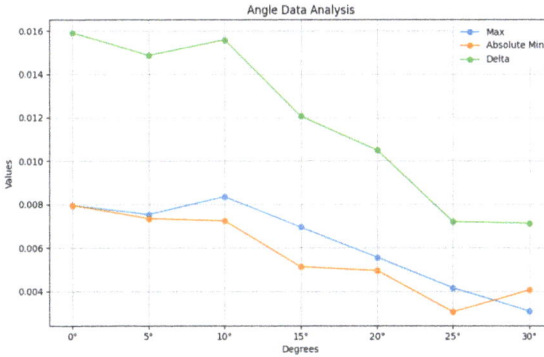

Fig. 9 Measured signal amplitude as a function of in-plane plate angle between RFSC LAMDA central-axis and PLB location, performed in a 50 mm radius from the centre of the first chiplet.

Fig. 9 shows the RFSC LAMDA signal amplitude from the first chiplet as a function of the distance from the leading edge of the sensor. The blue line depicts the maximum voltage of the PLB signal, the orange line depicts the absolute value of the minimum voltage (i.e. the absolute value of the maximum negative voltage) and the green line depicts the peak-to-peak voltage of the signal. As expected, the magnitude of the signal decreases as the distance from the leading edge of the sensor increases. Although the measured voltage decreases rapidly with distance (as 1/distance), the RFSC LAMDA was still able to detect the 6.6 µJ AE generated by a HB pencil lead PLB 200 mm away from the first chiplet, potentially further since the experiments were limited by the size of the plate.

The RFSC LAMDA signal amplitude as a function of angle to centre of the leading chiplet of the sensor is shown in Fig. 9, where again the blue line shows the maximum voltage of the signal, the orange line shows the absolute value of the minimum voltage, and the green line shows the peak-to-peak voltage. The amplitude of the signal is reduced by ~50% as the angle from the central-axis of the RFSC LAMDA increases from 0° to 30°. This decrease can be attributed to the

Structural Health Monitoring: 10APWSHM   Materials Research Forum LLC
Materials Research Proceedings 50 (2025) 233-243   https://doi.org/10.21741/9781644903513-27

fact that the RFSC material is highly directional, with a larger, negative piezoelectric coefficient in the 2-direction and a smaller, positive piezoelectric coefficient in the 1-direction. As the angle increases, the 1-direction component of the AE increases causing the measured voltage signal to decrease.

Conclusion

A new relaxor ferroelectric single crystal (RFSC) *linear array for modal decomposition and analysis* (LAMDA) sensor has been developed for in situ modal decomposition of Lamb waves generated by acoustic emission (AE) for potential use in air, space, land, and sea applications. RFSC LAMDA was tested using pencil lead breaks (PLB) to produce AEs on a 1.6 mm thick aluminium plate, and two dimensional (2D) plane strain model results were presented. The break-energy of PLBs for three different lead-types, 4H, HB and H, were measured, and found to be 8.5 µJ, 6.6 µJ, and 6.0 µJ, respectively. The HB lead type was selected for PLB AE experimentation based on its prevalent use in the literature. PLB experiments indicated that the amplitude of the AE signal detected by RFSC LAMDA decreases as 1/distance from the leading edge of the sensor (i.e. first RFSC chiplet, closest to the PLB location). The 6.6 µJ signal was still measured at a distance of 200 mm. Additionally, the signal amplitude decreases by 50% as the AE incident angle increases from 0° to 30° from the centre of the leading edge of the sensor. Modal decompositions were calculated using 2D *Fast Fourier Transforms* on both modelled and measured AE data, with good agreement found.

Acknowledgements

The authors would like to thank Mr. Daniel Bitton for his technical assistance. This work was supported by the ASCA Advanced Piezoelectric Materials and Applications program including DMTC Ltd., the DSTG Strategic Research Initiative in Advanced Materials and Sensors, and the DSTG Industry Experience Placement program.

References

[1] N. Rajic, C. Rosalie, S. van der Velden, L.F Rose, J. Smithard, W.K. Chiu, A novel high density piezoelectric sensing capability for in situ model decomposition of acoustic emissions, Proceedings of the 9th European workshop on Structural Health Monitoring. Manchester, UK, 10-13 July 2018.

[2] E. Baddiley, S.D. Moss, B. Vien, J. Gray, N. Rajic, C. Rosalie, D. Munk, C. Szydzik, A. Mitchell, W. Chiu, A single crystal Lamb wave sensing array, J. Acoust. Soc. Am. 154 (2023) A151. https://doi.org/10.1121/10.0023090

[3] C.U Grosse, M Ohtsu, D.G Aggelis, T Shiotani, Acoustic Emission Testing: Basics for research - applications in engineering, Springer Nature, 2021. https://doi.org/10.1007/978-3-030-67936-1

[4] N. Ghadarah, D. Ayre, A Review on Acoustic Emission Testing for Structural Health Monitoring of Polymer-Based Composites, Sensors. 23 (2023) 6945. https://doi.org/10.3390/s23156945

[5] Z. Su and L. Ye, Identification of Damage Using Lamb Waves, Lecture Notes in Applied and Computational Mechanics. 48 (2009) 15-58. https://doi.org/10.1007/978-1-84882-784-4_2

[6] M.J. Schipper, J. Gray, D. J. Munk, S. D. Moss, N. Rajic, C. Rosalie, J. Smithard, B. Vien, W. K. Chiu, C. Szydzik, A. Mitchell, Improving the performance of a lamb wave sensing array via relaxor ferroelectric single crystal transduction, Mater. Research Proc. 27 (2023) 84-94. https://doi.org/10.21741/9781644902455-11

[7] Y. He, S. Chen, D. Zhou, S. Huang, P. Wang, Shared Excitation Based Nonlinear Ultrasound and Vibrothermography Testing for CFRP Barely Visible Impact Damage Inspection. IEEE Trans. Indust. Info. 14 (2018) 5575-5584. https://doi.org/10.1109/TII.2018.2820816

[8] A. Wronkowicz-Katunin, A. Katunin, K. Dragan, Reconstruction of Barely Visible Impact Damage in Composite Structures Based on Non-Destructive Evaluation Results, Sensors. 19 (2019) 4629. https://doi.org/10.3390/s19214629

[9] D. Girolamo, L. Girolamo, F.G. Yuan, Automated laser-based barely visible impact damage detection in honeycomb sandwich composite structures, AIP Conf. Proc. 1650 (2015) 1392-1400. https://doi.org/10.1063/1.4914754

[10] S. Shan, L. Cheng, Mode-mixing-induced second harmonic A0 mode Lamb wave for local incipient damage inspection. Smart Mater. Struct. 29 (2020) 055020. https://doi.org/10.1088/1361-665X/ab7e37

[11] C.K. Lee, P.D. Wilcox, B.W. Drinkwater, J.J. Scholey, M.R. Wisnom, M.I. Friswell, Acoustic Emission during Fatigue Crack Growth in Aluminium Plates, Proc. ECNDT. 25-29 (2006) 1-8. https://doi.org/10.4028/0-87849-420-0.23

[12] D.N. Alleyne, P. Cawley, The interaction of Lamb waves with defects, IEEE T. Ultrason. Ferr. 39 (1992) 381-397. https://doi.org/10.1109/58.143172

[13] D. Alleyne, P. Cawley, A Two-Dimensional Fourier Transform Method for the Measurement for Propagating Multimode Signals. J. Acoust. Soc. Am. 89 (1991) 1159-1168. https://doi.org/10.1121/1.400530

[14] N. Rajic, C. Davis, A. Thomson, Acoustic-wave-mode separation using a distributed Bragg Grating Sensor, Smart Mater. Struct. 18 (2009) 125005. https://doi.org/10.1088/0964-1726/18/12/125005

[15] N. Rajic, C. Rosalie, B.S. Vien, S. van der Velden, L.F. Rose, J. Smithard and W.K. Chui, In situ wave number-frequency modal decomposition of acoustic emissions. Struct. Health Mon. 19 (2020) 2033-2050. https://doi.org/10.1177/1475921719885324

[16] X. Yao, B.S. Vien, N. Rajic, C. Rosalie, L.R.F. Rose, C. Davies, W.K. Chiu, Modal Decomposition of Acoustic Emissions from Pencil-Lead Breas in an Isotropic Thin Plate, Sensors. 23 (2023) 1988. https://doi.org/10.3390/s23041988

[17] R. Madarshahian, V. Soltangharaei, R. Anay, J.M. Caicedo, P. Ziehl, Hsu-Nielsen source acoustic emission data on a concrete block. Data in Brief. 23 (2019) 103813. https://doi.org/10.1016/j.dib.2019.103813

[18] M.G Sause, Investigation of pencil-lead breaks as acoustic emission sources, J. Acoust. Emission. 29 (2011) 184-196.

[19] E. Sun, S. Zhang, J. Luo, T.R. Shrout, W. Cao, Elastic, dielectric, and piezoelectric constants of Pb(In1/2Nb1/2)O3-P(Mg1/3Nb2/3)-PbTiO3 single crystal poled along [011]c, App. Phys. Lett. 97 (2010) 032902. https://doi.org/10.1063/1.3466906

[20] B. Pavlakovic, M. Lowe, D. Alleyne, P. Crawley, Disperse: A General Purpose Program for Creating Dispersion Curves, Rev. Prog. Quant. Non-destructive Eval. 16A (1997) 185-192. https://doi.org/10.1007/978-1-4615-5947-4_24

[21] J. Smithard, N. Rajic, S. van der Velden, P. Norman, C. Rosalie, S. Galea, H. Mei, B. Lin, V. Giurgiutiu, An Advanced Multi-Sensor Acousto-Ultrasonic Structural Health Monitoring System: Development and Aerospace Demonstration, Materials. 10 (2017) 832. https://doi.org/10.3390/ma10070832

[22] Information on COMSOL Multiphysics, Resolving time dependent waves, https://www.comsol.com/support/knowledgebase/1118 (accessed 18th January 2024).

Structural Health Monitoring: 10APWSHM   Materials Research Forum LLC
Materials Research Proceedings 50 (2025) 244-251   https://doi.org/10.21741/9781644903513-28

Efficient rebar dimension detection using prior-guided discrete global optimization full waveform inversion

Zihan Xia[1,a,*], Liyu Xie[1,b] and Songtao Xue[1,2,c]

[1] Department of Disaster Mitigation for Structures, Tongji University, Shanghai, China

[2] Department of Architecture, Tohoku Institute of Technology, Sendai, Japan

[a] xzhooo@tongji.edu.cn, [b] liyuxie@tongji.edu.cn, [c] xue@tongji.edu.cn

Keywords: FWI, Rebar Dimension Detection, Prior-Guided, Nonlinear Electromagnetic Responses

Abstract. In non-destructive rebar dimension detection in concrete using full waveform inversion (FWI), exhaustive search methods and global optimization algorithms struggle with nonlinear electromagnetic responses and computational complexity, hindering their integration into engineering detection systems. This paper proposes a prior-guided discrete global optimization full waveform inversion (PG-DGOFWI) method to improve the efficiency and accuracy of rebar dimension detection. The method first uses reliable prior information to convert rebar coordinates and dimensions from continuous to discrete arrays, reducing the search space. It then constructs an objective function from electromagnetic wave data and identifies parameters using a global optimization algorithm. Simulation using the gprMax compared the inversion results of exhaustive search, global optimization, and PG-DGOFWI methods. Experimental tests based on the simulation model showed that PG-DGOFWI significantly improves rebar dimension detection. This method uses prior information to guide the search, reducing computational complexity and avoiding local optima, thereby enhancing stability and reliability.

1. Introduction

Rebar is essential in modern construction for its critical role in seismic resistance, durability, crack prevention, and structural stability. It enhances load-bearing capacity and synergizes with concrete to ensure safety under various conditions[1]. In seismic zones, rebar's ductility and toughness absorb and dissipate energy, mitigating structural damage and collapse risks[2, 3]. Proper rebar design controls concrete shrinkage and expansion, preventing cracks and maintaining structural integrity and aesthetics[4, 5]. The performance of building structures is closely related to the functioning of the rebar, necessitating thorough inspection for structural safety assessments.

Current research on non-destructive rebar detection shows that, although these techniques perform well in terms of detection accuracy, they face challenges in seamlessly integrating with the engineering front-end during actual applications, mainly due to their reliance on calibration information.

Current studies indicate that while non-destructive rebar detection technologies offer high accuracy, they struggle with integration into practical engineering workflows due to their dependence on calibration data. R.D. Garg and S.S. Jain utilized Ground Penetrating Radar (GPR) technology, achieving high-resolution non-destructive rebar detection through precise antenna calibration and velocity adjustments, which effectively distinguish multilayered subsurface information [6]. In studies conducted in Canada, H. Rathod and R. Gupta obtained high-precision information about rebar at different depths and intervals within concrete structures by adjusting detector spacing and calibrating radar velocity[7]. D. Jeon and Y. Jeong integrated deep learning-based frequency-difference electrical resistance tomography technology, validating its high accuracy in rebar localization through calibration with cement mortar samples[8]. L. Zanzi and D.

Structural Health Monitoring: 10APWSHM   Materials Research Forum LLC
Materials Research Proceedings 50 (2025) 244-251   https://doi.org/10.21741/9781644903513-28

Arosio enhanced the sensitivity and precision of diameter detection by using dual-polarized GPR data and frequency-dependent calibration curves, ensuring reliable identification of rebars with varying diameters[9]. P. Asadi and M. Alvarez developed an automated image processing chain that employs computer vision to accurately locate and detect rebars on concrete bridge decks, significantly improving detection accuracy by matching image features with calibration information[10]. These studies highlight the superior performance of calibration-based rebar detection methods in diverse applications, advancing NDT technologies and offering robust technical support for engineering. However, their calibration dependency restricts their use in complex environments. To overcome this, this paper introduces a full waveform inversion (FWI) for rebar detection using GPR.

Traditional gradient-dependent FWI relies on the Helmholtz equation, often impractical for high-frequency real-world applications due to excessive theoretical assumptions. Numerical methods like finite element and finite difference modeling have advanced FWI applications, though they demand significant computational resources[11, 12]. Researchers have proposed global optimization methods to address these challenges, achieving accuracy without gradient dependency, as shown in geological and tunnel detection studies[13] [14]. Inspired by these advancements, this paper applies a non-gradient-dependent global optimization FWI method for rebar detection, utilizing GPR B-Scan data to simplify target parameters.

To improve computational efficiency, this paper proposes a prior-guided discrete global optimization FWI (PG-DGOFWI) method. By leveraging prior knowledge of rebar characteristics, PG-DGOFWI reduces unnecessary forward modeling calculations, enhancing convergence speed and accuracy in complex engineering environments. This method provides a more efficient and reliable approach for rebar detection, overcoming limitations of traditional calibration-dependent methods.

2. Detection Mechanism

2.1 GPR system

GPR primarily consists of two parts: the radar transmitter and the radar receiver, as shown in Figure 1.

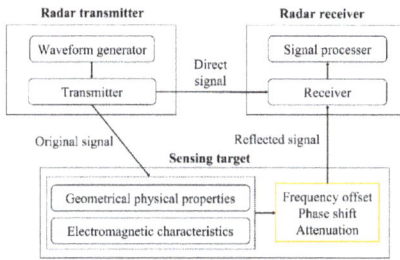

Figure 1. GPR system framework.

The radar transmitter modulates and generates high-frequency pulses within the waveform generator and transmits the signal into the structure under investigation. The geometric physical properties and electromagnetic characteristics of the structure affect the propagation of the electromagnetic waves within it, resulting in changes to the reflected signal's frequency, phase, and amplitude, among other attributes. Ultimately, the radar receiver captures the reflected signal, converting the collected high-frequency analog signal into a digital signal for processing.

Structural Health Monitoring: 10APWSHM   Materials Research Forum LLC
Materials Research Proceedings 50 (2025) 244-251   https://doi.org/10.21741/9781644903513-28

2.2 Prior-guided discrete global optimization full waveform inversion

To address these challenges in electromagnetic forward modeling, this paper proposes a prior-guided discrete global optimization full waveform inversion (PG-DGOFWI) method. PG-DGOFWI integrates prior-guided mechanisms with discrete global optimization strategies, thereby overcoming the issues of randomness and repetitive computations inherent in conventional global optimization approaches. By combining prior information with discrete optimization techniques, PG-DGOFWI enhances the computational efficiency and reliability of FWI.

The PG-DGOFWI method comprises three key steps, as shown in Figure 2.

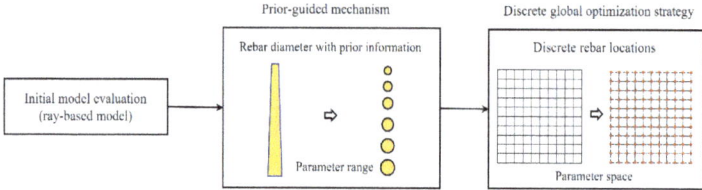

Figure 2. PG-DGOFWI method.

(1) *Initial model evaluation*: Establishing a model framework using ray-based methods to provide an initial model for inversion, which is similar as FWI.

(2) *Prior-guided mechanism*: Utilizing prior information to guide the optimization process, simultaneously narrowing the parameter search space and reducing ineffective random searches.

(3) *Discrete global optimization strategy*: Discretizing the continuous parameter space based on the prior-guided mechanism to direct the global optimization algorithm for parameter inversion, ensuring the coverage of possible parameter combinations while avoiding redundant computations.

Prior-guided mechanism

By incorporating relevant prior data, the prior-guided mechanism ensures that the optimization process is informed by reliable and contextual information, thereby enhancing both the accuracy and efficiency of rebar detection. A schematic diagram of the prior-guided mechanism is shown in Figure 3.

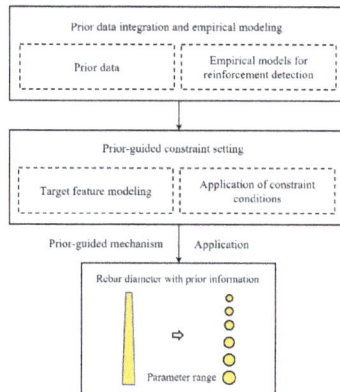

Figure 3. Schematic diagram of the prior-guided mechanism.

Structural Health Monitoring: 10APWSHM   Materials Research Forum LLC
Materials Research Proceedings 50 (2025) 244-251   https://doi.org/10.21741/9781644903513-28

Prior data integration and empirical modeling

To effectively implement the prior-guided mechanism, it is essential to systematically integrate and model the available prior data.

Prior data: In applications for rebar detection within structural health monitoring, prior information can include existing design blueprints, historical structural data, construction records, and previous inspection data. For example, when detecting the distribution of rebar in concrete structures, prior information may originate from building design diagrams or records of rebar layout during construction. Additionally, other inspection data, such as magnetic field strength and ultrasonic echo information, can also contribute to the comprehensive prior data set.

Empirical models for rebar detection: Long-term detection experience shows that the distribution of rebar within concrete often follows certain patterns, such as fixed spacing and arrangements within similar structural types. These empirical patterns can be applied through the prior-guided mechanism in the PG-DGOFWI method. For instance, if the variation range of rebar spacing within a structure is known, the optimization process can prioritize searches within these reasonable spacing ranges, avoiding unnecessary exploration of improbable parameter regions. This integration of empirical models allows the optimization algorithm to focus on plausible parameter spaces, reducing inefficient computations.

Prior-guided constraint setting

Target feature modeling: In electromagnetic wave detection, the material properties of rebar (such as conductivity and magnetic permeability) can be modeled using prior knowledge. For example, the electromagnetic characteristics of rebar can be represented as known parameters, and then possible physical feature models can be generated based on prior information. This constraint-based modeling process helps the optimization algorithm predefine the search range for parameters, thus reducing unnecessary searches and computations.

Application of constraint conditions: By converting target features into constraints within the optimization process, the search space can be significantly narrowed. For example, if prior data indicates that rebar is primarily distributed in certain specific areas, the optimization algorithm can focus its searches on these regions rather than randomly searching across the entire structure. Introducing such constraints not only enhances computational efficiency but also reduces the probability of false positives and omissions.

Discrete global optimization strategy

By designing global optimization algorithms tailored for discrete spaces, this approach ensures the efficient identification of global optima within complex, multi-dimensional parameter spaces. As illustrated in Figure 4, the discrete global optimization strategy employs a structured grid to systematically explore potential solutions, enhancing both accuracy and computational efficiency.

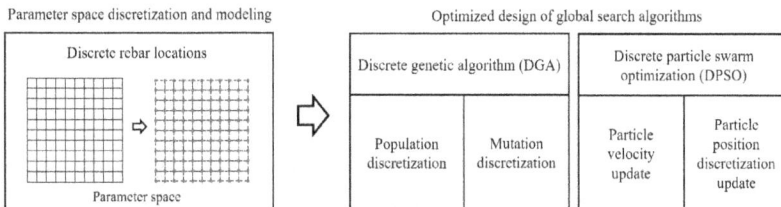

Figure 4. Discrete global optimization strategy

Structural Health Monitoring: 10APWSHM   Materials Research Forum LLC
Materials Research Proceedings 50 (2025) 244-251   https://doi.org/10.21741/9781644903513-28

Parameter space discretization and modeling

The continuous parameter space is converted into a set of discrete points through mathematical modeling and discretization algorithms.

Optimized design of global search algorithms

Discrete genetic algorithm (DGA): The innovation of DGA lies in setting the initial population using prior information, combined with adaptive mutation rates and crossover strategies, allowing the algorithm to converge more rapidly towards the global optimum. In rebar detection, DGA can utilize known constraints on rebar layout to reduce unnecessary searches, thereby improving computational efficiency.

Discrete particle swarm optimization (DPSO): Unlike traditional PSO in continuous spaces, DPSO incorporates discrete space search strategies, allowing efficient localization of potential optima within the discrete solution space. Specifically, by assigning priority to different parameters, DPSO can prioritize searches in areas where rebar is likely to be located, thus reducing optimization time.

3. Simulation

To validate the effectiveness of the PG-DGOFWI, this section conducts a systematic verification through simulation inversion experiments. The inversion model selected is a 2D model incorporating both tensile and compressive rebar. Numerical computations for both the forward and inversion processes are performed using gprMax. Subsequently, inversion analyses are carried out using the DGA and DPSO, respectively.

3.1 Model Construction

To demonstrate the efficacy of the PG-DGOFWI method, a two-dimensional concrete structural model with practical engineering relevance is first constructed. The model is based on a typical 2D cross-section of a concrete beam, measuring 0.3 meters in width and 0.4 meters in height. The concrete portion is modeled with uniformly distributed material properties, featuring a dielectric constant of 4.2 and a conductivity set to 0.

All rebars have a diameter of 14mm, selected based on standardized diameters commonly used in engineering to ensure the model's comparability and the engineering applicability of the simulation results.

The tensile rebars are positioned in the lower part of the beam cross-section, 50mm from the bottom edge. According to conventional design principles, the four tensile rebars are evenly distributed along the width of the beam, with a center-to-center spacing of 50mm to ensure uniform stress distribution. The compressive rebars are located in the upper part of the beam cross-section, 50mm from the top edge. Two compressive rebars are arranged with the same center-to-center spacing of 93mm to counteract compressive stresses in the upper region. Detailed parameter information is shown in Table 1.

Table 1. True model parameter

Param	ε_c	$l_1(mm)$	$l_2(mm)$	$\Delta l_t(mm)$	$\Delta l_s(mm)$	$d(mm)$	$h_1(mm)$	$h_2(mm)$
Value	4.2	50	50	50	93	14	50	50

As illustrated in Figure 5(a), the two-dimensional concrete structural model clearly depicts the specific arrangement of the tensile and compressive rebars.

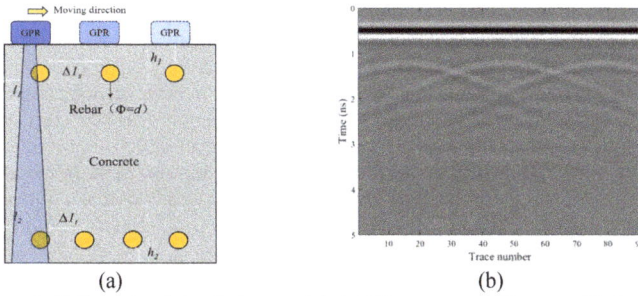

(a)   (b)

Figure 5. Schematic of simulation model. (a) 2D structural diagram, (b) Synthetic B-Scan Image.

The gprMax employs the finite-difference time-domain (FDTD) method for electromagnetic wave simulation, utilizing a wideband pulsed electromagnetic wave with a center frequency of 2.7 GHz as the excitation source. As illustrated in Figure 5(b), the generated B-Scan data clearly display the approximate distribution of tensile and compressive rebars within the two-dimensional concrete structure.

3.2 Application of the PG-DGOFWI

Initial model evaluation

This study employs a ray-based approach to analyze the features of the generated B-Scan data, thereby constructing the initial model for inversion and determining the necessary observation parameters.

The initial model encompasses the concrete region and the preliminary distribution of internal rebars. Based on the evaluation of the initial model, the following inversion parameters are determined, as shown in Table 1.

Table 2. Initial model of PG-DGOFWI

Param	ε_c	$l_1(mm)$	$l_2(mm)$	$\Delta l_t(mm)$	$\Delta l_s(mm)$	$d(mm)$	$h_1(mm)$	$h_2(mm)$
Value	3.5	30	30	60	100	8	40	40

Prior-guided mechanism

In practical engineering applications, the diameters, distribution patterns, and quantities of rebars typically exhibit regularity and standardization, while the electromagnetic parameters of concrete vary within specific ranges. By integrating prior information and applying empirical models, the parameter search space can be effectively reduced, providing a clear initial search direction for the optimization algorithm and enhancing the efficiency and accuracy of the inversion process.

This study adopts the following prior constraints:

(a) Rebar diameter constraints. Rebar diameters are categorized based on common engineering applications: **6mm, 8mm, and 10mm** for smaller components or auxiliary structures; **12mm, 14mm, 16mm, and 18mm** for general structural elements such as beams, slabs, and columns; **20mm, 22mm, 25mm, 28mm, and 32mm** for large load-bearing structures, including large beams, columns, bridges, and high-rise building cores; and **36mm and 40mm** for specialized structures or large infrastructure projects like bridge piers and dams.

Structural Health Monitoring: 10APWSHM   Materials Research Forum LLC
Materials Research Proceedings 50 (2025) 244-251   https://doi.org/10.21741/9781644903513-28

(b) Rebar distribution patterns. Tensile rebars are arranged in the lower part of the structure, and compressive rebars are placed in the upper part. Additionally, Devine et al. found that non-uniform spacing of rebars can reduce the load-bearing capacity of precast components[15]. Therefore, rebar spacing is assumed to be uniform in advance in the inversion process.

(c) Concrete electromagnetic parameter range. The relative dielectric constant of concrete is established within a range of 3 to 5, with incremental adjustments of 0.1.

By applying these prior constraints, the search space for the optimization algorithm is effectively narrowed, and the initial search direction is more clearly defined. This enhances the convergence speed and reliability of FWI.

Discrete global optimization strategy

In the PG-DGOFWI method, the DGA and DPSO are designed to adapt to the discrete parameter space and integrate with the prior-guided mechanism.

Discrete genetic algorithm
DGA generates the initial population by utilizing the parameter ranges and distribution patterns integrated through the prior-guided mechanism. In DGA, prior information is translated into step size constraints to ensure the discretization of parameter values. During population initialization, randomly generated variable values are discretized into integer multiples of the step size. Similarly, during the mutation process, mutated variable values are discretized into integer multiples of the step size. In each generation iteration, the current best fitness value and its corresponding parameter combination are recorded to ensure the final output of the global optimal solution.

Discrete particle swarm optimization
DPSO utilizes the prior-guided mechanism to generate initial particle positions, ensuring that particles are distributed within potential global optimal regions. Notably, the velocity update in DPSO is consistent with that of traditional PSO. However, during the position update process, the step size derived from the prior information is also used to control the discretization.

3.3 Discussion on simulation results
Finally, a systematic comparison and evaluation of the inversion results obtained using DGA and DPSO in FWI are conducted. The results, shown in Table 3, indicate that both DGA and DPSO, when combined with the prior-guided mechanism, can rapidly converge to the global optimal solution within the discrete parameter space.

Table 3. DGA and DPSO result

Param	ε_c	$l_1(mm)$	$l_2(mm)$	$\Delta l_t(mm)$	$\Delta l_s(mm)$	$d(mm)$	$h_1(mm)$	$h_2(mm)$
DGA	4.3	50	52	44	93	14	46	50
DPSO	4.2	50	50	48	95	14	52	50
True	4.2	50	50	50	93	14	50	50

4. Conclusion
The PG-DGOFWI leverages reliable a priori information to transform the search space for rebar coordinates and dimensions from continuous to discrete, effectively reducing the search range and computational complexity. Simulation results demonstrate that the PG-DGOFWI enhances the efficiency of rebar detection without compromising accuracy, while avoiding local optima and enhancing the stability and reliability of the inversion process.

References

[1] Kim S. Investigating Structural Stability and Constructability of Buildings Relative to the Lap Splice Position of Reinforcing Bars. Journal of The Korea Institute of Building Construction. 2023;23:315-26.

[2] Parra-Montesinos GJ. High-performance fiber-reinforced cement composites: an alternative for seismic design of structures. ACI Structural Journal. 2005;102:668. https://doi.org/10.14359/14662

[3] Zhang J, Liu X, Liu J, Zhang M, Cao W. Seismic performance and reparability assessment of recycled aggregate concrete columns with ultra-high-strength steel bars. Engineering Structures. 2023;277:115426. https://doi.org/10.1016/j.engstruct.2022.115426

[4] Abou-Zeid M, Fowler DW, Nawy EG, Allen JH, Halvorsen GT, Poston RW et al. Control of cracking in concrete structures. Rep ACI Comm. 2001;224:12-6.

[5] Liu J, Tian Q, Wang Y, Li H, Xu W. Evaluation method and mitigation strategies for shrinkage cracking of modern concrete. Engineering-Prc. 2021;7:348-57. https://doi.org/10.1016/j.eng.2021.01.006

[6] Bala D, Garg R, Jain S. Rebar detection using GPR: An emerging non-destructive QC approach. Int J Eng Res Appl(IJERA). 2011;1:2111-7.

[7] Rathod H, Debeck S, Gupta R, Chow B. Applicability of GPR and a rebar detector to obtain rebar information of existing concrete structures. Case Stud Constr Mat. 2019;11:e00240. https://doi.org/10.1016/j.cscm.2019.e00240

[8] Jeon D, Kim MK, Jeong Y, Oh JE, Moon J, Kim DJ et al. High-accuracy rebar position detection using deep learning-based frequency-difference electrical resistance tomography. Automat Constr. 2022;135:104116. https://doi.org/10.1016/j.autcon.2021.104116

[9] Zanzi L, Arosio D. Sensitivity and accuracy in rebar diameter measurements from dual-polarized GPR data. Constr Build Mater. 2013;48:1293-301. https://doi.org/10.1016/j.conbuildmat.2013.05.009

[10] Asadi P, Gindy M, Alvarez M, Asadi A. A computer vision based rebar detection chain for automatic processing of concrete bridge deck GPR data. Automat Constr. 2020;112:103106. https://doi.org/10.1016/j.autcon.2020.103106

[11] Warren C, Giannopoulos A, Giannakis I. gprMax: Open source software to simulate electromagnetic wave propagation for Ground Penetrating Radar. Comput Phys Commun. 2016;209:163-70. https://doi.org/10.1016/j.cpc.2016.08.020

[12] Giannopoulos A. Modelling ground penetrating radar by GprMax. Constr Build Mater. 2005;19:755-62. https://doi.org/10.1016/j.conbuildmat.2005.06.007

[13] Riedel C, Mahmoudi E, Trapp M, Lamert A, Hölter R, Zhao C et al. A hybrid exploration approach for the prediction of geological changes ahead of mechanized tunnel excavation. J Appl Geophys. 2022;203:104684. https://doi.org/10.1016/j.jappgeo.2022.104684

[14] Romdhane A, Grandjean G, Brossier R, Réjiba F, Operto S, Virieux J. Shallow-structure characterization by 2D elastic full-waveform inversion. Geophysics. 2011;76:R81-R93. https://doi.org/10.1190/1.3569798

[15] Devine RD, Barbachyn SM, Thrall AP, Kurama YC. Effect of tripping prefabricated rebar assemblies on bar spacing. J Constr Eng M. 2018;144:04018099. https://doi.org/10.1061/(ASCE)CO.1943-7862.0001559

Structural Health Monitoring: 10APWSHM   Materials Research Forum LLC
Materials Research Proceedings 50 (2025) 252-259   https://doi.org/10.21741/9781644903513-29

Improved reliability assessment of valve hall: Considering failure correlation

Xinzhu Qiao[1,a], Zhihang Xue[2,b], Qiang Xie[1,c] *

[1]Tongji University, Shanghai 200092, China

[2]State Grid Sichuan Electric Power Research Institute, Sichuan Province 610041, China

[a]2410032@tongji.edu.cn, [b]xuezhihang2@163.com, [c]qxie@tongji.edu.cn

Keywords: Valve Hall, Converter Station, Reliability Assessment, D-Vine Copula, Correlation

Abstract. Valve hall is the core system of converter stations and is critical for the operation of power systems. It is necessary to conduct seismic reliability assessments based on real response data. Traditional system reliability assessments are based on the response states of individual structures and the functional dependencies between them, typically assuming that the response states of different structures are completely independent. However, complex interactions exist between system components, meaning that such assumptions may lead to inaccurate assessment results. Copula method provides an effective tool for modeling the dependency structure between variables. By developing a detailed finite element model of the valve hall system, the actual response states of various equipment under seismic conditions are simulated. D-Vine Copula model is used to construct the dependency structure among the equipment response states, enabling a seismic reliability assessment of the valve hall system that accounts for equipment correlations. Comparisons reveal that reliability assessment methods based on the assumption of complete independence among components tend to underestimate the seismic performance of the valve hall system. This further confirms the necessity of considering the correlations between the response states of components in reliability assessments.

Introduction

Power system is one of the critical lifeline infrastructures, primarily consisting of power plants, transmission lines, substations, converter stations, and distribution lines [1,2]. As a vital part of a large and complex lifeline system, converter station plays a crucial role. If it sustains damage during a strong earthquake, it could lead to a collapse of the entire power network and negatively impact post-disaster rescue and reconstruction efforts [3,4,5]. The valve hall system, as the core structure within the converter station, is responsible for the conversion between AC and DC power. It is essential to conduct an accurate seismic reliability assessment to ensure the stable supply of electricity.

A system's reliability is directly determined by the operational states of its components. In existing studies, to simplify the analysis process, it is often assumed that the states of individual equipment are completely independent [6,7]. However, the applicability and accuracy of this assumption require further validation. Given the small spatial distribution of the valve hall system, it can be assumed that all equipment receives consistent external excitation inputs during an earthquake. Additionally, complex interactions exist between the components. These factors imply a degree of correlation among the failure states of the equipment, which can influence the reliability assessment of the system.

Scholars have conducted extensive research on the issue of failure dependencies within systems and have made significant progress. Gao et al. [8] considered the problem of strength degradation in mechanical transmission system components and calculated the time-varying reliability of the

Structural Health Monitoring: 10APWSHM	Materials Research Forum LLC
Materials Research Proceedings 50 (2025) 252-259	https://doi.org/10.21741/9781644903513-29

system using a dynamic fault correlation method. Yu et al. [9] proposed a function to describe redundancy systems and model the correlations between failure modes, further optimizing the system with this function. Yao et al. [10] addressed the issue of correlated failure modes in preventive maintenance and introduced a novel coefficient-based approach to characterize the interactions between failure modes.

Research on Copulas originated with Sklar, and Nelsen et al. [11] provided a systematic introduction to the definition, construction methods, and dependencies of Copula. Copula function allows the joint distribution of multiple random variables to be expressed using the marginal distributions of individual variables. This method has been applied across various engineering fields [12,13,14].

Wang et al. [15] used the Gumbel Copula function to calculate the actual operating state function of components considering failure dependencies and established a reliability allocation model, reducing product design and manufacturing costs. Navarro et al. [16] employed the Clayton Copula function to build a reliability function for the remaining life of multivariate systems. Yun et al. [17] introduced the Gumbel Copula function to describe the nonlinear dependencies between different parts and failure modes of the same part in gearbox systems, developing a reliability evaluation framework for high-speed train gearboxes. An et al. [18] utilized Copula theory for reliability modeling of mechanical series and parallel systems, effectively simplifying multivariate probability modeling and providing reasonable life predictions for mechanical products.

Due to the high cost, large size, and significant mass of the electrical equipment in the valve hall system, conducting seismic shaking table tests is challenging [19]. In this study, a valve hall system of an ±800kV converter station is modeled using the finite element analysis software ABAQUS to obtain the seismic responses of the equipment. These response values are treated as samples, and the D-Vine Copula function is employed for correlation analysis, enabling the seismic reliability assessment of the valve hall system. The results demonstrate the necessity of considering failure dependencies in the reliability evaluation of the valve hall system.

1. Failure Correlation Analysis

For series and parallel systems, the dependencies between the operating states of the components directly influence the system's overall performance. Consider a series system and a parallel system, each composed of n components, with the failure probabilities of individual components given by p_1, p_2, p_3, ..., p_n. Two scenarios are considered: complete independence and complete dependence. For the series system, the difference in system failure probabilities under these two scenarios is denoted as Δp, as shown in Eq. 1. For the parallel system, the difference is expressed in Eq. 2. Since the values of p_1, p_2, p_3, ..., p_n lie within the range [0, 1], Eq. 1$\geq$0, while Eq. 2$\leq$0. This indicates that assuming complete independence overestimates the failure probability of the series system and underestimates that of the parallel system. In reality, the dependencies between components fall between complete independence and complete dependence. To quantify the impact of these dependencies on the system's failure probability, it is necessary to accurately model the dependency structure between component states.

$$
\begin{aligned}
\Delta p &= 1 - \{p_1, p_2, \dots, p_n\}_{max} - (1 - p_1)(1 - p_2) \dots (1 - p_n) \\
&= (1 - p_{max}) - (1 - p_1)(1 - p_2) \dots (1 - p_n).
\end{aligned}
\tag{1}
$$

$$
\begin{aligned}
\Delta p &= p_1 \cdot p_2 \cdot \dots p_n - \{p_1, p_2, \dots, p_n\}_{min} \\
&= p_1 \cdot p_2 \cdot \dots p_n - p_{min}.
\end{aligned}
\tag{2}
$$

Structural Health Monitoring: 10APWSHM
Materials Research Proceedings 50 (2025) 252-259

Materials Research Forum LLC
https://doi.org/10.21741/9781644903513-29

2. Copula method

2.1 Copula function

Copula function is used to link a multivariate joint probability distribution with its one-dimensional marginal distributions. It is particularly suitable for scenarios where dependencies exist between variables, and the analytical form of the joint probability distribution is difficult to obtain directly. For a multivariate random variable $X = [x_1, x_2,..., x_n]$, according to Sklar's theorem, there exists a Copula function C that satisfies Eq. 3. Copula function describes the dependencies between variables from the perspective of their marginal distributions, separating the marginal distributions from their dependency structure. This simplifies the modeling of multivariate joint probability distributions.

$$F(x_1, x_2, ..., x_n) = C(F_1(x_1), F_2(x_2), ..., F_n(x_n)). \tag{3}$$

where $C(\cdot)$ denotes the Copula function, while $F_n(x_n)$ represents the marginal cumulative distribution functions of the variables.

2.2 D-Vine Copula

In the case of multivariate variables, directly estimating parameters for the Copula function can be quite challenging. A more common approach is to use Vine Copula functions. Vine Copula functions utilize a vine structure to layer and combine the probability distribution functions of pairs of variables, enabling the modeling of high-dimensional joint probability distributions. Depending on the characteristics of different vine structures, there are various types of Vine Copulas. D-Vine Copula is a type of Vine Copula characterized by a sequential distribution, as illustrated in Fig. 1. It is suitable for describing structures where variables exhibit adjacent relationships. In the valve hall system, the coupling effect between equipment is highly correlated with spatial distance, making D-Vine Copula function a suitable choice. The multivariate joint density probability function based on D-Vine Copula is represented in Eq. 4.

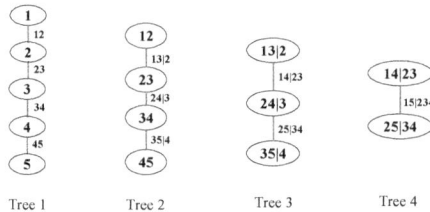

Fig. 1 *D-Vine Copula structure (five dimensions).*

$$f(x) = \prod_{k=1}^{n} f(x_k) \prod_{j=1}^{n-1} \prod_{i=1}^{n-j} c_{i,i+j|1:(i+j-1)} \left(F(x_i | x_{i+1}, ..., x_{i+j-1}), F(x_{i+j} | x_{i+1}, ..., x_{i+j-1}) \right) \tag{4}$$

where $c(\cdot)$ denotes the probability density function of Copula function. $f(\cdot)$ represents the marginal probability density distribution functions. $F(|)$ signifies the conditional distribution function.

Structural Health Monitoring: 10APWSHM   Materials Research Forum LLC
Materials Research Proceedings 50 (2025) 252-259   https://doi.org/10.21741/9781644903513-29

3. Correlation structure modeling

3.1 Response data acquisition

The valve hall system within the converter station is divided into two categories: high voltage valve halls and low voltage valve halls, totaling four valve halls arranged symmetrically. Each valve hall independently handles 25% of the power output without mutual interference. The main equipment within each valve hall includes valve towers, insulators, wall bushings, and connection fittings. Within a single valve hall, there exists a high degree of functional dependency among the internal equipment, meaning that the failure of any single component can compromise the overall functionality of the valve hall. Thus, the components within a single valve hall can be regarded as having a series relationship in terms of function. Given the substantial size and cost of the valve hall system, obtaining response data for each component through shaking table tests is impractical.

In this study, the finite element software Abaqus is utilized to model both the high voltage and low voltage valve halls. In the finite element model, B31 beam elements are used for numerical simulation of the beams, columns, and braces of the valve hall. S4R shell elements are employed to simulate components such as the valve tower equipment layer and mounting plates. Additional equipment is simplified into concentrated or distributed mass based on its mass distribution characteristics, and the connection conditions for components are set according to actual scenarios as rigid or pinned connections. The system's damping ratio is set at 2%. The finite element model and coordinate system are illustrated in Fig. 2.

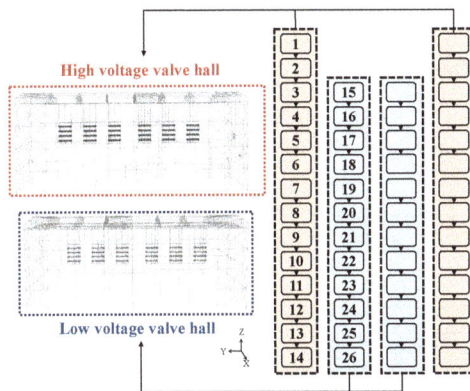

Fig. 2 *Finite element model and equivalent network of the valve hall.*

The design basis acceleration for the site where the valve hall is located is 0.4 g. Based on the demand response spectrum of the site, 30 sets of ground motion records containing three directional components are selected from the Pacific Earthquake Engineering Research Center (PEER) as input for ground motion. Through simulation calculations, the response values of each component in the valve hall system under seismic conditions are obtained. A preliminary analysis identifies 26 components in the high voltage and low voltage valve halls with non-zero probabilities of failure as the objects of analysis, as shown in Fig. 2.

3.2 Node structure and optimal Copula function

To effectively utilize D-Vine Copula for constructing the dependency structure of variables, it is essential to determine the node structure of D-Vine Copula. D-Vine Copula is a chain-dependent model, where the node structure of the k-th tree is entirely reliant on the node structure of the $(k-1)$-th tree. Thus, the overall node structure of the D-Vine Copula is fully determined by the first

Structural Health Monitoring: 10APWSHM   Materials Research Forum LLC
Materials Research Proceedings 50 (2025) 252-259   https://doi.org/10.21741/9781644903513-29

tree. In a Vine Copula model, different node structures dictate the pairing order of the variables, which directly affects both the modeling outcomes and computational efficiency.

For a D-Vine Copula model with a variable set $\{x_1, x_2, \ldots, x_n\}$, the optimal structure of the first tree should maximize the overall correlation among all variables. This problem can be equivalently described as finding a directed path that connects n vertices, containing n-1 edges. The weight of each edge corresponds to the strength of dependency between the vertices on either side. The objective is to determine a Hamiltonian path that maximizes the total weight of the path. Given the vast solution space, a classic heuristic algorithm, genetic algorithm is employed to solve this problem.

Considering the nonlinear correlation among variables, Kendall's rank correlation coefficient is utilized to quantify the dependency strength between nodes, as shown in Eq. 5. For any given path $P=\{x_{i_1}, x_{i_2}, \ldots, x_{i_n}\}$, a corresponding fitness indicator Q is defined, as detailed in Eq. 6. The goal is to maximize Q, using the order of variables as genetic coding, and applying selection, crossover, and mutation operations until the Q value converges to a stable state. The selection strategy employs roulette wheel selection, single-point crossover, and swap mutation. The flowchart of the genetic algorithm and the final node order obtained are presented in Fig. 3.

Once the node structure of the first tree is determined, the structures of subsequent trees are uniquely established. AIC criterion is utilized to select the optimal Copula function types for each connection edge in the D-Vine Copula, and the maximum likelihood estimation method is employed to determine the optimal parameters for each Copula function. Finally, the complete D-Vine Copula model is successfully constructed.

$$\tau = 2(N_c - N_d)/n(n-1) \tag{5}$$

where N_c represents the total number of concordant pairs in the dataset, and N_d represents the total number of discordant pairs in the dataset.

$$Q = \sum_{k=1}^{n-1} \tau\left(x_{i_k}, x_{i_{k+1}}\right) \tag{6}$$

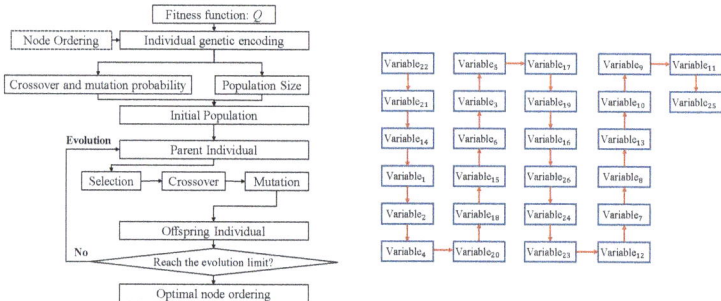

(a) Flowchart of genetic algorithm.   (b) Optimal node order.

Fig. 3 *Flowchart of genetic algorithm and optimal node order.*

4. Reliability assessment

Through finite element simulation analysis, the failure probabilities of 26 pieces of equipment under a PGA of 0.4 g can be obtained, as shown in Table 1. In traditional reliability analysis, the assumption is that the states of the equipment are independent of each other. In this case, the failure

Structural Health Monitoring: 10APWSHM   Materials Research Forum LLC
Materials Research Proceedings 50 (2025) 252-259   https://doi.org/10.21741/9781644903513-29

probabilities of each valve hall and the expected value of the overall power output can be directly determined using the probability multiplication rule. Once the correlation structure of equipment responses is established using D-Vine Copula model, a large number of correlated response samples can be generated based on this correlation structure. This enables a reliability assessment of the valve hall system that takes failure correlations into account. The comparison of the results from the two reliability assessment methods is shown in Fig. 4.

According to Fig. 4(a), under both independent and correlated states, the failure probability of the high voltage valve hall is higher than that of the low voltage valve hall, indicating worse seismic performance. This is due to the larger size of the equipment within the high voltage valve hall. Additionally, when considering the correlations in equipment failures, the failure probabilities of both valve halls decrease, by 22% and 8% respectively. Fig. 4(b) shows that the expected value of the power output of the valve hall system increases by 14.9% after considering correlations. In summary, assuming that the states of the equipment are completely independent significantly underestimates the seismic reliability of the valve hall system. It is necessary to conduct seismic performance assessments based on the consideration of the correlations in equipment states.

Table. 1 Equipment failure probability f.

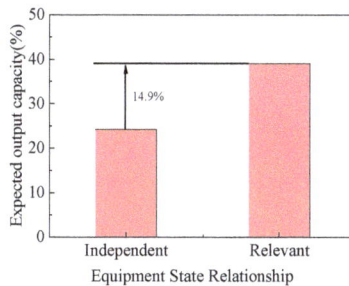

Equipment	1	2	3	4	5	6	7	8	9	10	11	12	13
f	0.06	0.00	0.13	0.19	0.00	0.13	0.00	0.09	0.19	0.16	0.31	0.31	0.03
Equipment	14	15	16	17	18	19	20	21	22	23	24	25	26
f	0.34	0.03	0.06	0.03	0.06	0.03	0.09	0.09	0.13	0.13	0.13	0.13	0.03

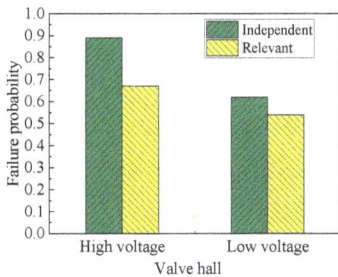

(a) Valve hall failure probability.   (b) Expected output capacity.
Fig. 4 Reliability assessment comparison.

Summary

This paper utilizes D-Vine Copula model to construct the correlation structure between the response states of equipment in the valve hall system under earthquake and conducts a reliability assessment. The following conclusions are drawn.

1. The correlation between equipment states affects system reliability. For serial systems, assuming that the states of equipment are completely independent tends to overestimate the system's failure probability, whereas the opposite is true for parallel systems.

2. D-Copula function can effectively model the correlation structure of the response states of various pieces of equipment in the valve hall system.

Structural Health Monitoring: 10APWSHM   Materials Research Forum LLC
Materials Research Proceedings 50 (2025) 252-259   https://doi.org/10.21741/9781644903513-29

3. Assuming that equipment states are completely independent significantly underestimates the seismic reliability of the valve hall system.

References

[1] SCHIFF A J. Northridge earthquake : lifeline performance and post-earthquake response1995[M]//ASCE, 1995, 16.

[2] XIE Q, ZHU R. Damage to electric power grid infrastructure caused by natural disasters in China[J]. IEEE Power and Energy Magazine, 2011, 9(2):28-36. https://doi.org/10.1109/MPE.2010.939947

[3] MOHAMED A, MOUSTAFA K M M. Structural performance of porcelain and polymer post insulators in high voltage electrical switches [J]. Journal of Performance of Constructed Facilities, 2016, 30(5). https://doi.org/10.1061/(ASCE)CF.1943-5509.0000848

[4] He C, Liu R, He Z. Seismic vulnerability assessment on porcelain electrical equipment based on Kriging model[J]. Structures, 2023, 55: 1692-1703. https://doi.org/10.1016/j.istruc.2023.06.134

[5] Li J, Wang T, Shang Q. Probability-based seismic reliability assessment method for substation systems[J]. Earthquake Engineering & Structural Dynamics, 2019, 48(3): 328-346. https://doi.org/10.1002/eqe.3138

[6] Liang H, Xie Q. System Vulnerability Analysis Simulation Model for Substation Subjected to Earthquakes[J]. IEEE Transactions on Power Delivery, 2021: 1-1.

[7] Liu X, Zheng S, Wu X, et al. Research on a seismic connectivity reliability model of power systems based on the quasi-Monte Carlo method[J]. Reliability Engineering & System Safety, 2021, 215: 107888. https://doi.org/10.1016/j.ress.2021.107888

[8] Gao P, Xie L Y, Pan J. Reliability and availability models of belt drive systems considering failure dependence[J]. Chinese Journal of Mechanical Engineering, 2019, 32(2): 133-144. https://doi.org/10.1186/s10033-019-0342-x

[9] Yu H, Chu C, Chatelet E. Availability optimization of a redundant system through dependency modeling[J]. Applied Mathematical Modelling, 2014, 38(19-20): 4574-4585. https://doi.org/10.1016/j.apm.2014.03.006

[10] Yao Y Z, Meng C, Wang C. Optimal preventive maintenance policies for multi-unit system considering failure interactions[J]. The International Journal of Advanced Manufacturing Technology, 2013, 19(12): 2976-2981.

[11] Nelsen R B. Copulas and Association[C].Advances in Probability Distributions with Given Marginals, Dor- drecht, 1991 https://doi.org/10.1007/978-94-011-3466-8_3

[12] Wen J, Li X, Xue J. Feasibility evaluation of Copula theory for substation equipment with multiple nonlinear-related seismic response indexes[J]. Reliability Engineering & System Safety, 2024, 247: 110132. https://doi.org/10.1016/j.ress.2024.110132

[13] Wen J, Li X, Zhu Y. Improved seismic risk evaluation for high-voltage switchgear equipment: A copula-based framework considering joint failure modes[J]. Earthquake Engineering & Structural Dynamics, 2024, 53(2): 694-716. https://doi.org/10.1002/eqe.4041

[14] Lyu M Z, Fei Z J, Feng D C. Copula-based cloud analysis for seismic fragility and its application to nuclear power plant structures[J]. Engineering Structures, 2024, 305: 117754. https://doi.org/10.1016/j.engstruct.2024.117754

[15] Wang H, Zhang Y M, Yang Z. A reliability allocation method of CNC lathes based on copula failure correlation model[J]. Chinese Journal of Mechanical Engineering, 2018, 31(1): 1-9. https://doi.org/10.1186/s10033-018-0303-9

[16] Navarro J, Durante F. Copula-based representations for the reliability of the residual lifetimes of coherent systems with dependent components[J]. Journal of Multivariate Analysis, 2017, 158: 87-102. https://doi.org/10.1016/j.jmva.2017.04.003

[17] Liu Y, Chen Y. Dynamic Reliability Evaluation of High-speed Train Gearbox Based on Copula Function[J]. IEEE Access, 2022, 10: 51792-51803. https://doi.org/10.1109/ACCESS.2022.3174043

[18] An H, Yin H, He F. Analysis and application of mechanical system reliability model based on copula function[J]. Polish Maritime Research, 2016, 23: 187-191. https://doi.org/10.1515/pomr-2016-0064

[19] Liang H, Xie Q, He C. Seismic Performance and Aseismic Measures of ±800 kV EHV Wall Casing-Valve Hall System[C]//2019 IEEE 3rd Conference on Energy Internet and Energy System Integration (EI2). IEEE, 2019: 2246-2251. https://doi.org/10.1109/EI247390.2019.9062139

Structural Health Monitoring: 10APWSHM   Materials Research Forum LLC
Materials Research Proceedings 50 (2025) 260-268   https://doi.org/10.21741/9781644903513-30

Testbed assessment of wave propagation analysis:
Towards scum characterisation in covered anaerobic lagoon

Dat Nha BUI[1,*], Thomas KUEN[2], Shouxun LU[1], L.R. Francis ROSE[1],
Wing Kong CHIU[1]

[1]Department of Mechanical & Aerospace Engineering, Monash University, Clayton, VIC 3008,
Australia

[2]Melbourne Water Corporation, 990 La Trobe Street, Docklands, VIC 3008, Australia

*Dat.Bui1@monash.edu

Keywords: Scum, Scum Assessment, Wave Propagation Analysis, Dispersion
Characteristics, Anaerobic Lagoon, Wastewater, Wastewater Treatment, Large-Scale
Floating Geomembranes

Abstract. The assessment of the state and extent of scum conditions under floating covers is crucial for operational management and structural health monitoring of wastewater treatment plants. Advancing beyond our previous research that employed a single accelerometer to examine the covered scum's frequency response, we propose a more comprehensive approach using wave propagation analysis with an array of sensors to gain a deeper understanding of the complex cover-scum system. Traditional wave propagation analysis applications typically target either large length scale, low-frequency scenarios as seen in seismic surface waves analysis, or small length scale, high-frequency scenarios as observed in ultrasonic guided waves analysis. However, in anticipation of a mid-scale of sub-metre depth and sub-hundred hertz frequency range for early-stage scum in covered anaerobic lagoons, we established a water bladder testbed to evaluate the adaptability of the multi-sensor approach. In this study, we deployed 16 accelerometers on the bladder filled with water to three distinct water levels and used a durometer to gauge the shore hardness. We applied impulse excitations, recorded the wavefields and examined dispersion diagrams. Our findings showed that the wave was dispersive across all tested shore hardness levels, with the phase velocity decreasing as the frequency increased. Notably, a higher shore hardness corresponded to a higher phase velocity which can be clearly differentiated on the dispersion diagrams. By selecting different subsets of accelerometers, we further explore the impact of sensor array length and source offset on the quality of the analysis. Our results confirmed that the sensor array should be sufficiently close to the excitation source to capture high-frequency component information, which attenuates rapidly over distance. For a given number of sensors, optimising the sensor pitch is important to maintain an optimal balance between wavenumber resolution and maximum resolvable wavenumber. These insights could inform the design and deployment of sensor arrays to characterise varying hardness levels of scum. This research offers potential for monitoring and assessing the state of this scum and enhancing our understanding and management of large-scale wastewater treatment lagoons covered with floating covers.

Introduction

The assessment of the state and extent of scum conditions under floating covers is crucial for operational management and structural health monitoring of wastewater treatment plants. Floating covers, often made from high-density polyethylene (HDPE), are used to create an airtight seal over anaerobic lagoons, facilitating the anaerobic digestion of sewage and the collection of methane-rich biogas [1]. The accumulation of scum, which consists of fibrous materials, fats, floating solids, and buoyed sludge, can impede biogas collection and compromise the structural integrity of the

Structural Health Monitoring: 10APWSHM   Materials Research Forum LLC
Materials Research Proceedings 50 (2025) 263-271   https://doi.org/10.21741/9781644903513-30

covers [2]. Building on our previous research that employed a single accelerometer to examine the covered scum's frequency response [2], we propose a more comprehensive approach using wave propagation analysis with an array of sensors to gain a deeper understanding of the complex cover-scum system.

Wave propagation analysis using a linear array of sensors is not a new technique. In seismology, the analysis of surface waves such as Rayleigh and Love waves has been used for decades to assess subsurface properties and study earthquakes [3, 4]. Among various methods, the multi-channel analysis of surface waves, introduced in 1999 by Park, Miller [5], has become one of the most widely used techniques [4, 6, 7]. In structural health monitoring and non-destructive evaluation, the analysis of guided waves, particularly Lamb waves, has been extensively explored to detect damages in structures [8-10]. Comprehensive guidelines and textbooks providing both practical information and theoretical details around the multi-channel analysis of surface waves [4, 11] or guided waves [8] have been developed. With this available knowledge, we aim to adapt the above approaches into scum condition assessments.

However, several practical challenges must be addressed before these approaches can be adapted and deployed in the field. The first challenge is the mismatch in frequency range and length scale of interest. The multi-channel analysis of surface waves for seismological applications typically targets large length scales, ranging from several metres up to several kilometres, at very low frequencies of five hertz down to millihertz [11]. In contrast, the Lamb wave ultrasonic method operates in the kilohertz to megahertz frequency range and is typically used at a length scale of millimetres or sub-millimetres [8, 10]. For characterising early-stage scum in covered anaerobic lagoons, informed by our previous study, the frequency range of interest is around ten to one hundred hertz [2], while the length scale is of one metre or less, guided by the depth of early-stage scum. Consequently, a sensing system targeting the right bandwidths to appropriately capture the wavefields is needed. Other challenges include the deployability and portability of the instrument system. The potential explosive atmosphere on top of covered anaerobic lagoons limits the power levels that can be used. The massive size of these lagoons makes the availability of a host computer impractical. Bonding the sensors onto the cover is also a challenge as techniques like using adhesive may affect the integrity of the cover. Under such conditions, sensing techniques using scanning laser Doppler vibrometers or PZT (lead zirconate titanate) sensor arrays as in Lamb wave ultrasonics or seismic geophones may become irrelevant.

To address these challenges, we deliberately designed and assembled an array of 16 analogue MEMS accelerometers along with necessary hardware creating a low-cost, low-power, highly portable instrument system. In this study, we established a water bladder testbed to evaluate the adaptability of the wave propagation approach using this custom-built system to assess the state of scum particularly using wavenumber-frequency analysis. By varying the firmness of the bladder with different levels of filled water, we could simulate the changes in the scum hardness under the cover. We also studied the effects of important parameters such as the number of sensors, sensor pitch and source offset on the quality of analysis, allowing for strategic deployment in the field to obtain optimal results.

Materials and Methods

Instrumentation

In this study, we assembled an array of 16 triaxial analogue MEMS accelerometers and tested its capability to resolve propagating waves. The sensing chips of these MEMS accelerometers are of the ADXL356C type, acquired prepopulated on EVAL-ADXL356CZ evaluation printed circuit boards (PCBs) provided by the same manufacturer of the chips (Figure 1a). These PCBs were also prepopulated with antialiasing filters of 50 Hz cut-off frequency on all x, y and z channels externally to the ADXL chip by the manufacturer, however were modified with the removal of the

Structural Health Monitoring: 10APWSHM   Materials Research Forum LLC
Materials Research Proceedings 50 (2025) 263-271   https://doi.org/10.21741/9781644903513-30

filter capacitors. This modification allows the signals to be output directly from the base chip at its native bandwidth of 1.5 kHz limited by the internal antialiasing filters.

The ADXL356C chips can operate at two different full-scale ranges: ±10 g ("low range") and ±40 g ("high range"). Accordingly, three of the sensors were set to high range, while the remaining 13 were set at low range.

The PCBs were attached to 3-metre thin cables and securely fastened in custom-built aluminium casings (Figure 1b). Plastic suction cups were tightly threaded onto the sensor assemblies' casings so that they can easily be attached to and removed from the top surface of the water bladder or any flat surface under test. A custom-made power supply and switch box was manufactured to simultaneously supply regulated DC voltage to the 16 assembled sensors and interface with the data acquisition device.

To record the signals, a 16-channel DI-710-UHS data acquisition device was utilised. This device, with its capability of recording data directly onto a memory card in standalone mode (i.e., without a host computer), offers great portability. In the experiments, it was set at its maximum throughput of 14,400 samples per second, corresponding to a 900 Hz per channel sampling rate.

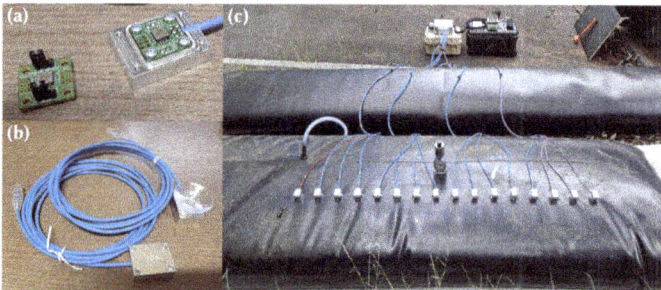

Figure 1. (a) EVAL-ADXL356CZ evaluation boards, (b) the fully assembled sensors and (c) the sensor array deployed on the water bladder.

Experiment setup

A rainwater bladder filled with water was used as a testbed for the wave propagation analysis. A schematic of the experiment setup is shown in Figure 2. The water bladder, manufactured by Fleximake, was made of PVC (polyvinyl chloride) reinforced with high-tenacity 1000 denier fabric. When fully deflated, the bladder's flat measurement is 1 m × 4 m. When fully inflated, its height is approximately 0.45 m. When the bladder was filled with water, any trapped air was entirely released through a breather hole at the top before closing the attached valve.

Impulse excitations were generated by striking a mallet on a hardwood board placed against one end of the bladder, acting as a large aperture for the excitations. An array of 16 accelerometers was attached to the top surface of the bladder (Figure 2a) to measure the wavefield in terms of vertical accelerations in the space-time domain, which was recorded by a 16-channel data acquisition device. The sensor pitch dx was selected as 100 mm providing a wavenumber resolution of 0.625 m^{-1} and a Nyquist wavenumber of 5 m^{-1} (Figure 3a), or a wavelength range of 0.2 m to 1.6 m. As suggested in the guidelines [4], the distance from the hardwood board (i.e. the source of excitations) to the first sensor of the array x_0 was selected as 400 mm (Figure 3a), which is four times the sensor pitch, to avoid near-field effects while ensuring a good signal-to-noise ratio at the last sensor of the array.

The experiments were conducted at three water levels, referred to as "low, moderate" and "high". A shore durometer (an instrument used to measure material hardness by assessing

Structural Health Monitoring: 10APWSHM
Materials Research Proceedings 50 (2025) 263-271

Materials Research Forum LLC
https://doi.org/10.21741/9781644903513-30

resistance to indentation) with a type "A" indenter was used to measure the corresponding shore hardness values at the top surface while the bladder remained stationary. It is worthwhile to note that gauging the internal pressure of the bladder or measuring the volume of filled water could be more direct measurements; however, measuring shore hardness with a durometer is simple yet effective enough to quantitatively differentiate the levels of water. Subsequently, the durometer was removed, and impulse excitations were applied four times at each water level, with sufficient time intervals between each mallet blow to ensure that one blow did not influence the next. Waves propagating across the bladder were recorded by the accelerometer array.

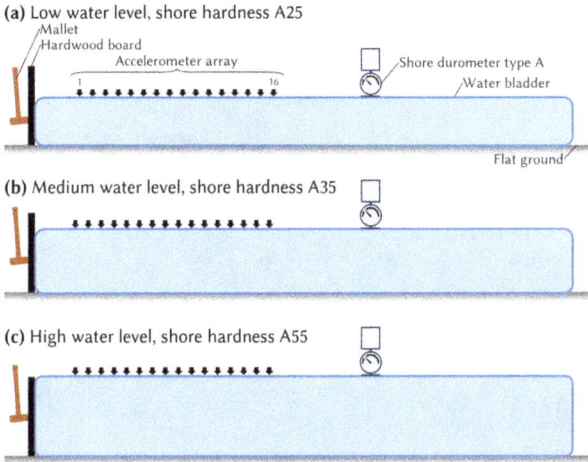

Figure 2. Experiment setup.

Data processing
Standard discrete two-dimensional Fourier transformation (2D FFT) is used to transform wavefield data sampled in the time-space domain in terms of acceleration $a(t_m, x_n)$ into frequency-wavenumber domain $A(f_p, \tilde{v}_q)$, which is given in the below equation:

$$A(f_p, \tilde{v}_q) = \mathcal{F}_{2D}[a(t_m, x_n)] = \sum_{n=1}^{N} \sum_{m=1}^{M} a(t_m, x_n) \exp\left(-\mathrm{j}mp\frac{2\pi}{M} + \mathrm{j}nq\frac{2\pi}{N}\right).$$

Here, t_m [s] and x_n [m] represent the m-th sampling time and n-th sampling location in space, respectively. M and N denote the total numbers of samples in time and space. Additionally, f_p [Hz] and $\tilde{v}_q$ [m^{-1}] are the p-th and q-th discretised frequency and wavenumber (not to be confused with angular wavenumber $k = 2\pi\tilde{v}$ [rad·m^{-1}]) respectively.

To reduce leakage, a hamming window is applied along the spatial dimension while an exponential window is applied along the temporal dimension. The attenuation of the exponential window is set to be 2% at the end of the data length.

$A(f_p, \tilde{v}_q)$ is the dispersion spectrum of the wavefield in the frequency-wavenumber space. This spectrum can be mapped into phase velocity-frequency space simply by using the relation $c_p = f/\tilde{v}$. Here, c_p is the phase velocity (where non-italic subscript 'p' stands for "phase" and should not be confused with the discrete frequency index 'p'). For each point $(f_p, \tilde{v}_q)$, its corresponding phase velocity $c_p(p, q) = f_p/\tilde{v}_q$ is calculated; then, it is mapped to point $(c_p(p, q), f_p)$ in the new space:

$$A(f_p/\tilde{v}_q, f_p) = A(f_p, \tilde{v}_q).$$

Structural Health Monitoring: 10APWSHM   Materials Research Forum LLC
Materials Research Proceedings 50 (2025) 263-271   https://doi.org/10.21741/9781644903513-30

The averaging of a set of L data records (e.g., a set of each data record with respect to every mallet strike collected for the same water level) is done in the frequency-wavenumber domain after each dispersion spectrum $A_l(f_p, \tilde{v}_q)$ of l-th record is normalised by its maximum value:

$$\bar{A}(f_p, \tilde{v}_q) = \frac{1}{L} \sum_{l=1}^{L} \frac{A_l(f_p, \tilde{v}_q)}{A_{l,max}}, \text{ where } A_{l,max} = \max_{p,q} A_l(f_p, \tilde{v}_q).$$

To further investigate the effects of the source offset and sensor pitch on the resolution of the dispersion characteristics, signals from different subarrays of the accelerometers, as detailed in Table 1 and illustrated on Figure 3, were extracted from the recorded wavefields and analysed. The values of effective sensor number, equivalent source offset and sensor pitch of each subarray are provided in Table 1.

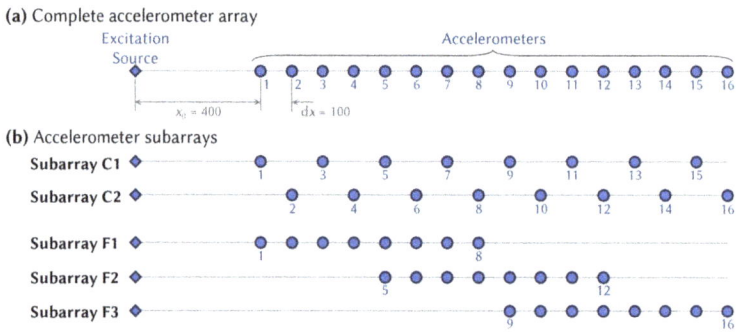

Figure 3. Complete accelerometer array (a) and extracted subarrays (b).

Table 1. Subarrays of accelerometers and their effective configurations.

Subarray	Selected Sensors	Number of Sensors N_x	Source Offset x_0 [mm]	Sensor Pitch dx [mm]	Wavenumber Resolution $d\tilde{v}$ [m^{-1}]	Nyquist Wavenumber $\tilde{v}_{Nyq}$ [m^{-1}]
Complete array	[1...16]	16	400	100	0.625	5
C1	[1, 3, 5, 7, 9, 11, 13, 15]	8	400	200	0.625	2.5
C2	[2, 4, 6, 8, 10, 12, 14, 16]	8	500	200	0.265	2.5
F1	[1...8]	8	400	100	1.25	5
F2	[5...11]	8	800	100	1.25	5
F3	[9...16]	8	1200	100	1.25	5

Results

Table 2 summarises the durometer readings at each tested water level. Apparently, the bladder becomes firmer with more water filled into it.

Table 2. Shore hardnesses measured on the water bladder top surface at tested water levels.

Water Level	Shore Hardness
Low	A25
Moderate	A35
High	A55

16-accelerometer array

Figure 4 shows the waveforms, or traces, of vertical acceleration recorded at each accelerometer. Averaged dispersion spectra of the three tested water levels are visualised in Figure 5 in the

Structural Health Monitoring: 10APWSHM   Materials Research Forum LLC
Materials Research Proceedings 50 (2025) 263-271   https://doi.org/10.21741/9781644903513-30

wavenumber-frequency space and mapped into frequency-phase velocity space in Figure 6. From the averaged spectrum associated with each water level, a frequency-dependent ridge was also traced.

It can be seen on Figure 4 that the attenuation over distance is stronger, i.e. the signals reaching the downstream sensors are weaker, at higher shore hardness. The higher the water lever, the shorter the duration the disturbance propagates across all the sensors indicating a faster travelling wave. The traces oscillate more rapidly with higher water level indicating that the signals contain higher frequency components. This observation is further confirmed by the dispersion spectra (Figure 5), where the high intensity regions range from 10 to 30 Hz, 10 to 40 Hz and 40 to 80 Hz respectively for low, moderate and high-water level.

In all tested cases, the waves exhibit dispersive behaviour, with phase velocity decreasing as frequency increases (Figure 6). A higher shore hardness corresponds to a higher phase velocity which can be clearly differentiated on the dispersion diagrams. Specifically, at the lowest shore hardness level (A25), the phase velocity decreases from 15 m/s at 10 Hz to 10 m/s at 30 Hz. At the moderate shore hardness level (A35), the phase velocity ranges from 25 m/s at 15 Hz down to 15 m/s at 40 Hz. For the highest shore hardness level (A55), the phase velocity decreases from 80 m/s at 40 Hz to 20 m/s at 80 Hz.

8-accelerometer subarrays
Figure 7 shows the dispersion spectra calculated using data extracted from different subarrays of accelerometers as detailed in Table 1 at the three tested water levels. All these subarrays have 8 evenly spaced accelerometers, half the number of sensors in the complete array.

Subarrays F1, F2 and F3 ("Fs") used a sensor pitch of 100 mm, the same as in the original array, while subarrays C1 and C2 ("Cs") used a coarser pitch of 200 mm. This configuration allows the Cs subarrays to achieve the same wavenumber resolution of 0.625 m^{-1} as the original array, with the trade-off of halving the Nyquist wavenumber $\tilde{v}_{Nyq}$ from 5 m^{-1} to 2.5 m^{-1}. However, the aliased information in the range of $\tilde{v}_{Nyq}$ to $2\tilde{v}_{Nyq}$ (2.5 m^{-1} to 5 m^{-1}) could be fully recovered from the range $-\tilde{v}_{Nyq}$ to 0 by unwrapping the spectra without introducing additional noise [11]. Consequently, for all water levels, the spectra calculated from the Cs subarrays (Figure 7) appeared as similar to those from the complete 16-sensor array (Figure 5). In contrast, the Fs subarrays retain the maximum resolvable wavenumber but provide a poorer wavenumber resolution of 1.25 m^{-1}. Their spectra appeared less distinctive (Figure 7) due to this loss in wavenumber resolution.

The source offset was significantly increased from 400 mm to 800 mm and finally to 1200 mm from the F1 to F2 and F3 subarrays. With this increase, the high-frequency components on the spectra appear weaker for all water levels, especially in the case of high water level. This observation is less pronounced when comparing C2 with C1, as the change in the source offset is smaller, only from 400 mm to 500 mm. At the high water level, the high-intensity lobe (around a wavenumber of 3 m^{-1} and frequency of 70 to 80 Hz) on the spectrum calculated from C1 becomes much weaker in that from C2. At the low and moderate water levels, only a subtle change in the spectra between C1 and C2 across their high-frequency components can be observed.

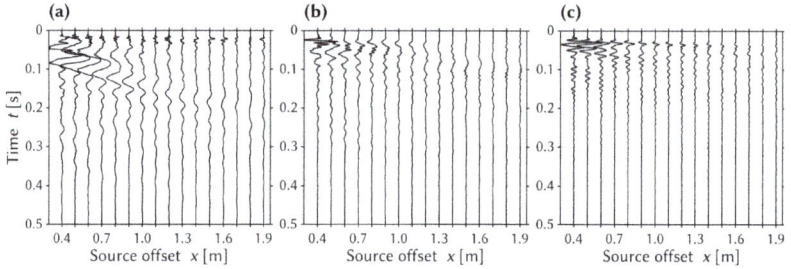

Figure 4. Wavefields recorded at (a) low, (b) moderate and (c) high water level.

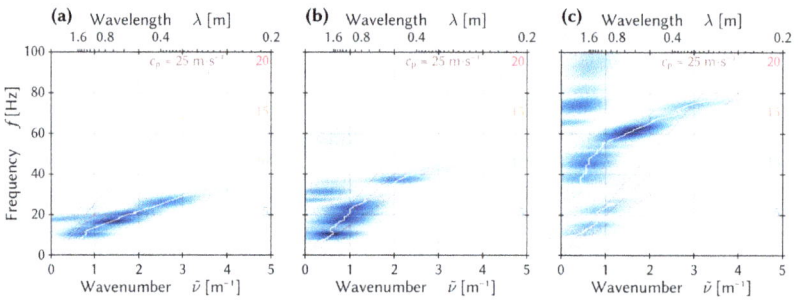

Figure 5. Dispersion spectra in wavenumber-frequency space corresponding to (a) low,
(b) moderate and (c) high water level. White traces represent the ridges of the spectra traced
along the frequency dimension; diagonal lines are constant phase velocity lines.

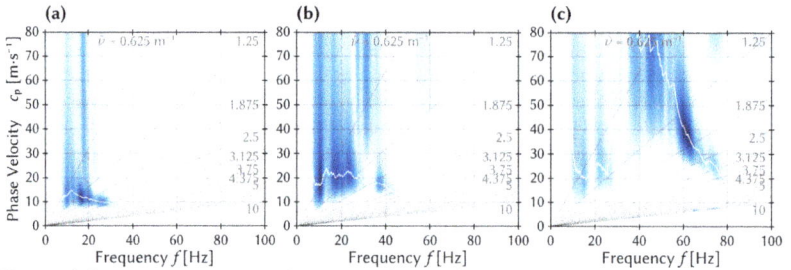

Figure 6. Dispersion spectra in frequency-phase velocity space corresponding to (a) low,
(b) moderate and (c) high water level. White traces represent the ridges of the spectra traced
along the frequency dimension; diagonal lines are constant wavenumber lines.

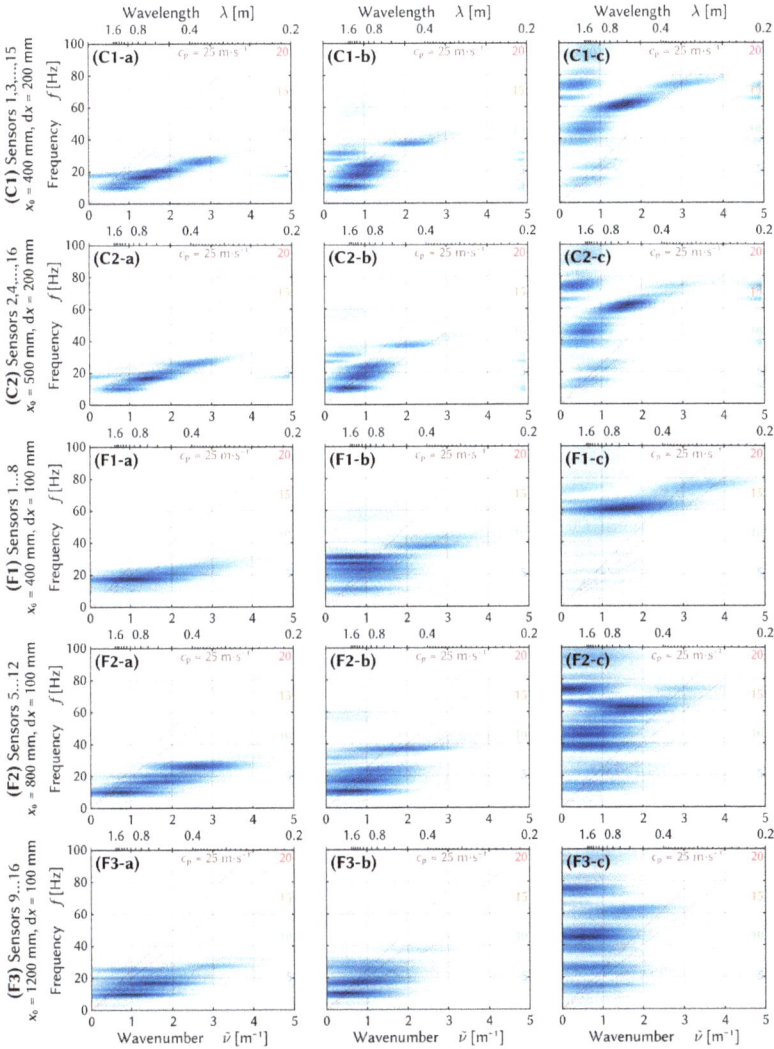

Figure 7. Frequency-wavenumber spectra calculated using data extracted from subarrays of sensors 1,3,...,15 (C1); 2,4,...,16 (C2); 1...8 (F1); 5...12 (F2) and 9...16 (F3) corresponding to (a) low, (b) moderate and (c) high water level.

Summary and discussion

This study confirmed the functionality of our low-cost, custom-built linear accelerometer array for sampling wavefields in the space-time domain for wavenumber-frequency analysis. Our findings showed that the wave was dispersive across all tested shore hardness levels, with the phase velocity decreasing as the frequency increased. Firmer water bladder conditions corresponded to higher phase velocities which can be clearly differentiated on the dispersion diagrams. This suggests that dispersion diagrams can differentiate scum of different hardness, with phase velocity serving as a quantitative measure.

We also gained insights for optimal sensor deployment by tuning key parameters: number of sensors, sensor spacing, and source offset. With a fixed number of sensors, increasing sensor pitch improves wavenumber resolution but also decreases the cut-off wavenumber. In our experiments, the Cs subarrays with only half the number of sensors delivered spectra as clear as the original array. This was achieved by doubling the sensor spacing and unwrapping the spectra up to twice Nyquist wavenumber. This confirms that the maximum resolvable wavenumber is twice the Nyquist wavenumber if sampling wavefields of only positive travelling waves. Additionally, source offset should be optimised to avoid near-field effects while ensuring that high-frequency components still reach the last sensor with a good signal-to-noise ratio.

These insights allow for strategic field deployment of the sensor array, paving the way for our future work to characterise scum of varying hardnesses in the covered anaerobic lagoons using the wavenumber-frequency analysis. This research offers potential for enhancing our understanding and management of large-scale covered wastewater treatment lagoons.

Acknowledgements

Gratitude is expressed to the Monash Electrical and Computer Systems Engineering Workshop for their invaluable consultation and assistance in the development and manufacturing of the sensing system. The in-kind contributions from Melbourne Water are gratefully acknowledged. Financial support from the Monash Graduate Scholarship is also deeply appreciated.

References

[1] DeGarie, C.J., et al., *Floating geomembrane covers for odour control and biogas collection and utilization in municipal lagoons.* Water Science and Technology, 2000. **42**(10-11): p. 291-298.

[2] Bui, D.N., et al. *A Mechanical method of classifying the state of solid matter beneath a floating cover over an anaerobic lagoon.* in *Materials Research Proceedings.* 2023.

[3] Maranò, S., D. Fäh, and Y.M. Lu, *Sensor placement for the analysis of seismic surface waves: sources of error, design criterion and array design algorithms.* Geophysical Journal International, 2014. **197**(3): p. 1566-1581.

[4] Foti, S., et al., *Guidelines for the good practice of surface wave analysis: a product of the InterPACIFIC project.* Bulletin of Earthquake Engineering, 2018. **16**(6): p. 2367-2420.

[5] Park, C.B., R.D. Miller, and J. Xia, *Multichannel analysis of surface waves.* GEOPHYSICS, 1999. **64**(3): p. 800-808.

[6] Rahimi, S., C.M. Wood, and D.P. Teague, *Performance of Different Transformation Techniques for MASW Data Processing Considering Various Site Conditions, Near-Field Effects, and Modal Separation.* Surveys in Geophysics, 2021. **42**(5): p. 1197-1225.

[7] Ali, A., et al., *Multi-channel analysis of surface waves (MASW) using dispersion and iterative inversion techniques: implications for cavity detection and geotechnical site investigation.* Bulletin of Engineering Geology and the Environment, 2021. **80**(12): p. 9217-9235.

[8] Rose, J.L., *Ultrasonic Guided Waves in Solid Media.* 2014, Cambridge: Cambridge University Press.

[9] Yu, L., C.A.C. Leckey, and Z. Tian, *Study on crack scattering in aluminum plates with Lamb wave frequency–wavenumber analysis.* Smart Materials and Structures, 2013. **22**(6): p. 065019.

[10] Tian, Z., et al., *Guided wave imaging for detection and evaluation of impact-induced delamination in composites.* Smart Materials and Structures, 2015. **24**(10): p. 105019.

[11] Foti, S., et al., *Surface Wave Methods for Near-Surface Site Characterization.* 2014.

Structural Health Monitoring: 10APWSHM
Materials Research Proceedings 50 (2025) 269-276

Materials Research Forum LLC
https://doi.org/10.21741/9781644903513-31

Accurate diagnosis of bone degradation-induced prosthesis loosening using harmonic vibration analysis: An experimental study with a simplified model

Qingsong Zhou[1,a*], L. R. Francis Rose[1,b], Benjamin Steven Vien[1,c], Peter R. Ebeling[2,d], Matthias Russ[3,e], Mark Fitzgerald[4,f] and Wing Kong Chiu[1,g]

[1]Department of Mechanical & Aerospace Engineering, Monash University, Clayton, VIC, 3800, Australia

[2]Department of Medicine, School of Clinical Sciences, Monash University, Clayton, VIC, 3800, Australia

[3]The Alfred Hospital, Melbourne, VIC 3004, Australia

[4]The National Trauma Research Institute, Commercial Road, Melbourne, VIC, 3004, Australia

[a]Qingsong.Zhou@monash.edu, [b]louis.rose@monash.edu, [c]ben.vien@monash.edu
[d]Peter.Ebeling@monash.edu, [e]matthias.russ@russorthopaedics.com.au,
[f]M.Fitzgerald@alfred.org.au, [g]wing.kong.chiu@monash.edu

Keywords: Implant Loosening Monitoring, Vibration Analysis, Harmonic, Transfemoral Osseo-integrated Prosthesis

Abstract. The Osseointegration Prosthetic Limb is a transfemoral implant that attaches the prosthesis directly to the bone, providing greater comfort than a traditional socket. However, transferring load to an implant can induce stress shielding and bone resorption, potentially leading to loosening and pain. This experimental study investigates the nonlinear vibration response of a simplified femur-implant model as the basis for a non-invasive strategy to monitor implant loosening. Three femur-implant assemblies were constructed with prescribed levels of interference fit between the bone and implant. To mimic bone degradation due to stress shielding, bone thickness was reduced by material removal, which led to a reduction of contact stability. In the vibration analysis, a static longitudinal gait force was applied at the distal end of the implant to preload it with a load representative of the compression experienced during walking. Subsequently, the model was excited at the resonant frequency of the first torsional vibration mode to assess interfacial contact. Slippage at the interface occurs when the interfacial shear stress exceeds a prescribed critical threshold, leading to the generation of higher-order harmonics in the vibration response. The evolution of these harmonics was analysed in relation to the magnitude of torsional excitation and contact stability. Results suggest that the presence of the higher-order harmonic serves as a reliable indicator of interfacial slippage, and the corresponding load level provides a quantitative index of the implant's load-bearing capacity. These findings demonstrate the feasibility of using nonlinear vibration analysis, as a non-invasive means of assessing the criticality of the bone-implant interface.

Introduction

Osseointegrated prostheses, which are anchored directly to the bone, offer several advantages over conventional socket prostheses. These advantages include improved patient mobility, increased comfort, and improved osseo-perception, which is the ability to sense the prosthetic limb as part of the body [1]. The implant achieves initial mechanical stability through a press-fit technique, where the implant is oversized to ensure a secure fit within the bone remnant. This secure fit provides the essential contact stability required during the early stages of osseointegration [2].

Structural Health Monitoring: 10APWSHM   Materials Research Forum LLC
Materials Research Proceedings 50 (2025) 269-276   https://doi.org/10.21741/9781644903513-31

However, a significant clinical challenge associated with Osseo-integrated prostheses is implant loosening, with stress shielding being a major contributing factor [3-5]. Stress shielding occurs when the presence of an implant alters the natural load transfer pattern within the surrounding bone, resulting in reduced mechanical stimuli for adaptive remodelling. As a result, the bone weakens progressively over time, which can ultimately lead to aseptic implant loosening [6]. This process poses a considerable challenge to the long-term stability and success of osseointegrated prostheses. Periprosthetic cortical thinning [7] is a typical type of bone resorption. It is the resorption around the implant and will raise concerns about outbreak fracture or aseptic loosening. From a vibrational standpoint, cortical thinning not only affects the interfacial stability between the bone and implant but also modifies the modal mass and modal stiffness of the bone, thereby influencing the vibration characteristics of the implant-bone assembly.

Recent advancements have highlighted vibration analysis as a promising diagnostic approach for monitoring implant integrity [8]. The percutaneous nature of these implants allows the integration of external sensors to provide real-time data on the implant's vibration characteristics. Previous studies by Rosenstein, *et al.* [9], Li, *et al.* [10] and have shown that the presence of harmonics in the vibration response can indicate implant loosening in total hip arthroplasty. For example, applying sinusoidal excitation to the knee and analysing the vibration response at the greater trochanter revealed that distortion in the output waveform and the presence of harmonics can broadly categorize the implant status as loose, tight, or moderately loosened [11].

When the implant loses its secure connection with the femoral cavity, harmonics are generated due to the nonlinear interactions at the bone-implant interface. In our previous research [12], we investigated the mechanisms of harmonic generation in an axisymmetric transfemoral implant model in detail. Our experimental results demonstrated that nonlinear slip-stick dynamics at the interface produced odd harmonics when the implant was excited at its torsional modal frequency. However, this method has limited sensitivity in detecting early stages of loosening, as the bone-implant interface remains tightly secured, preventing activation of the nonlinear slip-stick dynamics with low excitation levels. A more sensitive approach is needed to closely monitor implant health, enabling healthcare professionals to optimize rehabilitation plans in a timely manner.

This paper presents an innovative approach for quantitatively evaluating implant loosening caused by cortical thinning. A series of experiments using press-fit bone-implant models were conducted to investigate contact nonlinearity at the bone-implant interface. Instead of replicating the exact mechanical properties of real bones and implants, simplified bone models, each simulating different degrees of cortical thinning, were prepared. The approach involves applying different levels of axial preload, followed by excitation of the torsional vibration mode to induce rotational motion of the implant. The results suggest that the presence of the higher-order harmonics reliably indicates interfacial slippage. Additionally, the corresponding preload level at which harmonics emerge serves as a quantitative index of the load-bearing capacity of the implant, which is directly related to its safety for routine activities such as walking and running.

Experiment Method:

Implant Model:

Three pairs of simplified press-fit constructs were manufactured (Figure 1.c). In each pair, a polyether ether ketone (PEEK) tube (density of 1.31 g/cm³, tensile modulus of 4.2 GPa, compression modulus of 3.4 GPa) represented the bone with an outer diameter of 40mm, an inner diameter of 25mm and a length of 300mm. Three aluminium 2011-T3 rods, each with a diameter of 25 mm and a length of 175 mm, were used to represent the implants. All pairs were initially machined to achieve an identical interference fit of 0.01 mm in the overlap region.

Structural Health Monitoring: 10APWSHM   Materials Research Forum LLC
Materials Research Proceedings 50 (2025) 269-276   https://doi.org/10.21741/9781644903513-31

To simulate varying degrees of cortical thinning, the outer diameter of the tubes was reduced at the distal end. Model A retained an outer diameter of 40 mm, Model B was reduced to 35 mm, and Model C to 30 mm, thereby simulating cortical thinning by removing the outer material of the tubes. Subsequently, the aluminium rods were press-fitted into the PEEK tubes with an insertion depth of 10 mm. This configuration allowed for the analysis of contact variations resulting from different levels of simulated cortical thinning.

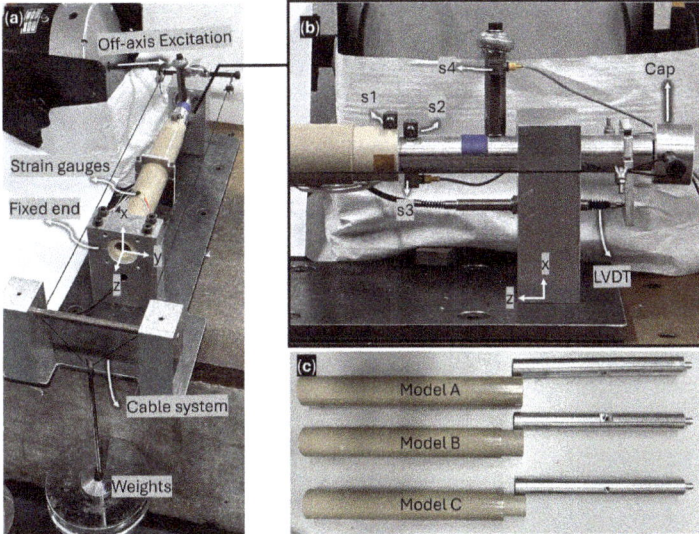

Figure 1 Experiment setup. (a) Test rig set up. (b) Zoom view of the experiment setup showing placement of accelerometers and LVDT. (c) Implant and bone models.

Test Rig Setup:
The bone-implant model was cantilevered at the proximal end of the bone. A cable system was designed to apply an axial compression load to the distal end of the implant. To minimise external influences on the bone-implant structure, a high-damping rubber band was placed inside the cap to isolate the cable system from the model. The load capacity of the implants was validated through push-in tests, during which a progressively increasing compressive load was applied to each implant via a cable system until slippage onset.

The axial compressive force was calculated from the strain measurements recorded by strain gauges attached to the outer surface of the tube. Additionally, a Linear Variable Differential Transformer (LVDT) was positioned to track the implant's inward movement during loading. The load-bearing capacity of the implants corresponds to the axial load at the point of slippage.

Vibration analysis procedure:
The vibration test consisted of two parts. The first part aimed to identify the natural frequency of the torsional vibration mode of the models. A shaker applied broad-band frequency excitation using a random signal. The vibration response in the y-direction was recorded by accelerometers s2 and s3 (refer to Figure 1), placed diametrically opposite each other on the implant to distinguish torsional modes from bending modes. A phase angle of 180 degrees between s2 and s3 indicates the sensors are moving in opposite directions, showing rotational motion of the implant, whereas a 0-degree phase angle indicates they are moving in phase, which corresponds to bending motion.

Accelerometer s4 was mounted on the link arm between the shaker and the implant to measure the excitation signal, serving as a reference for the shaker's excitation amplitude. The natural frequency of the implant was identified by a 90-degree phase shift in the frequency response function (FRF) between s4 and s2.

In the second part, the implant was preloaded with static weights to simulate loading conditions experienced during walking. The shaker then applied a ramped excitation at the natural frequency of the first torsional mode. Accelerometer s1 was attached to the bone adjacent to s2 to assist in validating interfacial slippages. When bone and implant rotate in synchrony, the time histories of acceleration s1 and s2 will remain linear. However, if the implant slips, relative displacement will occur, reflected as distortions in time history, and the emergence of higher-order harmonics in the frequency domain. The auto-power spectrum of s2 was calculated, with spectral magnitudes normalised by the magnitude of the first harmonic at the excitation frequency.

Results

The results of the implant push-in test, including the loads and displacements, are presented in Figure 2. The axial displacement of the implant, as recorded by LVDT, exhibited a notable increase during instances of macro slippage. The load-bearing capacity until macro slippage occurs was determined to be 362 N, and 238 N for Implants B, and C, respectively. No macro slippage occurred in the test of implant A. According to the well-known Lamé's equations [13], a reduction in the outer diameter of the peak tube will result in a corresponding decrease in contact pressure and, consequently, in the load-bearing capacity of the implant. The results of the push-in test demonstrated that cortical thinning will effectively reduce the load capacity of the implant.

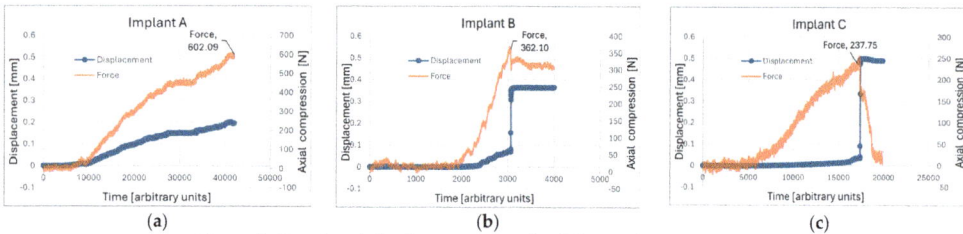

Figure 2. Load and displacement result of the implant press-in test.

To ascertain the natural frequency of the torsional modes, the frequency response function (FRF) and the cross-spectra of the vibration response were calculated. The torsional natural frequency for all implant constructs was identified within the range of 1200-1400 Hz (see Figure 3). In the phase spectrum of the FRF, a phase angle of 90 degrees indicated the natural frequency, while a phase angle of 180 degrees in the cross-spectrum denoted rotational motion.

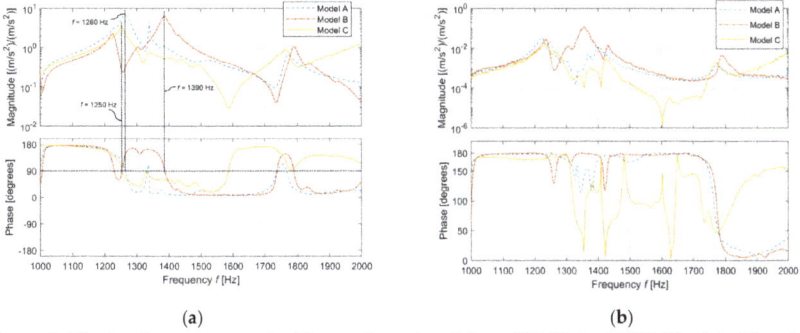

(a)   (b)

Figure 3. The implants were excited by random signal from 300 Hz to 10000 Hz. (a): Magnitude spectrum and phase spectrum of the frequency response function between acceleration s4 and s2. (b) Magnitude spectrum and phase spectrum of the cross-spectrums between acceleration s2 and s3.

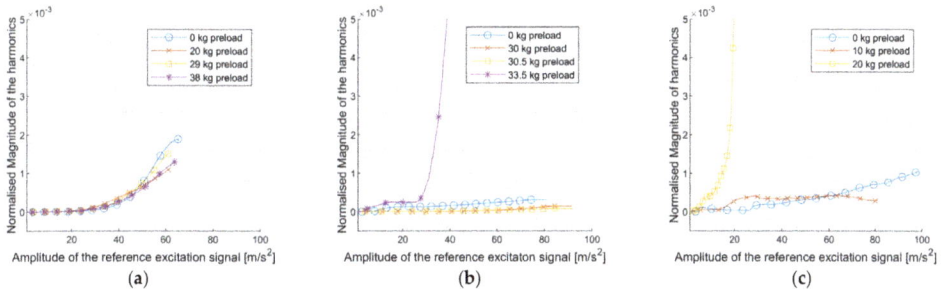

(a)   (b)   (c)

Figure 4. Changes in the magnitude of the sum of the second and third harmonics in the normalised auto power spectrum of acceleration at s2. Results are shown for: (a) model A, (b) model B and (c) model C.

Subsequently, the bone-implant models were excited at the frequency corresponding to their first torsional mode using a ramped sinusoidal input with preload weights. For Implant B, a significant increase in harmonic magnitude was observed with a preload of 33.5 kg, while Implant C demonstrated a notable rise at 20 kg (Figure 4). In the case of lower preload conditions, the harmonic magnitude for both implants remained below the threshold value of 1×10^{-3}. It could be proposed that a threshold of 1×10^{-3} may be established to differentiate between significant and non-significant harmonic activity, with values exceeding this threshold indicative of notable harmonics. In contrast, the normalised harmonic magnitude of implant A reached 1.5×10-3, which exceeded the limit. As the maximum preload level of 38 kg was lower than the load capacity of model A, the interfacial slippages were not anticipated during the vibration test. The higher harmonics were likely originated from the soft beeswax for affixing the accelerometers, which was replaced with super glue in Implants B and C.

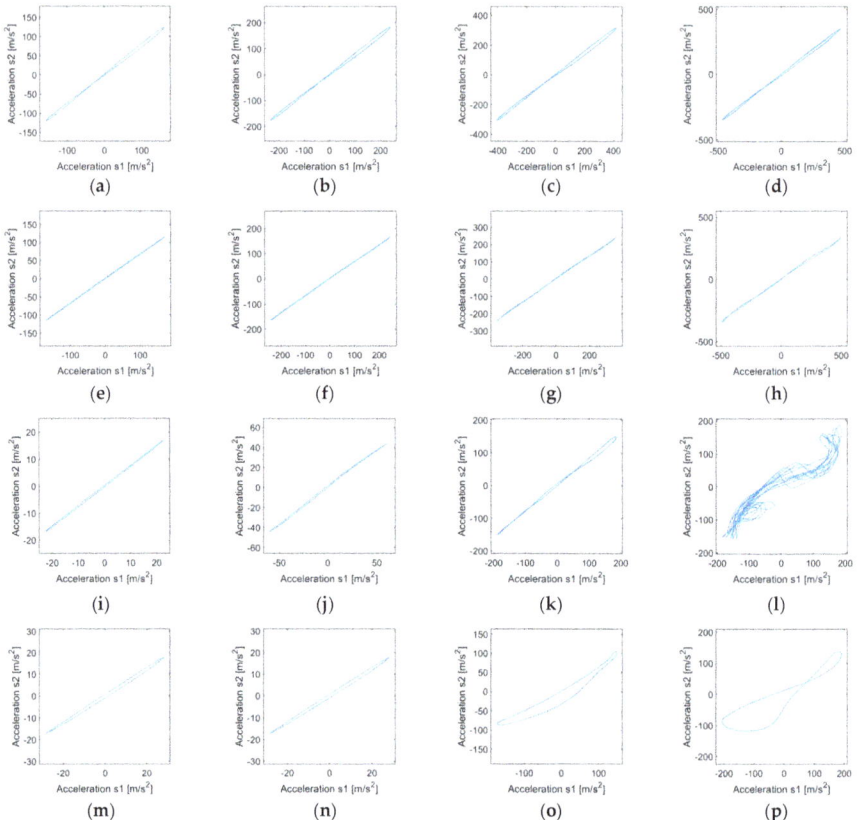

Figure 5. Interface motion profile represented as acceleration s1 plotted against s2. Model C with 10 kg preload at (a) 24 m/s², (b) 34 m/s², (c) 58 m/s², (d) 80 m/s². Model B with 30 kg preload at: (e) 24 m/s², (f) 35 m/s², (g) 56 m/s², (h) 84 m/s². Model C with 20 kg preload at: (i) 3 mm/s², (j) 11 mm/s², (k) 23 mm/s², (l) 33 m/s². Model B with 33.5 kg preload at: (m) 5 mm/s², (n) 27 m/s², (o) 42 m/s², (p) 57 m/s²

The interface motion profile, represented by the time history of acceleration s1 (located on the hollow cylinder) plotted against acceleration s2 (located on the implant), was illustrated in Figure 5 for the purpose of facilitating the examination of interfacial slippage. If the harmonics magnitude displayed in Figure 4 remains below the specified threshold, it can be observed that the bone and implant adhere to each other, exhibiting a linear relationship between s1 and s2 (Figure 5.a-d and Figure 5.e-h). In contrast, for implant B, which was preloaded with 33.5 kg and exhibited a significant increase in harmonic magnitude above the threshold during the test, the relationship between s1 and s2 initially remained linear. However, as the vibration excitation level increased, the relative motion between s1 and s2 becomes non-linear, indicating that s1 and s2 are no longer moving in synchrony (Figure 5.i-l and Figure 5.m-p). Similar pattern was also evident for Model C with 20 kg preload. The distortion observed in the interface profile confirms the association between the emergence of higher-order harmonics and changes in the interface.

Structural Health Monitoring: 10APWSHM
Materials Research Proceedings 50 (2025) 269-276

Materials Research Forum LLC
https://doi.org/10.21741/9781644903513-31

Discussion and conclusion.

The early-stage detection of implant loosening remains challenging, as the implant typically remains firmly bonded to the bone cavity during this phase, and the reduction in osteointegration is limited. The experimental results suggest that the vibration excitation of the shaker can provide only a limited energy input to the implant system. Consequently, it will be difficult to use vibrational excitation along to detect interfacial slippage for the diagnose of the early loosening, even when driving the implant construct at the resonant frequency. Without any preload, none of the implant models exhibited a notable harmonic component in the vibration response.

The preload, representing the compression experienced by the implant during walking, was applied to bring the implant close to its slip threshold, thereby promoting the activation of interfacial stick-slip dynamics as the nonlinear source responsible for generating harmonics under torsional vibration excitation. This preload establishes a mean shear stress while the subsequent sinusoidal excitation induces a subtle rotational motion of the implant, generating a small shear traction tangential to the bone-implant interface. The principal stress at the onset of slippage corresponds to the friction limit of the overlap, which is directly correlated to the implant's load-bearing capacity.

It is suggested that the load capacity of the implant can be monitored in real-time during daily activities, allowing for continuous and sensitive detection before severe loosening occurs. This is a significant benefit because it enables the evaluation of the implant's safety based on the load range experienced during typical activities, which are influenced by the patient's personal characteristics, such as weight. Consequently, assessing the safety of an implant involves determining whether it can withstand the loads encountered during normal walking or running.

In conclusion, an experiment was conducted to investigate the vibration response of the bone-implant construct for the early detection of implant loosening. The implant construct was subjected to a preload of static weight and subsequently excited by vibration at the frequency corresponding to the first torsional mode. Slippage at the interface occurs when the interfacial shear stress exceeds the prescribed critical value, resulting in the generation of higher-order harmonics in the vibration response. The findings indicated that the axial preload, promoted the activation of interfacial stick-slip dynamics as the nonlinear source responsible for generating harmonics under torsional vibration excitation. The result suggests that monitoring the implant's load capacity during routine activities, such as walking, can offer a more practical and sensitive method for detecting early indications of implant failure.

References

[1] Overmann, A.; Forsberg, J. The state of the art of osseointegration for limb prosthesis. Biomed Eng Lett 2020, 10, 5-16. https://doi.org/10.1007/s13534-019-00133-9

[2] Prochor, P.; Sajewicz, E. The Influence of Geometry of Implants for Direct Skeletal Attachment of Limb Prosthesis on Rehabilitation Program and Stress-Shielding Intensity. Biomed Res Int 2019, 2019, 6067952. https://doi.org/10.1155/2019/6067952

[3] Wiese, A.; DiGuglielmo, A.; Mellet, J.; Rebillon, M.; Mandayam, S.; Brewer, E.; Austin, L. A Radial Basis Function Technique for the Early Detection and Measurement of Hip Implant Loosening. 2020; pp. 1-4. https://doi.org/10.1109/SAS48726.2020.9220036

[4] Patil, N.; Goodman, S.B. 7 - Wear particles and osteolysis. In Orthopaedic Bone Cements, Deb, S., Ed.; Woodhead Publishing: 2008; pp. 140-163. https://doi.org/10.1533/9781845695170.1.140

[5] Parlee, L.; Kagan, R.; Doung, Y.C.; Hayden, J.B.; Gundle, K.R. Compressive osseointegration for endoprosthetic reconstruction. Orthop Rev (Pavia) 2020, 12. https://doi.org/10.4081/or.2020.8646

[6] Frost, H.M. A 2003 update of bone physiology and Wolff's Law for clinicians. The Angle Orthodontist 2004, 74, 3-15.

[7] Nebergall, A.; Bragdon, C.; Antonellis, A.; Kärrholm, J.; Brånemark, R.; Malchau, H. Stable fixation of an osseointegated implant system for above-the-knee amputees: titel RSA and radiographic evaluation of migration and bone remodeling in 55 cases. Acta orthopaedica 2012, 83, 121-128. https://doi.org/10.3109/17453674.2012.678799

[8] Cachão, J.H.; Soares dos Santos, M.P.; Bernardo, R.; Ramos, A.; Bader, R.; Ferreira, J.A.F.; Torres Marques, A.; Simões, J.A.O. Altering the Course of Technologies to Monitor Loosening States of Endoprosthetic Implants. Sensors 2020, 20, 104. https://doi.org/10.3390/s20010104

[9] Rosenstein, A.D.; McCoy, G.F.; Bulstrode, C.J.; McLardy-Smith, P.D.; Cunningham, J.L.; Turner-Smith, A.R. The differentiation of loose and secure femoral implants in total hip replacement using a vibrational technique: an anatomical and pilot clinical study. Proc Inst Mech Eng H 1989, 203, 77-81. https://doi.org/10.1243/PIME_PROC_1989_203_014_01

[10] Li, P.L.; Jones, N.B.; Gregg, P.J. Vibration analysis in the detection of total hip prosthetic loosening. Med Eng Phys 1996, 18, 596-600. https://doi.org/10.1016/1350-4533(96)00004-5

[11] Georgiou, A.P.; Cunningham, J.L. Accurate diagnosis of hip prosthesis loosening using a vibrational technique. Clin Biomech (Bristol, Avon) 2001, 16, 315-323. https://doi.org/10.1016/S0268-0033(01)00002-X

[12] Zhou, Q.; Rose, L.R.F.; Ebeling, P.; Russ, M.; Fitzgerald, M.; Chiu, W.K. Harmonic Vibration Analysis in a Simplified Model for Monitoring Transfemoral Implant Loosening. Sensors 2024, 24, 6453. https://doi.org/10.3390/s24196453

[13] Imaninejad, M.; Subhash, G. Proportional loading of thick-walled cylinders. International Journal of Pressure Vessels and Piping 2005, 82, 129-135. https://doi.org/10.1016/j.ijpvp.2004.07.013

Structural Health Monitoring: 10APWSHM   Materials Research Forum LLC
Materials Research Proceedings 50 (2025) 277-284   https://doi.org/10.21741/9781644903513-32

Seismic control of structures considering frequency dependency of inerter-based dynamic vibration absorbers

XIE Ruihong[1, a] and IKAGO Kohju[1, b *]

[1]Tohoku University, Sendai, Japan

[a]xie.ruihong.q8@dc.tohoku.ac.jp, [b]koju.ikago.e8@ tohoku.ac.jp

Keywords: Inerter, Dynamic Vibration Absorber, Frequency Dependency, Multi-Modal Response, Complex Modal Analysis

Abstract. An inerter is a mechanical element that generates inertial forces proportional to the relative accelerations of its two terminals. The apparent mass of the inerter can be thousands of times larger than its physical mass, which enables to realize the lightweight modification of conventional dynamic vibration absorbers. Thus, the inerter has recently drawn widespread attention of civil engineering researchers and various inerter-based dynamic vibration absorbers (IDVAs) have been developed. In the optimization of IDVAs, previous studies have mainly focused on the response control performance at the targeted resonant frequency neglecting the influence of IDVAs on other frequency regions. It is acceptable for the response control of single-degree-of-freedom structures, but for multi-degree-of-freedom (MDOF) structures, it could weaken the validity of the optimization results. Compared to a linear viscous damper (LVD), the energy dissipation efficiency of the IDVA is significantly amplified at the targeted resonant frequency, whereas at other frequencies, it declines to i) almost the same value as the LVD or ii) nearly zero. It implies that the IDVA has a strong frequency-dependent energy dissipation capability and the frequency dependency of the IDVA is totally different from that of the LVD. This paper investigates the influence of the frequency dependency on the performance of two typical IDVAs, the tuned viscous mass damper (TVMD) and the tuned inerter damper (TID). Through complex modal analysis and numerical simulations, it can be found that due to the frequency dependency, the TVMD differs significantly from the TID in mitigating inter-story drifts and floor accelerations of MDOF structures.

Introduction

A dynamic vibration absorber (DVA) is a device that effectively absorbs energy by adding a secondary vibratory system that resonates with the controlled structure, allowing for large amplitude oscillations resulting in enhanced damping. When applied to civil structures, a DVA is often referred to as a tuned mass damper (TMD), with one of the earliest examples being the Sydney Tower in Australia during the 1970s. TMDs are considered effective for controlling wind-induced vibrations, but they have been thought to be barely effective against seismic motions [1]. In Japan, efforts have been made to increase the effective mass ratio of TMDs to about 5% to address the challenges caused by long-period ground motions.

By replacing the effective mass of a TMD with an inerter, it is possible to realize a DVA with a large mass ratio without weight penalty [2,3]. Notable examples of DVAs using inerters (IDVAs) include the tuned viscous mass damper (TVMD) [4], tuned inerter damper (TID) [5], and tuned mass damper inerter (TMDI) [6].

Among DVAs utilizing inerters, the TVMD and TID are representative examples, differing in the arrangement of their dampers. The damper arrangement in TID is the same as that in traditional TMDs, arranged in parallel to the spring. In contrast, in the TVMD, the damper is arranged in parallel to the inerter. This paper investigates the impact of this topological difference on the frequency response characteristics and clarifies the frequency performance of the structural control capabilities of TVMD and TID.

Equations of Motion

The analytical model of a multi-story structure incorporated with dampers is shown in Fig. 1. The equations of motion can be expressed as

$$\mathbf{M}_p\ddot{\mathbf{x}} + \mathbf{C}_p\dot{\mathbf{x}} + \mathbf{K}_p\mathbf{x} + \mathbf{f} = -\mathbf{M}_p\mathbf{r}a_g \tag{1}$$

where $\mathbf{M}_p$, $\mathbf{K}_p$, and $\mathbf{C}_p$ are the mass, stiffness, and inherent damping matrices of the primary structure; $\mathbf{x}$, $\mathbf{f}$, and $\mathbf{r} = [1,1, ...,1]^T$ are the vectors of the displacement of the primary structure relative to the ground, the controlling force of dampers, and the influence coefficients; a_g is the ground acceleration. Controlling force of different dampers at the jth story can be expressed as

$$f_j^{\text{TVMD}} = m_{d,j}\ddot{x}_{d,j} + c_{d,j}\dot{x}_{d,j} = k_{b,j}\left(x_j - x_{j-1} - x_{d,j}\right)$$
$$f_j^{\text{TID}} = m_{d,j}\ddot{x}_{d,j} = k_{b,j}\left(x_j - x_{j-1} - x_{d,j}\right) + c_{b,j}\left(\dot{x}_j - \dot{x}_{j-1} - \dot{x}_{d,j}\right) \tag{2}$$
$$f_j^{\text{LVD}} = c_{a,j}\left(\dot{x}_j - \dot{x}_{j-1}\right)$$

where $k_{b,j}$ is the stiffness of the spring; $m_{d,j}$ and $x_{d,j}$ are the inertance and deformation of the inerter; $c_{d,j}$ and $c_{b,j}$ are the damping coefficients of the dashpots arranged in parallel with the inerter and the spring; $c_{a,j}$ is the damping coefficient of the LVD. It should be noted that $x_0 \equiv 0$ for j equals 1.

In the frequency domain, Eqs. 1 and 2 can be rewritten as

$$\left(-\omega^2\mathbf{M}_p + i\omega\mathbf{C}_p + \mathbf{K}_p\right)\mathbf{X}\left(i\omega\right) + \mathbf{F}\left(i\omega\right) = -\mathbf{M}_p\mathbf{r}A_g\left(i\omega\right) \tag{3}$$

$$Q_j^{\text{TVMD}}(i\omega) = \frac{F_j^{\text{TVMD}}(i\omega)}{[X_j(i\omega) - X_{j-1}(i\omega)]} = \frac{k_{b,j}(-\omega^2 m_{d,j} + i\omega c_{d,j})}{-\omega^2 m_{d,j} + i\omega c_{d,j} + k_{b,j}}$$

$$Q_j^{\text{TID}}(i\omega) = \frac{F_j^{\text{TID}}(i\omega)}{[X_j(i\omega) - X_{j-1}(i\omega)]} = \frac{-\omega^2 m_{d,j}(i\omega c_{b,j} + k_{b,j})}{-\omega^2 m_{d,j} + i\omega c_{b,j} + k_{b,j}} \tag{4}$$

$$Q_j^{\text{LVD}}(i\omega) = \frac{F_j^{\text{LVD}}(i\omega)}{[X_j(i\omega) - X_{j-1}(i\omega)]} = i\omega c_{a,j}$$

where $\mathbf{X}(i\omega)$, $\mathbf{F}(i\omega)$, and $A_g(i\omega)$ are the Fourier transforms of $\mathbf{x}$, $\mathbf{f}$, and a_g; i and ω are the imaginary unit and excitation angular frequency; $Q_j^{\text{TVMD}}(i\omega)$, $Q_j^{\text{TVMD}}(i\omega)$, and $Q_j^{\text{TVMD}}(i\omega)$ represent the dynamic stiffness of the TVMD, TID, and LVD at the jth story.

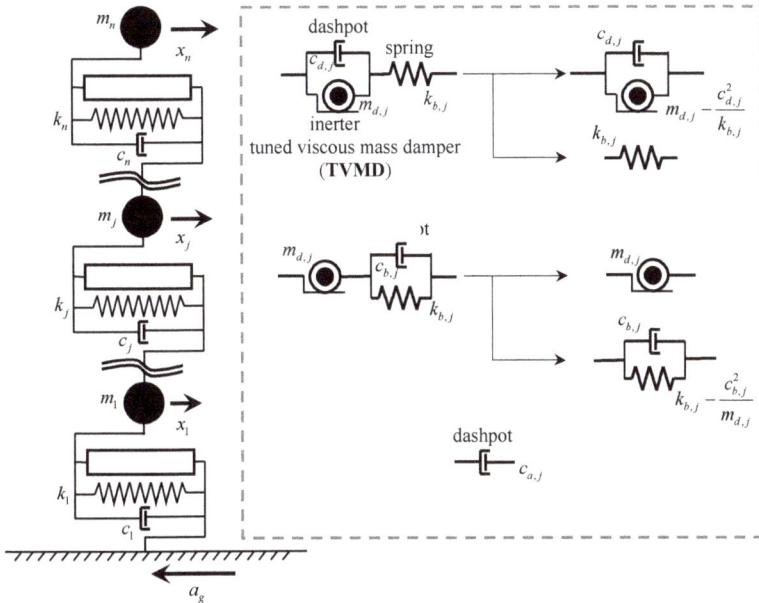

Fig. 1 Analytical model of a multi-story structure incorporated with dampers

Frequency Dependency

For effective response control, the IDVA should be tuned at a proper frequency ω_s [4,5]. Fig. 2 compares the storage and loss stiffness (i.e., the real and imaginary parts of the dynamic stiffness) of the three dampers with identical damping coefficients for a single-story structure and Table 1 lists the nondimensional parameters of the dampers [7]. As shown in Fig.2, the loss stiffness of the TVMD and TID is significantly larger than that of the LVD around the tuned frequency ω_s, illustrating the damping amplification effect of the IDVA [4,8].

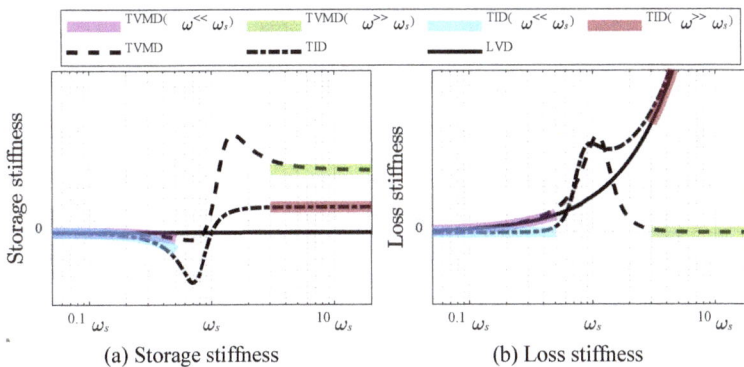

(a) Storage stiffness   (b) Loss stiffness
Fig. 2 Comparison of dynamic stiffness

Structural Health Monitoring: 10APWSHM   Materials Research Forum LLC
Materials Research Proceedings 50 (2025) 277-284   https://doi.org/10.21741/9781644903513-32

Table 1. Parameters of dampers in SDOF structure [7]

Damper	$\dfrac{m_{d,1}}{m_1}$	$\dfrac{k_{b,1}}{k_1}$	$\dfrac{c_{d,1} \text{ or } c_{b,1} \text{ or } c_{a,1}}{2\sqrt{m_1 k_1}}$
TVMD	0.26	0.35	0.10
TID	0.40	0.24	0.10
LVD	—	—	0.10

It should be noted that the damping amplification effect is strongly dependent on frequency. If the excitation frequency is far away from the tuned frequency, the dynamic stiffness of the TVMD and TID can be approximate as

$$Q_j^{\text{TVMD}}(i\omega) \approx \begin{cases} -\omega^2 \left(m_{d,j} - \dfrac{c_{d,j}^2}{k_{b,j}} \right) + i\omega c_{d,j}, & \omega \ll \omega_s \\ k_{b,j}, & \omega \gg \omega_s \end{cases} \tag{5}$$

$$Q_j^{\text{TID}}(i\omega) \approx \begin{cases} -\omega^2 m_{d,j}, & \omega \ll \omega_s \\ \left(k_{b,j} - \dfrac{c_{b,j}^2}{m_{d,j}} \right) + i\omega c_{b,j}, & \omega \gg \omega_s \end{cases} \tag{6}$$

and their configurations can be further simplified as shown in Fig. 1. When $\omega \ll \omega_s$, the dynamic stiffness of the inerter is much lower than that of the spring and thereby the deformation of the spring in the TVMD and TID can be ignored. Consequently, the TVMD and TID reduce to a viscous mass damper (i.e., an inerter connected with a dashpot in parallel) and an inerter, respectively. As for $\omega \gg \omega_s$, the dynamic stiffness of the inerter is relatively high to hinder its deformation, respectively making the TVMD and TID reduce to a spring and a Kelvin-Voigt element (i.e., a spring connected with a dashpot in parallel). It can be observed in Figs. 1 and 2 that the energy dissipation efficiency of the IDVA is no longer amplified and even vanishes outside the tuned frequency.

Analytical Example
To investigate the effect of the frequency dependency of IDVAs on a specific mode and other non-targeted modes, the complex modal analysis was conducted on a 10-story benchmark shear building model presented by JSSI [9]. The mass and stiffness distributions of the benchmark structure are listed in Table 2, and consequently the first three natural frequencies $\omega^{(1)}$, $\omega^{(2)}$, and $\omega^{(3)}$ can be obtained as 3.12, 8.29, and 13.61 rad/s, respectively. The inherent damping distribution is proportional to that of the stiffness and assumed to be 2% of the critical damping, i.e., $c_j = 2 \times 0.02 \times k_j / \omega^{(1)}$.

Table 2. Structural information about 10-story benchmark building

Story No.	Mass (ton)	Stiffness (kN/m)	Story No.	Mass (ton)	Stiffness (kN/m)
1	700	279960	6	667	291890
2	628	383550	7	660	244790
3	680	383020	8	656	220250
4	676	382260	9	649	180110
5	670	306160	10	875	158550

Structural Health Monitoring: 10APWSHM   Materials Research Forum LLC
Materials Research Proceedings 50 (2025) 277-284   https://doi.org/10.21741/9781644903513-32

Four designs of the IDVAs are discussed in this study, namely TVMD-$\omega^{(1)}$-a, TVMD-$\omega^{(2)}$-a, TID-$\omega^{(1)}$-d, and TID-$\omega^{(2)}$-d. -$\omega^{(1)}$ and -$\omega^{(2)}$ represent the tuned frequency; -d and -a represent the displacement and the absolute acceleration control designs. For example, TVMD-$\omega^{(1)}$-a represents the control design of the TVMD tuned to $\omega^{(1)}$ to minimize the absolute acceleration amplification factor. The distributions of the inertance of the inerters are proportional to the structural stiffness, and the total damping coefficients in the four designs are identical and equal to $2\times0.06\times\Sigma k_j/\omega^{(1)}$. More details about the designing method of the IDVAs installed to MDOF structures can be found in the previous study [7]. In addition, the parameter of the LVD is set to $c_{a,j} = 2\times0.06\times k_j/\omega^{(1)}$ for comparison.

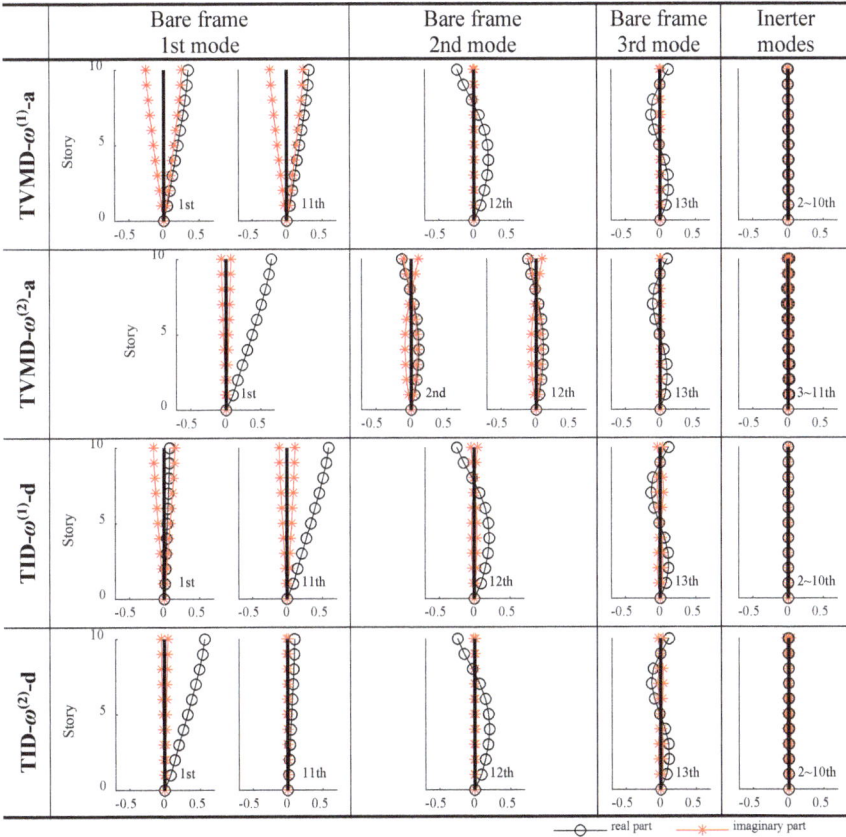

Fig. 3 Participation mode vectors of 10-story benchmark structure with different IDVAs

Complex Modal Analysis

The participation mode vectors and modal characteristics of the 10-story benchmark structure with different designs of the IDVAs are shown in Figs. 3 and 4, respectively. Due to the small components of the participation mode vectors, the inerter modes are insignificant and thereby the corresponding modal characteristics are ignored in Fig. 4. The modal characteristics of the bare

Structural Health Monitoring: 10APWSHM   Materials Research Forum LLC
Materials Research Proceedings 50 (2025) 277-284   https://doi.org/10.21741/9781644903513-32

frame and the frame with LVDs are added in Fig. 4 and their mode numbers are labeled with brackets.

At lower modes (the 1st and 11th modes), damping ratios of TVMD-$\omega^{(1)}$-a and TID-$\omega^{(1)}$-d are amplified relative to that of the LVD, while damping ratios TVMD-$\omega^{(2)}$-a and TID-$\omega^{(2)}$-d are nearly the same as those of the LVD and bare frame, respectively. Although TID-$\omega^{(2)}$-d has a large damping ratio at the 11th mode, its influence to the structural response is neglectable because the corresponding participation mode vectors are minor. At higher modes (the 2nd, 12th, and the latter modes), the amplified damping ratios of the TID approach to the LVD but those of the TVMD tend towards the bare frame. It can be generally explained by the frequency dependency of IDVAs as shown in Figs. 1 and 2. For example, the TVMD behaves like the spring to increase structural stiffness when the modal frequency is much larger than the tuned frequency so that the damping ratios of the frame with TVMDs are even less than those of the bare frame in higher modes.

Both TVMD and TID exhibit the damping amplification effect at the tuned mode. The main difference between them is that the TVMD provides effective damping to lower modes relative to the tuned mode, whereas the TID provides effective damping to higher modes.

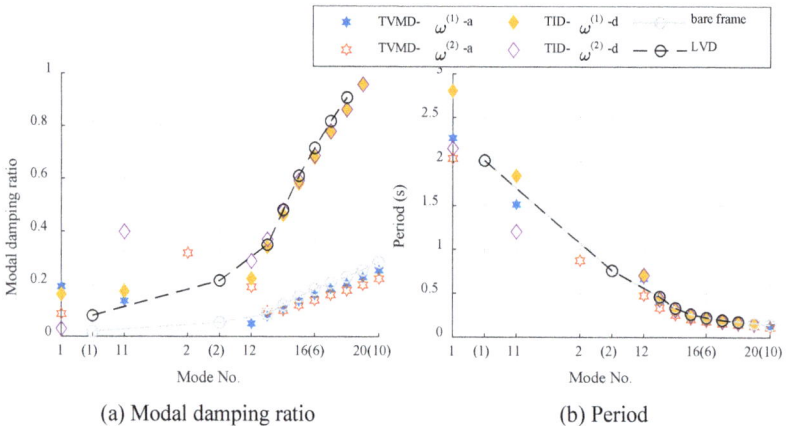

(a) Modal damping ratio   (b) Period

Fig. 4 Comparison of modal characteristics

Time-history analysis

Two ground motion records, the historic East-West component of the 1952 Kern County Earthquake recorded at Taft (Taft 1952 E-W) and the long-period East-West component of the 1999 Chi-Chi Earhquake recorded at ILA056 (ILA056 1999 E-W), were used in the numerical simulations. Peak ground velocities (PGVs) of the two records are scaled to 0.5 m/s and the response spectra are shown in Fig. 5. Taft 1952 E-W and ILA056 1999 E-W are dominated by high- and low-frequency components, respectively.

Fig. 6 compares the maximum inter-story drift and floor response acceleration of the 10-story benchmark building under Taft 1952 E-W and ILA056 1999 E-W records. TVMD-$\omega^{(1)}$-a performs the best in mitigating inter-story drift yielded by ILA056 1999 E-W but the worst in mitigating floor response acceleration yielded by Taft 1952 E-W. Due to the frequency dependency, TVMD-$\omega^{(1)}$-a can effectively suppress the response (mostly displacements) induced by lower-frequency components relative to the tuned frequency rather than the response (mostly accelerations) induced by higher-frequency components. TID-$\omega^{(2)}$-d also exhibits opposite seismic performance under the two ground motions, i.e., the smallest floor response acceleration yielded by Taft 1952 E-W but the largest inter-story drift yielded by ILA056 1999 E-W. The reason is contrary to that of TVMD-

Structural Health Monitoring: 10APWSHM   Materials Research Forum LLC
Materials Research Proceedings 50 (2025) 277-284   https://doi.org/10.21741/9781644903513-32

$\omega^{(1)}$-a. Since TVMD-$\omega^{(2)}$-a, as well as TID-$\omega^{(1)}$-d, can effectively control the response in a wider range of frequency, both inter-story drift and floor response acceleration are well controlled.

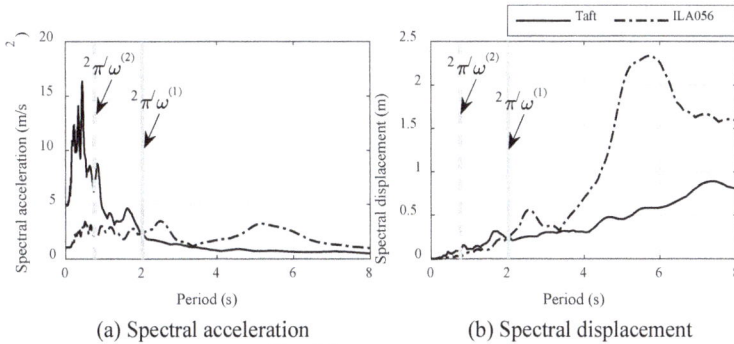

(a) Spectral acceleration   (b) Spectral displacement

Fig. 5 Response spectra

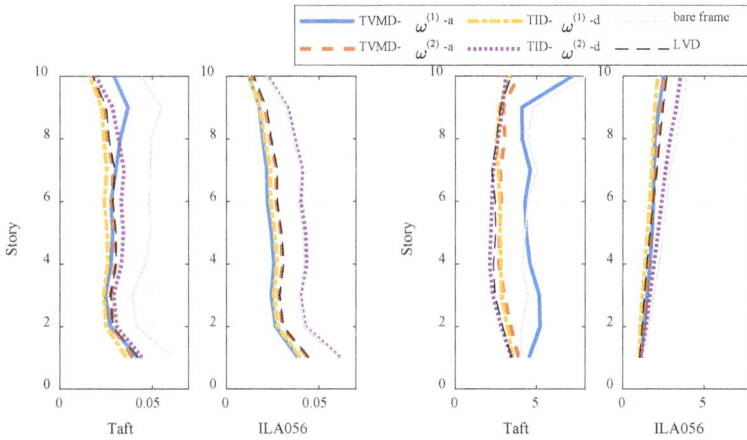

(a) Inter-story drift (m)   (b) Floor response acceleration (m/s^2)

Fig. 6 Maximum seismic responses (Taft 1952 E-W and ILA056 1999 E-W)

Conclusions

One of the attractive features of IDVAs is the damping amplification effect that contributes to dissipating more vibration energy than conventional LVDs. Nevertheless, the amplification does not always exist and is strongly dependent on excitation frequency, which has significant influence on the seismic performance of IDVAs. This study examined the frequency dependency of two typical IDVAs, the TVMD and TID. According to the dynamic stiffness, the energy dissipation in both TVMD and TID is amplified at around the tuned frequency. When the excitation frequency is much lower than the tuned frequency, the TVMD and TID reduce to a viscous mass damper and an inerter, respectively. As for the excitation frequency is much higher, the TVMD and TID reduce to a spring and a Kelvin-Voigt element. Therefore, the TVMD cannot effectively control the higher modes relative to the tuned modes in an MDOF structure, while the TID is on the contrary. Since the inter-story drift is mainly from lower modes and the floor response acceleration is mainly from higher modes, the TVMD performs better in displacement control but worse in acceleration control than the TID in MDOF structures.

References

[1] R.J. McNamara, Tuned mass dampers for buildings, J. Struct. Div., 103.9 (1977) 1785-1798. https://doi.org/10.1061/JSDEAG.0004721

[2] M.C. Smith, Synthesis of mechanical networks: the inerter. IEEE Trans. Automat. Contr. 47 (2002) 1648-62. https://doi.org/10.1109/TAC.2002.803532

[3] R. Ma, K. Bi, and H. Hao, Inerter-based structural vibration control: A state-of-the-art review, Eng. Struct. 243 (2021) 112655. https://doi.org/10.1016/j.engstruct.2021.112655

[4] K. Ikago, K. Saito, and N. Inoue, Seismic control of single-degree-of-freedom structure using tuned viscous mass damper, Earthq. Eng. Struct. Dyn. 41.3 (2012) 453-474. https://doi.org/10.1002/eqe.1138

[5] I.F. Lazar, S.A. Neild, and D.J. Wagg, Using an inerter-based device for structural vibration suppression, Earthq. Eng. Struct. Dyn. 43.8 (2014) 1129-1147. https://doi.org/10.1002/eqe.2390

[6] L. Marian and A. Giaralis, Optimal design of a novel tuned mass-damper-inerter (TMDI) passive vibration control configuration for stochastically support-excited structural systems, Probabilistic Eng. Mech. 38 (2014) 156-164. https://doi.org/10.1016/j.probengmech.2014.03.007

[7] R. Xie and K. Ikago, Device topology optimization for an inerter-based structural dynamic vibration absorber, Front. Built Environ. 10(2024), 1508190. https://doi.org/10.3389/fbuil.2024.1508190

[8] R. Zhang, Z. Zhao, C. Pan, K. Ikago, and S. Xue, Damping enhancement principle of inerter system, Struct. Control Health Monit. 27.5 (2020) e2523. https://doi.org/10.1002/stc.2523

[9] JSSI, Manual for Design and Fabrication for Passive Vibration Control of Structures, third ed, JSSI, Tokyo, 2013. (in Japanese)

Structural Health Monitoring: 10APWSHM   Materials Research Forum LLC
Materials Research Proceedings 50 (2025) 285-291   https://doi.org/10.21741/9781644903513-33

Assessing structural safety under repeated earthquakes: Integration of an energy-based function in NSmos®

Hiroki Kazama[1, a], Hirotaka Imaeda[1, b *], Hiroki Kameda[1, c] and Nicolas Giron[1, d]

[1]NIKKEN SEKKEI LTD, Structural Design Section, Engineering Department, 2-18-3 Iidabashi, Chiyoda-ku, Tokyo, Japan

[a]hiroki.kazama@nikken.jp, [b]imaedah@nikken.jp, [c]kameda.hiroki@nikken.jp, [d]giron.nicolas@nikken.jp

Keywords: Structural Health Monitoring, Repeated Earthquakes, Structural Safety Assessment, Energy

Abstract. The authors have developed and integrated a new function into an existing operational monitoring system (NSmos®) to assess structural safety of a steel structure during repeated earthquakes, such as the 2016 Kumamoto earthquakes. Using data from the system's accelerometers, the acceleration and displacement of each story of a building are estimated. The acceleration is then multiplied by the pre-inputted story weight to estimate the earthquake-induced shear force. Based on the resulting restoring force, the earthquake's induced hysteretic energy is calculated and added to that of previous earthquakes. This cumulative energy is then compared to the building's energy-dissipation capacity, which is determined from the specifications of its structural members in accordance with energy-based design theory. Finally, the accuracy of this approach is verified through computational analysis.

Introduction

Recent seismic events, such as the 2016 Kumamoto earthquakes, have highlighted the need for more advanced methods to assess and manage the structural safety of buildings subjected to repeated earthquake forces. Traditional monitoring systems typically evaluate the impact of individual seismic events, which may not fully capture the cumulative effects of multiple earthquakes over the life span of a building. In response to this, the authors have developed an enhanced function for NSmos® (Nikken Sekkei monitoring system), an operational monitoring system designed to evaluate structural integrity and safety based on real-time data.

The original system focused on single-event analysis, lacking the ability to account for the cumulative damage that can occur when a building experiences multiple seismic events. To address this limitation, the newly integrated function expands the system's capabilities to include the assessment of cumulative damage for steel structure, using the energy-based design theory.

The proposed methodology is verified using computational analysis to a specific case study—the Nikken Sekkei Tokyo Building—where the modified system is applied to assess the building's cumulative structural performance under varying seismic intensities, providing valuable insights into the practicality and reliability of the approach.

Brief presentation of NSmos®

NSmos® is a system designed to evaluate the level of vibration and degree of damage sustained by a building following an earthquake. Developed by structural engineers at Nikken Sekkei, an architectural firm based in Japan, the system uses accelerometers strategically distributed throughout the building. By recording real-time data on building vibrations during an earthquake, it assesses the building's damage and determines whether evacuation is necessary. This rapid evaluation of a building's safety and damage level provides building managers with essential information regarding evacuation requirements or the feasibility of continuing operations.

Since March 2014, NSmos® has been implemented in approximately 70 buildings, including high-rise structures, seismically isolated buildings, as well as both existing and newly constructed buildings. Detailed explanations about the system can be found in [1]. Over the past decade, NSmos® has yielded valuable insights for both structural engineers and building managers. However, it is primarily focused on individual earthquake events and does not account for cumulative damage over a building's lifespan. While it contributes to damage assessment, it lacks information regarding the need for retrofitting and the potential for increased damage in future seismic events.

To address these limitations, a new function, discussed in the following section, has been introduced.

An energy-based function to assessing structural safety under repeated earthquakes

The evaluation of the degree of damage in NSmos® is based solely on the maximum estimated story drift and the allowable drift determined by a structural engineer. This straightforward approach provides results that are consistent with experimental data and ensures robustness. To incorporate the effects of cumulative damage from repeated earthquakes, the following energy-based method based on [2] is employed. The procedure described below is applied to each story.

First, the hysteretic energy E_H of an earthquake is estimated. The acceleration recorded by the sensors distributed throughout the building is integrated twice to calculate the inter-story drift δ, which is then multiplied by the mass m specified by a structural engineer to obtain the shear force Q. By integrating the hysteretic loop derived from the drift and shear force, E_H is determined.

Next, the distribution of the dissipated energies between the hysteretic elements, such as the moment-resisting frame (E_f) and braces (E_v) is calculated. An equivalent bilinear model based on computational results is used, and the displacements calculated in the first step are applied to estimate the pseudo-dissipated energy of each element as E_f' and E_v'. The dissipated energy is then calculated as $E_f = E_f'/(E_f'+E_v')$ and $E_v = E_v'/(E_f'+E_v')$.

Following this, the cumulative plastic deformation demand η of a story is calculated based on the dissipated energy of hysteretic elements. This value is then summed with the deformation demand from previous earthquakes and compared to the cumulative plastic deformation capacity $_r\eta_u$. The cumulative plastic deformation demand η is given by equation (1) and the cumulative plastic deformation capacity $_r\eta_u$ is given by equation (2)

$$\eta = E/(2 \times_r Q_y \times_r \delta_y). \tag{1}$$

$$_r\eta_u = _m\delta_y/_r\delta_y \times _m\eta_u \times a_p \times a_b + a_d. \tag{2}$$

Where E is the previously calculated dissipated hysteretic energy. $_rQ_y$ and $_r\delta_y$ are the yielding shear force and drift of the equivalent bilinear model, respectively. $_m\delta_y$ is the member displacement at story yield (where $_m\delta_y/_r\delta_y = 1/3$). $a_p = 1.5$ is the energy dissipation factor for the connection panel. The factor a_b, which accounts for the Bauschinger effect, is 5/3 for columns and 2.0 for girders. $_m\eta_u$ represents the cumulative plastic deformation capacity for columns and girders, and $a_d = 2.0$ accounts for the additional deformation capacity between the ultimate state and failure. For other hysteretic elements besides the moment-resisting frame, the capacity can be estimated using an appropriate fatigue curve. An example is given for the buckling restrained braces in the next section.

Finally, the demand-to-capacity ratio $\eta/(_r\eta_u - \Sigma\eta_{previous})$ is calculated. Based on this ratio, messages are displayed to building users after an earthquake. For instance, in the case of a moment-resisting frame, if the ratio is less than 10%, the message "No problem" is displayed; if it is between 10% and 50%, the message "The same scale earthquake may occur XX times with no problem" is shown; and if it exceeds 50%, the message "In the case of a similar earthquake, building failure may occur due to cumulative damage" is provided.

The overall process is illustrated in Figure 1. Additionally, for seismically isolated buildings, the evaluation includes the fatigue of steel dampers.

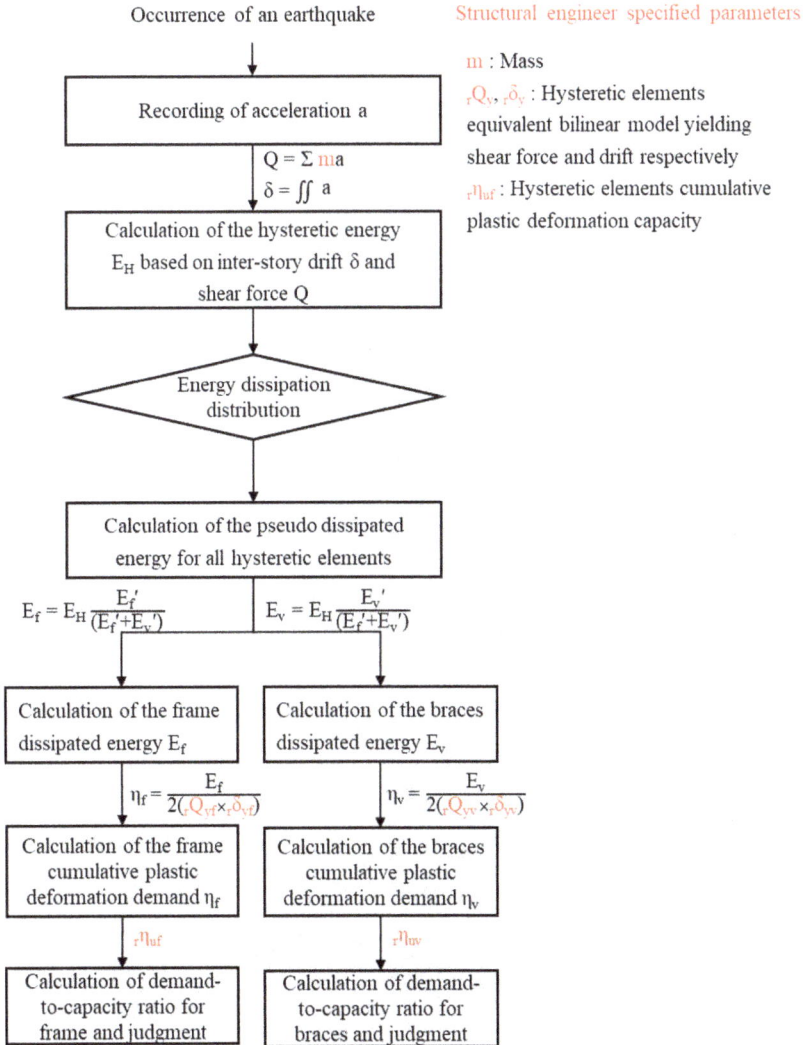

Figure 1. Flow chart of the proposed method.

Case study and performance evaluation

The building studied in references [3,4] is the Nikken Sekkei Tokyo Building, which serves as a test case for the proposed methodology. This building was designed with a high seismic resistance

Structural Health Monitoring: 10APWSHM | Materials Research Forum LLC
Materials Research Proceedings 50 (2025) 285-291 | https://doi.org/10.21741/9781644903513-33

and energy dissipation structure to maintain the functionality of its core operations even during a major earthquake (Fig. 2). Seismic energy is absorbed by low-yield-point steel (yield strength = 100 N/mm²) buckling-restrained braces and viscous damping walls, ensuring that the structure typically remains undamaged during significant seismic events. The primary structural system is a steel moment-resisting frame, with the columns constructed as concrete-filled tubes. The energy dissipation braces are designed to absorb energy during moderate to severe earthquakes, while the viscous damping walls address small to moderate earthquakes and strong winds.

For the purpose of this study, a modified version of the building is used, consisting solely of the moment-resisting frame and buckling-restrained braces, excluding the viscous damping walls. As for the seismic input, the 1995 Kobe earthquake record from the Japan Meteorological Agency [5] is utilized, with its intensity scaled by factors of 2.0, 2.5, and 3.0 to simulate varying levels of damage.

To evaluate the dissipated pseudo-energy of the hysteretic elements, equivalent bilinear models are calculated as follows. The initial stiffness K_i is set equal to that from the pushover analysis. The area under the pushover curve E_u is calculated from 0 to the critical drift, where δ_u is the ultimate inter-story drift and in this case is taken as 1/50 of the floor height. The yield drift $_r\delta_y$ is then determined using the formula $_r\delta_y = \delta_u - \sqrt{(\delta_u^2 - 2 \times E_u / K_i)}$, and the yield shear load Q_u is calculated as $Q_u = 2 \times E_u / (2 \times \delta_u - _r\delta_y)$. This ensures that the area under the pushover curve and the equivalent bilinear model is consistent. The input energy is calculated by determining the area of the hysteretic loop for each floor. The results for the 3rd story of the building are shown in fig. 3.

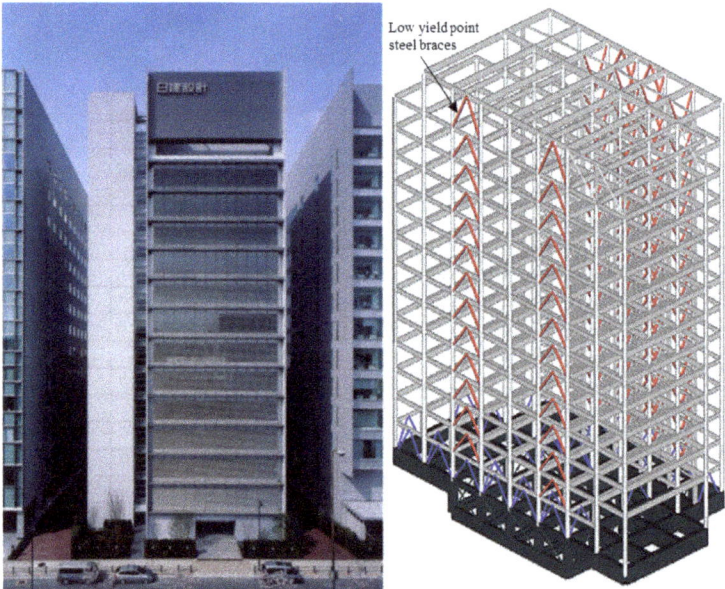

Figure 2. Nikken Sekkei Tokyo building external view and schematic structural diagram.

The same approach is used to calculate the pseudo-dissipated energy for both the frame and braces. For the input energy, the shear force is computed by multiplying the acceleration at each story with its mass and summing over the upper floors. For the dissipated pseudo-energy, the shear force is derived using the recorded displacement and the pre-defined bilinear restoring force. The calculation results of the proposed method are summarized in Table 1, showing a strong agreement between the total hysteretic energy and its distribution among the hysteretic elements with direct computational analysis results. As the bilinear models are calculated for a drift angle of 1/50, the agreement is better for large intensity earthquakes.

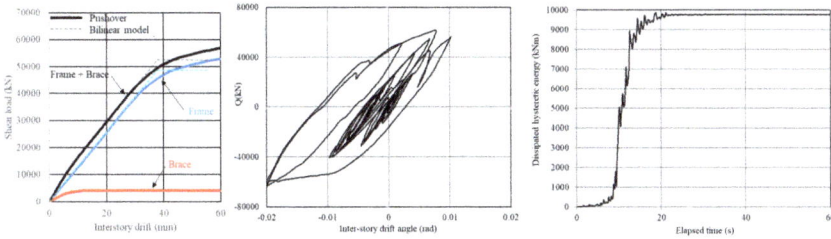

Figure 3. Third floor, Pushover results and equivalent bilinear models (left), hysteretic loop (middle) and corresponding dissipated hysteretic energy (right) during JMA Kobe×3

Table 1. Comparison of dissipated energy distribution

Floor	JMA Kobe ×2.0					JMA Kobe ×2.5					JMA Kobe ×3.0				
	Computational analysis			Proposed method		Computational analysis			Proposed method		Computational analysis			Proposed method	
	Maximum drift angle (rad)	Hysteretic Energy (kNm)	Frame bearing ratio	Hysteretic Energy (kNm)	Frame bearing ratio	Maximum drift angle (rad)	Hysteretic Energy (kNm)	Frame bearing ratio	Hysteretic Energy (kNm)	Frame bearing ratio	Maximum drift angle (rad)	Hysteretic Energy (kNm)	Frame bearing ratio	Hysteretic Energy (kNm)	Frame bearing ratio
14	1/160	7	52%	222	91%	1/144	37	45%	273	75%	1/138	72	36%	317	32%
13	1/125	127	10%	195	0%	1/111	224	20%	238	6%	1/105	349	28%	242	27%
12	1/109	300	20%	314	0%	1/94	587	41%	488	19%	1/88	906	48%	719	48%
11	1/94	609	38%	641	0%	1/78	1,171	54%	1,092	37%	1/72	1,788	60%	1,617	56%
10	1/85	869	43%	1,031	3%	1/67	1,717	59%	1,847	45%	1/59	2,682	65%	2,795	62%
9	1/82	1,347	36%	1,503	3%	1/61	2,570	52%	2,690	37%	1/51	4,050	60%	4,126	53%
8	1/84	1,456	41%	1,763	10%	1/59	2,870	58%	3,188	46%	1/46	4,799	66%	4,997	61%
7	1/82	1,709	44%	2,114	28%	1/56	3,331	59%	3,815	52%	1/42	5,574	65%	6,075	65%
6	1/81	2,245	42%	2,567	5%	1/55	4,076	55%	4,591	41%	1/41	6,400	62%	7,141	53%
5	1/79	2,875	48%	3,097	16%	1/56	5,145	60%	5,321	46%	1/42	7,696	66%	7,961	58%
4	1/82	3,575	55%	3,642	22%	1/61	5,731	62%	5,918	50%	1/47	8,025	67%	8,342	61%
3	1/79	3,635	52%	3,892	36%	1/62	5,572	59%	6,125	52%	1/52	7,608	63%	8,507	63%
2	1/90	1,809	15%	2,909	54%	1/75	2,624	24%	4,504	65%	1/65	3,568	32%	6,330	70%
sum	1/79	20,564	44%	23,891	23%	1/55	35,654	55%	40,089	48%	1/41	53,518	62%	59,169	60%

For the cumulative damage evaluation, the previously presented formulas are applied and compared with the beam end fracture verification method for super high-rise steel buildings due to repeated deformation, as proposed by Japan's Ministry of Land, Infrastructure, and Tourism [6]. The results are summarized in Table 2, demonstrating good agreement between the computational analysis results and the proposed method's estimates. While some adjustments may be necessary, the method already provides valuable insights into the structural health of the building and the potential need for evacuation in the event of aftershocks.

Table 2. Comparison of damage level.

Floor	JMA Kobe ×2.0				JMA Kobe ×2.5				JMA Kobe ×3.0			
	Frame		Brace		Frame		Brace		Frame		Brace	
	Computational analysis[1]	Proposed method[2]	Computational analysis[2]	Proposed method[2]	Computational analysis[1]	Proposed method[2]	Computational analysis[2]	Proposed method[2]	Computational analysis[1]	Proposed method[2]	Computational analysis[2]	Proposed method[2]
14	0%	3%	0%	0%	0%	3%	0%	0%	0%	1%	0%	0%
13	0%	0%	0%	0%	0%	0%	0%	0%	0%	1%	1%	0%
12	0%	0%	1%	0%	1%	1%	1%	0%	2%	2%	1%	0%
11	1%	0%	1%	1%	3%	2%	1%	1%	5%	5%	2%	1%
10	1%	0%	1%	1%	5%	4%	2%	1%	9%	8%	2%	1%
9	2%	0%	1%	1%	7%	4%	2%	1%	15%	9%	3%	1%
8	3%	1%	1%	1%	9%	6%	2%	1%	25%	13%	3%	1%
7	4%	3%	2%	1%	11%	9%	2%	1%	35%	18%	3%	2%
6	4%	0%	2%	1%	11%	7%	2%	1%	30%	14%	3%	2%
5	5%	2%	2%	2%	12%	9%	3%	2%	29%	18%	3%	2%
4	6%	3%	2%	2%	11%	11%	3%	2%	24%	19%	3%	2%
3	7%	5%	2%	2%	12%	11%	3%	2%	22%	19%	4%	3%
2	1%	8%	2%	1%	1%	15%	3%	2%	2%	23%	3%	2%

*1 Beam end failure probability with C = 8 as in []
*2 Plastic cumulative deformation demand over capacity

Calculations example

Subsequent section include examples of the calculations for the deformation capacity of various building elements.

The building's design prioritizes strong columns and weak beams, so to evaluate the cumulative plastic deformation capacity of each story, the weakest girder is used to determine $_m\eta_u$. For the third story, the cross-section considered is H-900×300×19×25 (H×B×t_w×t_f) , with a yield strength of 325 N/mm². The cumulative plastic deformation capacity $_r\eta_u$ is calculated as follows:

$_r\eta_u = _m\delta_y/_r\delta_y \times _m\eta_u \times a_p \times a_b + a_d = 1.3 \times 5.21 \times 1.5 \times 2 + 2 = 7.21$

$_m\eta_u = (s-1)/S \times (E/E_{st} \times (s-1) + 2(\varepsilon_p/\varepsilon_y))$
$\quad = (1.20-1)/1.20 \times (60 \times (1.20-1) + 2 \times 10) = 5.21$

where $1/s = 0.2868/\alpha_f + 0.0588/\alpha_w + 0.7730 = 0.0164 + 0.0457 + 0.7730 = 0.836$
$\alpha_f = (E/\sigma_{yf}) \times (t_f/b)^2 = (205000/325) \times (25/150)^2 = 17.5$
$\alpha_w = (E/\sigma_{yw}) \times (t_w/d_e)^2 = (205000/325) \times (19/425)^2 = 1.26$
$d_e = H/2 - t_f = 900/2 - 25 = 425$, $b = B/2 = 300/2 = 150$
$E/E_{st} = 60$, $\varepsilon_p/\varepsilon_y = 10$

For the buckling-restrained braces, a low yield point of 100N/mm² is assumed, with a brace angle of 56 degrees to the horizontal. The yielding segment of the brace is considered to be 30% of the total brace length from floor to floor. A maximum inter-story drift angle of 1/50 is used to calculate the strain maximum amplitude. The cumulative plastic deformation capacity η_s is expressed as:

$\eta_s = 2 \times (\Delta\varepsilon_t/\varepsilon_y - 2) \times N_f \times \cos(56)^2 = 2 \times (5.54/0.09 - 2) \times 59.4 \times \cos(56)^2$
$\quad = 7841 \times \cos(56)^2 = 2450 \rightarrow 2400$
$N_f = (\Delta\varepsilon_t/20.48)^{-1/0.49} = 59.4$
$\Delta\varepsilon_t = 2/(50 \times 0.30) \times \sin(56) \times \cos(56) = 5.54\%$
$\varepsilon_y = 100/205000 \times \cos(56)/0.30 = 0.09\%$

The parameters inputted in NSmos® for the case study are summarized in Table 3.

Table 3. Model parameters.

Floor	Mass m (kN)	Floor height h (mm)	X direction						Y direction					
			Frame			Brace			Frame			Brace		
			Stiffness (kN/mm)	Yield drift (mm)	Deformation capacity	Stiffness (kN/mm)	Yield drift (mm)	Deformation capacity	Stiffness (kN/mm)	Yield drift (mm)	Deformation capacity	Stiffness (kN/mm)	Yield drift (mm)	Deformation capacity
14	9,629	5,900	263	44.4	6.96	26	39.0	2,200	349	23.8	6.96	197	18.5	2,600
13	12,126	4,450	526	38.5	6.96	78	18.9	2,300	669	21.5	6.96	361	16.4	2,800
12	10,772	4,000	700	38.0	6.96	122	14.0	2,400	836	23.4	6.96	497	12.9	2,800
11	10,579	4,000	770	41.6	6.96	148	11.4	2,400	869	28.1	6.96	563	11.5	2,800
10	10,591	4,000	824	44.5	6.96	171	9.9	2,400	908	30.1	6.96	627	10.1	2,800
9	10,544	4,000	852	45.5	7.21	277	11.1	2,400	908	31.7	7.21	791	9.6	2,800
8	10,650	4,000	950	41.7	7.21	276	10.9	2,400	1,035	30.9	7.21	816	9.4	2,800
7	10,800	4,000	996	38.9	7.21	306	9.5	2,400	1,058	31.7	7.21	904	8.4	2,800
6	10,811	4,000	1,028	42.9	7.21	406	9.8	2,400	1,055	31.5	7.21	1,127	7.9	2,800
5	10,841	4,000	1,093	40.9	7.21	449	8.7	2,400	1,101	33.0	7.21	1,263	7.2	2,800
4	10,934	4,000	1,196	39.9	7.21	499	7.9	2,400	1,197	32.2	7.21	1,413	6.5	2,800
3	11,060	4,000	1,284	39.1	7.21	592	6.6	2,400	1,230	32.7	7.21	1,693	5.4	2,800
2	11,023	4,000	1,569	30.2	6.84	688	5.3	2,400	1,470	29.0	6.84	1,989	4.6	2,800
1	9,791	4,000	1,289	0.7	-	7,075	11.2	600	1,299	1.1	-	6,970	10.2	750

Summary

The integration of a cumulative damage assessment function for steel structures into the NSmos® monitoring system provides a significant advancement in evaluating building safety under repeated earthquake events. By utilizing accelerometer data to calculate earthquake induced hysteretic energy and comparing it with the building's energy dissipation capacity, the system offers a comprehensive approach to monitoring structural health over time. The methodology has been validated through computational analysis, demonstrating its reliability and effectiveness. This enhanced capability not only improves decision-making regarding building evacuation and maintenance but also contributes valuable insights for designing resilient structures capable of withstanding multiple earthquakes. Future developments may focus on refining the system's accuracy and expanding its applicability to a wider range of building types and seismic scenarios.

References

[1] N. Giron, H. Imaeda, Y. Sakai, T. Ishizaki, K. Hirai and H. Harada, NSmos® A structural health monitoring system developed and operated by structural engineers, 8h WCSCM, Orlando, 2022

[2] Technical standards for seismic calculation methods based on energy balance (Japanese), The Building Center of Japan, 2005

[3] H. Imaeda, H. Harada and T. Shinohara, Assessment of the Damage Estimation System in Nikken Sekkei Tokyo Building, 6th APWSHM, Hobart, 2017 pp400-407. https://doi.org/10.1016/j.proeng.2017.04.501

[4] H. Harada, T. Shinohara and K. Sakakibara, A Study on Dynamic Behavior of Nikken Sekkei Tokyo Building Equipped with Energy Dissipation Systems when Struck by The 2011 Great East Japan Earthquake, 15 WCEE, Lisboa, 2012

[5] Japan Meteorological Agency, 1995 South Hyogo Earthquake, accessed 16 October 2024, [https://www.data.jma.go.jp/eqev/data/kyoshin/jishin/hyogo_nanbu/dat/H1171931.csv],Japanese

[6] Building research institute, technical materials and data related to long-period earthquake motion countermeasures, accessed 16 October 2024, [https://www.kenken.go.jp/japanese/contents/topics/lpe/51.pdf], Japanese

Structural Health Monitoring: 10APWSHM
Materials Research Proceedings 50 (2025) 292-298

Materials Research Forum LLC
https://doi.org/10.21741/9781644903513-34

Enhanced seismic isolation system by using a yoke-type inerter

Li Zhang[1,a], Songtao Xue[1,2,b], Ruifu Zhang[1,c] and Liyu Xie[1,d] *

[1]Department of Disaster Mitigation for Structures, Tongji University, Shanghai 200092, China

[2]Department of Architecture, Tohoku Institute of Technology, Sendai 982-8577, Japan

[a]zhangli24@tongji.edu.cn, [b]xue@tongji.edu.cn, [c]zhangruifu@tongji.edu.cn,
[d]liyuxie@tongji.edu.cn

Keywords: Yoke-Type Inerter, Nonlinear Inertia, Performance Improvement

Abstract. Previous studies have shown that inserting an inerter whose force-acceleration relationship is linear into the isolation layer of a base-isolated structure can effectively improve its seismic performance. A nonlinear strategy has the potential to provide a better control effect than its linear counterpart. This study proposed a yoke-type nonlinear inerter device by employing the Scotch yoke mechanism. The mechanical model of the proposed nonlinear inerter is built to illustrate the nonlinear inertial behavior using the Euler-Lagrange method. Then, The yoke-type inerter is incorporated into the isolation layer of a base-isolated structure to improve seismic performance. The relative displacement of isolation layer and responses of the superstructures are checked under different seismic events. The nonlinear inertial behavior of the yoke-type inerter is indicated by the proposed mechanical model. The relative displacement of the isolation layer and responses of the superstructures for the base-isolated structure with yoke-type inerter is smaller than that of the base-isolated structure with a linear inerter whose inertance is the same as the linearized yoke-type inerter, especially for the rare earthquake. It is said that the proposed yoke-type inerter can be seen as a candidate of the nonlinear inerter and can be used to improve the seismic performances of base-isolated structures.

Introduction

The inerter is a recently developed inertia element. Different with the traditional mass element whose inertia force dependent on the absolute acceleration, the output force of inerter is proportional to its relative acceleration [1]. The coefficient of the proportional constant is named as inertance or apparent mass, which is usually hundreds or thousands of times larger than the actual gravitational mass of the inerter itself using some specific implements (e.g., ball screw mechanism [2], hydraulic mechanism [3] and so on). The development of inerter technique facilitates the generation of nonlinear inertia.

Analogy to the mechanical behaviors of nonlinear spring and damping elements based on their relative deformations, the nonlinear inerter that generates the nonlinear inertia force can provide an alternative approach to realize adaptive control. In addition, owing to the apparent mass effect, the nonlinear inerter will not substantially increase the additional weight of the structure as well. Previously, the authors [4] proposed a yoke-type inerter that can generate nonlinear inertia force. However, the corresponding experimental validation is absent on this device. The adaptive control effect of the yoke-type inerter is unclear as well.

This study investigates the adaptive seismic response mitigation of the isolation structure equipped with yoke-type inerter under muti-level earthquakes. The operational principle of the yoke-type inerter is described and its mechanical model is established. Analytical solutions of an isolator equipped with yoke-type inerter are derived under base excitations to investigate its adaptive control potential. The seismic response a base-isolated structure with yoke-type inerter is

Structural Health Monitoring: 10APWSHM   Materials Research Forum LLC
Materials Research Proceedings 50 (2025) 292-298   https://doi.org/10.21741/9781644903513-34

checked under muti-level earthquakes. Comparative analyses are conducted to illustrate the adaptive seismic response mitigation of the yoke-type inerter.

Basic Theory of Yoke-type Inerter

Fig. 1 shows the representation of the proposed yoke-type inerter. Given a relative deformation between the two nodes of this device, the translational motion of the shaft drives the pin moving up and down within the yoke ring. Then, the pin drives the rotation of the flywheel. As a consequence, the yoke-type inerter generates the apparent mass that is larger than its physical mass. Since the relationship between the rotation angle acceleration of the flywheel and translational motion acceleration is nonlinear, the yoke-type generates nonlinear inertia force.

In Fig. 1, the radius of the flywheel is denoted as R. The distance between the pin and the center of the flywheel is r. $x(t)$ denotes the relative displacement of the yoke-type inerter with the vertical axis of the flywheel as the initial motion position, which can be expressed as $x_r(t) - x_l(t)$, where $x_r(t)$ and $x_l(t)$ are the absolute displacements of the right and left connections, respectively.

(a)

(b)

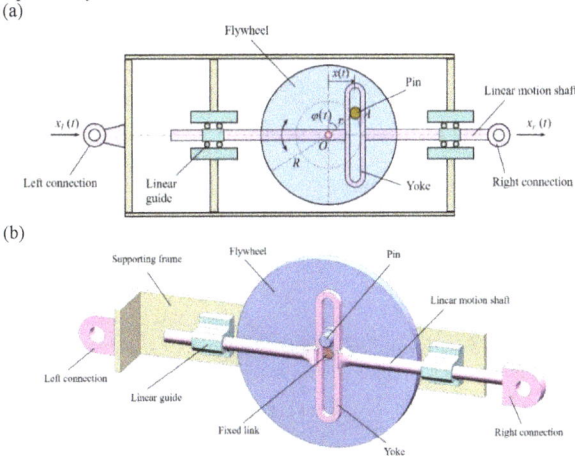

Fig. 1: Representation of the yoke-type inerter. (a) Schematic diagram and (b) three-dimensional model

The formulation of the inertia force of yoke-type inerter can be obtained by the Euler-Lagrange method, which can be expressed as

$$F_I = \frac{m_i R^2}{2} \cdot \left(\frac{1}{r^2 - x^2} \ddot{x} + \frac{x}{\left[r^2 - x^2\right]^2} \dot{x}^2 \right) \tag{1}$$

where m_i is the physical mass of the flywheel. For the sake of simplicity, the derivation process of the inertia force F_I is not presented here. The detailed solution process of F_I can be found in [4].

By considering the nonnegligible friction of the inerter devices as Coulomb friction, the mechanical model of the yoke-type inerter is shown in Fig. 2. The schematic diagrams of force-displacement and force-acceleration relationships for the inertia force of the inerter under specified

Structural Health Monitoring: 10APWSHM
Materials Research Proceedings 50 (2025) 292-298

Materials Research Forum LLC
https://doi.org/10.21741/9781644903513-34

harmonic excitation are shown in Fig. 2(b) and Fig. 2(c), respectively. The nonlinear inertia behaviors of the yoke-type inerter are shown clearly. For comparison, the linear counterparts of inertia behaviors are shown in Fig. 2(b) and Fig. 2(c) as well. In Fig. 2(b), the slope indicates the linearized negative stiffness $-m_{in,L}\omega_e^2$ of the inerter, where $m_{in,L}$ denotes the linearized inertance; ω_e is the frequency of harmonic excitation. In Fig. 2(b), the slope represents the linearized inertance $m_{in,L}$, which can be calculated as follows:

$$m_{in,L} = \frac{m_i R^2}{2r^2} \tag{2}$$

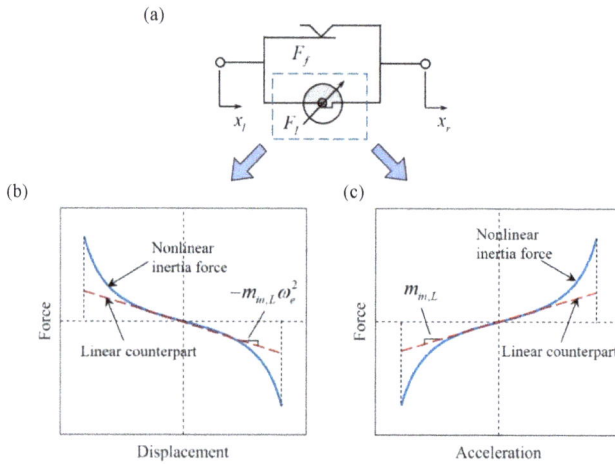

Fig. 2: Mechanical model and nonlinear inertial behaviors of the yoke-type inerter. (a) Mechanical model, (b) force versus displacement and (c) force versus acceleration

Evaluation of An Isolator with Yoke-type Inerter

A vibration isolator equipped with yoke-type inerter under base-excitation is proposed here to check its performance and understand the structural control principle of the inerter. Fig. 3 shows the mechanical model of the vibration isolator with yoke-type inerter. The friction of the yoke-type inerter is ignored to not loss focus, and also the friction is quite small compared to the inertia force. Base on dynamic equilibrium principle, the equation of motion of the isolator with yoke-type inerter is built as follows:

$$m\ddot{y}_r + c\dot{y}_r + ky_r + \frac{1}{2}m_i R^2 \left(\frac{1}{r^2 - y_r^2}\ddot{y}_r + \frac{y_r}{\left[r^2 - y_r^2\right]^2}\dot{y}_r^2 \right) = -m\ddot{q}_e = mq_0\omega^2 \cos\omega t \tag{3}$$

where $y_r = y_d - q_e$ denotes the relative displacement of the isolation mass to the base; y_d and q_e are the absolute displacements of the isolation mass and base, respectively; m, c and k are the mass, damping and stiffness of the isolator, respectively.

Structural Health Monitoring: 10APWSHM   Materials Research Forum LLC
Materials Research Proceedings 50 (2025) 292-298   https://doi.org/10.21741/9781644903513-34

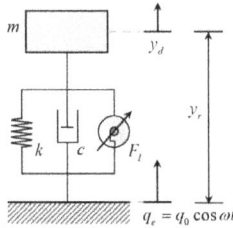

Fig. 3: A vibration isolator equipped with yoke-type inerter under base-excitation

The following dimensionless parameters are adopted to rewrite Eq. (3), which are defined as:

$$\omega_0 = \sqrt{k/m},\ \xi = c/2m\omega_0,\ \mu = m_i/m,\ \gamma = R/r,$$

$$\Omega = \omega/\omega_0,\ \tau = \omega_0 t,\ Y_r = \frac{y_r}{r},\ Q_0 = \frac{q_0}{r},\ \mu_{in} = \frac{\mu\gamma^2}{2} \tag{4}$$

where ω_0 and ξ denote the natural frequency and damping ratio of the isolator, respectively; μ is the ratio of the flywheel mass to the isolator mass; γ is the nondimensional radius of the flywheel; Ω, τ, Y_r and Q_0 are the nondimensional frequency of the excitation, the nondimensional time, the nondimensional displacement response and the nondimensional excitation amplitude, respectively; μ_{in} is the nominal inertance-to-mass ratio.

Substituting Eq. (4) into Eq. (3), a dimensionless form equation of motion is rewrite as

$$Y_r'' + 2\xi Y_r' + Y_r + \mu_{in}\left(\frac{1}{1-Y_r^2}Y_r'' + \frac{Y_r}{\left(1-Y_r^2\right)^2}Y_r'^2\right) = Q_0\Omega^2\cos\Omega\tau \tag{5}$$

where Y_r' and Y_r'' are the first and second order differentials of Y_r with respect to τ, respectively.

For obtaining analytical solutions, the nonlinear terms in Eq. (5) are approximately expanded using Taylor series, expressed as follows:

$$\frac{1}{1-Y_r^2} \approx 1 + Y^2 + Y^4,\ \ \frac{Y_r}{\left(1-Y_r^2\right)^2} \approx Y + 2Y^3 + 3Y^5 \tag{6}$$

Then, the frequency-response relationship of the isolator with yoke-type inerter at steady state can be obtained using averaging method as follows:

$$\left[a_r - a_r\Omega^2 - \mu_{in}\Omega^2 a_r\left(1 + \frac{1}{2}a_r^2 + \frac{3}{8}a_r^4 - \frac{15}{64}a_r^6\right)\right]^2 + 4a_r^2\xi^2\Omega^2 = Q_0^2\Omega^4 \tag{7}$$

The displacement transmissibility of the isolator with yoke-type inerter is adopted to assess its performance, which is defined as follows:

$$TR_D = \frac{|Y_d|}{Q_0} = \frac{\sqrt{a_r^2 + Q_0^2 + 2a_r^2\left[\frac{1}{\Omega^2} - 1 - \mu_{in}\left(1 + \frac{1}{2}a_r^2 + \frac{3}{8}a_r^4 - \frac{15}{64}a_r^6\right)\right]}}{Q_0} \tag{8}$$

Using Eqs. (7) and (8), the frequency-displacement relationship and displacement transmissibility curves are plotted in Fig. 4(a) and Fig. 4(b) under excitations with various amplitudes by setting $\mu_{in} = 2$, $\xi = 0.02$, respectively. In Fig. 4(a), the frequency-displacement curve bends towards the low frequency region with increasing Q_0, which indicates a soften characteristic of the isolator with yoke-type inerter. The resonance frequency for the peak displacement decreases with the increase of Q_0. This is different with linear isolator whose resonance frequency is independent to the excitation amplitude. A varying resonance frequency also indicates that the isolator with yoke-type inerter can operate under excitation with broadband frequency.

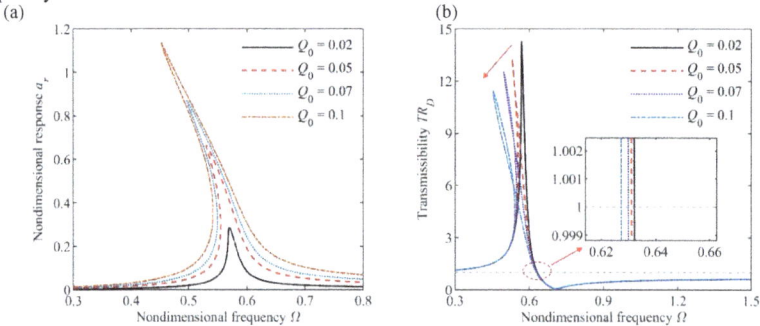

Fig. 4: Isolation performances of the isolator equipped with yoke-type inerter under base-excitation with various amplitudes. (a) Frequency-displacement relationship curves and (b) Displacement transmissibility curves

In Fig. 4(b), the peak of displacement transmissibility for the isolator with yoke-type inerter decreases with the increase of Q_0, whereas the displacement transmissibility for a linear isolator is independent to the excitation amplitude. This indicates a better displacement isolation effect of the isolator under excitation with lager amplitude. In addition, the subfigure in Fig. 4(b) indicates the isolator with yoke-type inerter can achieve effective displacement isolation within a broader frequency band that the transmissibility is less than 1. As a consequence, an adaptive control is said to be achieved by adding yoke-type inerter to the vibration isolator.

Seismic performance of a Base-isolated Structure with Yoke-type Inerter
Fig. 5 shows the mechanical model of a base-isolated structure with yoke-type inerter, which has a single-degree-of-freedom superstructure. The mass m_t, damping c_t and stiffness k_t of the superstructure are 21015 ton, 6340.9 kN/(m/s) and 1.196×10^6 kN/m, respectively. For the isolation layer, the mass m_b of the isolation floor is 6115 ton, the stiffness k_b of the bearing is 2.678×10^5 kN/m. The related isolation period $T_b = 2\pi / \sqrt{k_b / (m_b + m_t)}$ can be calculated as 2 s. The damping coefficient c_b of the isolation layer is 1.585×10^4 kN/(m/s). The acceleration of the ground motion is denoted as $\ddot{u}_g$. u_t and u_b are displacement of the superstructure and isolation floor relative to the ground, respectively. For the base-isolated structure with yoke-type inerter, the equations of motion can be expressed as follows:

Structural Health Monitoring: 10APWSHM
Materials Research Proceedings 50 (2025) 292-298

Materials Research Forum LLC
https://doi.org/10.21741/9781644903513-34

$$\begin{cases} m_t\ddot{u}_t + c_t\left(\dot{u}_t - \dot{u}_b\right) + k_t\left(u_t - u_b\right) = -m_t\ddot{u}_g \\ m_b\ddot{u}_b + c_b\dot{u}_b + k_bu_b + \dfrac{m_iR^2}{2}\left(\dfrac{1}{r^2 - u_b^2}\ddot{u}_b + \dfrac{u_b}{\left(r^2 - u_b^2\right)^2}\dot{u}_b^2\right) - c_t\left(\dot{u}_t - \dot{u}_b\right) - k_t\left(u_t - u_b\right) = -m_b\ddot{u}_g \end{cases}$$

$$(9)$$

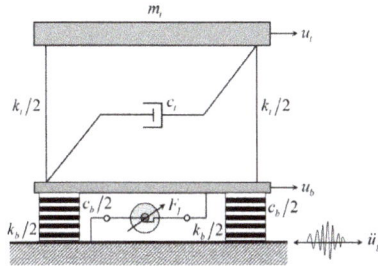

Fig. 5: Base-isolated structure equipped with yoke-type inerter under seismic excitation

By setting the nominal inertance-to-mass ratio μ_{in} equal to 2 for the additional yoke-type inerter, the physical mass of the flywheel m_i is 840.6 ton. The value of r is 0.1 m. The nondimensional radius γ of the flywheel is 10. The mass ratio $\mu_{t,i}$ of the flywheel to the mass of superstructure is 4%. The seismic record of Coalinga wave is adopted as the ground motion excitation. The fourth-order Runge-Kutta method is adopted to solve Eq. (9). The peak ground accelerations (PGA) of the seismic records are adjusted as 70 gal and 400 gal for frequent and rare earthquakes, respectively. The isolation displacement u_b are calculated under different earthquake intensities for Coalinga wave, as shown in Fig. 6. The inerter with inertance of the linear counterpart of yoke-type inerter expressed by Eq. (2) is also incorporated to base-isolated structure as comparison. In this study, the base-isolated structure without inerter, with linear inerter and yoke-type inerter are denoted as ST0, ST1 and ST2, respectively. The isolation displacements of all these structures are shown in Fig. 6.

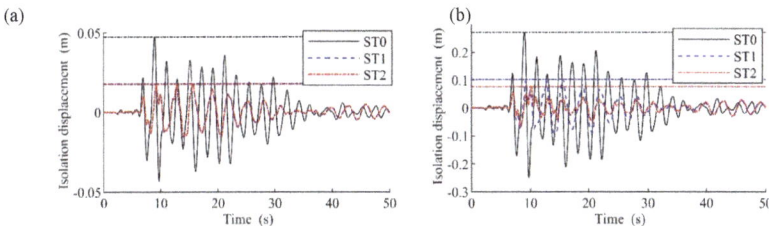

Fig. 6: Isolation displacement responses of the base-isolated structure with yoke-type inerter under various earthquake intensities. (a) Coalinga wave with PGA = 70 gal, (b) Coalinga wave with PGA = 200 gal, (c) Loma Prieta wave with PGA = 70 gal and (d) Loma Prieta wave with PGA = 200 gal

It can be seen in Fig. 6 that the isolation displacements for both the base-isolated structures with linearized inerter and with yoke-type inerter are less than the isolation displacement of base-isolated structure without inerter. This indicates adding both linear and nonlinear inerter are

Structural Health Monitoring: 10APWSHM   Materials Research Forum LLC
Materials Research Proceedings 50 (2025) 292-298   https://doi.org/10.21741/9781644903513-34

beneficial for reducing isolation displacement. For the Coalinga wave excitation, the isolation displacements are almost the same under frequent earthquake level with PGA = 70gal. When the value of PGA grows to 400 gal for the rare earthquake level, the isolation displacements for the structure with yoke-type inerter are clearly less than the isolation displacements for the structure with linearized inerter under Coalinga wave excitation. This indicated the adaptive seismic response mitigation effect of the base-isolated structure with yoke-type inerter is achieved successfully.

Conclusions

This study proposed a yoke-type inerter for adaptive seismic response mitigation. Theoretical analysis and numerical simulation are conducted involving the proposed yoke-type inerter technology in isolation system. Unlike the linear control method, the resonance peak frequency and transmissibility of the isolator with yoke-type inerter decrease with the increase of excitation amplitude, which indicates the adaptive control potential of the yoke-type inerter under excitations with various amplitudes. The base-isolated structure with yoke-type inerter show almost the same seismic mitigation as its linear counterpart under frequent earthquakes, while a better seismic mitigation is clearly shown under rare earthquakes for the structure with yoke-type inerter compared to its linear counterpart.

References

[1] C. Pan, R. Zhang, H. Luo, C. Li, H. Shen, Demand-based optimal design of oscillator with parallel-layout viscous inerter damper, Structural Control & Health Monitoring, 25 (2018) e2051. https://doi.org/10.1002/stc.2051

[2] K. Ikago, K. Saito, N. Inoue, Seismic control of single-degree-of-freedom structure using tuned viscous mass damper, Earthquake Engineering & Structural Dynamics, 41 (2012) 453-474. https://doi.org/10.1002/eqe.1138

[3] S. Kawamata, Development of a vibration control system of structures by means of mass pumps, Institute of Industrial Science, University of Tokyo, Tokyo, Japan, 1973.

[4] L. Zhang, R. Zhang, L. Xie, S. Xue, Dynamics and isolation performance of a vibration isolator with a yoke-type nonlinear inerter, International Journal of Mechanical Sciences, 254 (2023) 108447. https://doi.org/10.1016/j.ijmecsci.2023.108447

Structural Health Monitoring: 10APWSHM   Materials Research Forum LLC
Materials Research Proceedings 50 (2025) 299-306   https://doi.org/10.21741/9781644903513-35

Analog demodulation circuit for structural health monitoring based on nonlinear wave modulation

Takashi Tanaka[1,a] *, Shinsei Nogami[2,b] and Yasunori Oura[1,c]

[1]The University of Shiga Prefecture, 2500 Hassaka, Hikone, Shiga, Japan

[2]Graduate School of Engineering, The University of Shiga Prefecture, 2500 Hassaka, Hikone, Shiga, Japan

[a]tanaka.ta@mech.usp.ac.jp, [b]etz22snogami@ec.usp.ac.jp, [c]oura@mech.usp.ac.jp

Keywords: Diagnostics, Structural Health Monitoring, Nonlinear Wave Modulation, Demodulation, Complex Analog Circuit

Abstract. The purpose of this study is the development of analog demodulation system for a structural health monitoring system based on nonlinear wave modulation. When the structure with contact-type failure, such as tips of fatigue cracks, delamination of composite materials and bolt loosening is vibrating at low-frequency, the amplitude and phase modulation of ultrasonic waves or vibrations occurs and this phenomenon is called nonlinear wave modulation or vibro-acoustic modulation. The demodulation process is necessary to evaluate the failure development and localize the failure location with a simple and inexpensive signal processing system. The authors proposed a demodulation system using a complex analog filter. This complex analog circuit, which operates as Hilbert filter, has the characteristic of widening the operating frequency range as Hilbert filter by increasing the number of filter stages; however, the gain, which is ratio of output voltage to input voltage decreases with the increase of the number of filter stages. In this paper, numerical investigation for design of this complex analog circuit was performed. First, the basic performance of this complex analog circuit was introduced. Next, the design result of this complex analog circuit for specific condition was shown. Moreover, evaluation terms for the design of this complex analog circuit using performance characteristics was proposed.

Introduction

The deterioration of infra-structures is a serious problem in many countries. To prevent a serious accident by the deterioration, structural health monitoring, which is monitoring system attached to the structure, is proposed. By this system, the continuously monitoring and diagnosis of the structural integrity can be realized.

As one of detection method of contact-type failures, such as tips of fatigue cracks, delamination of composite materials and bolt loosening, by using ultrasonic waves or vibrations, the detection method of contact-type failures based on nonlinear phenomenon called nonlinear wave modulation or vibro-acoustic modulation [1-4]. When low-frequency disturbances from wind, construction or traffic is applied to the structure, the contact interface of failure is tapping and clapping. As this result, the amplitude and phase of ultrasonic waves or vibrations modulates in synchronization with low-frequency disturbances. To evaluate failure development [5] and localize failure [6], the demodulation process is necessary. In previous studies [2, 5-6], the demodulation process using digital signal processing was performed. For demodulation of these ultrasonic signals, the measurement system with high-sampling frequency and long-time record and the digital signal processor, that can process at high speeds, is necessary; however, this system is expensive and the data capacity is large.

The authors focused on the fact that the demodulated signal is low-frequency. By the realization of demodulation using analog system, the measurement system with low-sampling frequency and

Structural Health Monitoring: 10APWSHM   Materials Research Forum LLC
Materials Research Proceedings 50 (2025) 299-306   https://doi.org/10.21741/9781644903513-35

the digital signal processor, that process at low speeds, can be used to build the structural health monitoring system of contact-type failures based on nonlinear wave modulation.

The authors propose a demodulation system using complex analog circuit and analog calculation circuit. In this demodulation system, the complex analog circuit called Resistance-Capacitance (RC) polyphase filter was used. In general, polyphase filters used in the field of communications were used to generate of quadrature four signals from two or four input signals with reversed phase [7]. As one of simplest analog polyphase filter, RC polyphase network circuit was proposed [8]. The RC polyphase network circuit consists of only resistors and capacitors; therefore, this circuit is passive circuit.

Here, to demodulate ultrasonic signals measured by vibration sensors, the sensor signal is usually one signal; therefore, the active analog circuit for generate the phase-reversal signal of the sensor signal is necessary. Because the power source of the structural health monitoring system is popularly limited, the lower the power consumption, the better. Accordingly, the authors developed the RC polyphase network circuit for generation of quadrature two signals from one input signal. This complex analog circuit with multi-stages operates as Hilbert filter within a specific frequency range [9] has the characteristic of widening the operating frequency range as Hilbert filter by increasing the number of filter stages; however, the gain, which is ratio of output voltage to input voltage decreases with the increase of the number of filter stages. For optimal design of this complex analog circuit, the evaluation method of performance characteristics, which depends on specific systems, is necessary.

In this paper, numerical investigations for design of this complex analog circuit are performed. First, the basic performance of this complex analog circuit is introduced. Next, the design result of this complex analog circuit for specific condition is shown. Moreover, an evaluation method for the design of the stage number of this complex analog circuit using performance characteristics is proposed.

Detection Method of Contact-Type Failure Based on Nonlinear Wave Modulation
In this section, the overview of structural health monitoring system of contact-type failure is introduced. Moreover, the structure of the proposed analog demodulation system is explained.

The overview diagram of nonlinear wave modulation is shown in Fig. 1(a). When the structure with contact-type failure is vibrating due to environmental disturbance, the contact surfaces of the failure is tapping and clapping in synchronization with this vibration. In this condition, the amplitude and phase of ultrasonic waves or vibrations excited to the structure are modulated.

In previous studies to evaluate the development of contact-type failures and estimate failure location, the digital demodulation processing of ultrasonic signals shown in Fig. 1(b) is used. First, ultrasonic signals are measured using A/D converter as digital signals. Second, the complex signals of measured signals are obtained by Hilbert transformation. Third, the time variation of amplitude and phase of this complex signals are calculated. To calculate the time variation of phase, the average value of time variation of amplitude or phase is decrease from the time variation of amplitude or phase. The time variation signals obtained by this processing is demodulated signals.

Two problems of this processing as a structural health monitoring system is explained. One of them is that the A/D converter with high speeds and the digital signal processor with high performance is necessary to measure and process ultrasonic signals correctly. Another of them is that the large data capacity of digital modulated ultrasonic signals is necessary for this demodulation processing. By these problems, the demodulation system is expensive. When the structural health monitoring of infra-structures using ultrasonic waves or vibrations is considered,

Materials Research Forum LLC
https://doi.org/10.21741/9781644903513-35

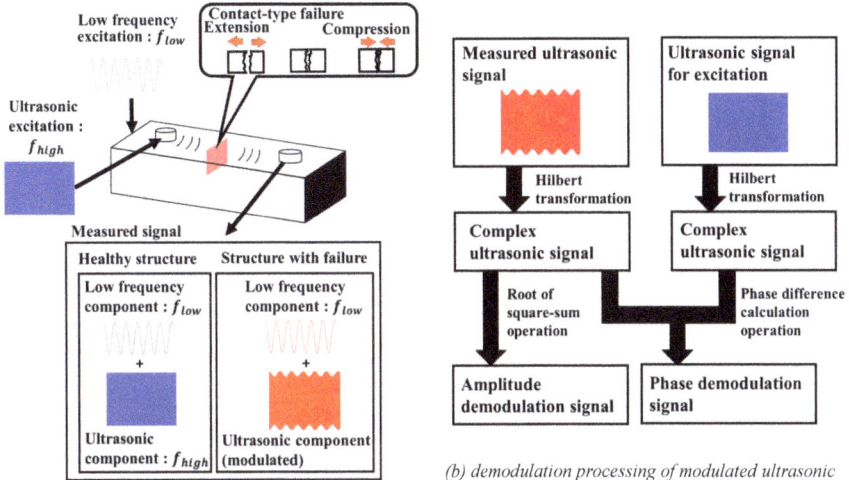

(a) Overview diagram of nonlinear wave modulation

(b) demodulation processing of modulated ultrasonic signals

Fig. 1 Structural health monitoring system of contact-type failures based on nonlinear wave modulation

mounting monitoring units at various locations of infra-structures is necessary. The initial cost of the system installation should be lowered.

Complex Analog Circuit for Demodulation System
In this section, the complex analog circuit for this demodulation system is explained. First, a classical RC poly-phase filter is introduced. Next, the idea of this complex analog circuit is explained. Finally, the design method of this analog circuit is explained.

First, a RC poly-phase filter, which was studied and put into practical use in the field of information and communication technology, is introduced [8]. This RC polyphase filter shown in Fig. 2 consists of resistors and capacitors and is passive circuit. The two signals, of which phases is reversal, input to this filter. The phase difference between output signals ($V_{output\ 1} - V_{output\ 2}$, $V_{output\ 2} - V_{output\ 3}$, $V_{output\ 3} - V_{output\ 4}$) is 90 degrees. As this result, it was confirmed that this poly-

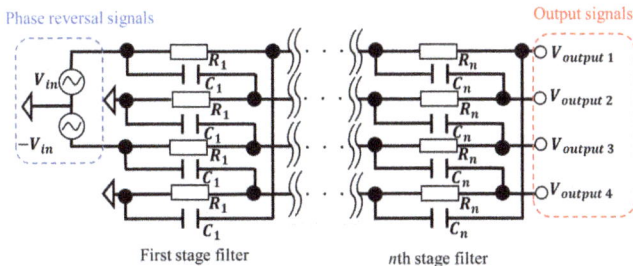

Fig. 2 Classical multi-stage poly-phase filter with two input signals and four output signals.

phase filter can regard as Hilbert filter.

Structural Health Monitoring: 10APWSHM   Materials Research Forum LLC
Materials Research Proceedings 50 (2025) 299-306   https://doi.org/10.21741/9781644903513-35

Here, the problem of this poly-phase filter to use Hilbert filter in demodulation system for structural health monitoring of contact-type failure based on nonlinear wave modulation is described. In the structural health monitoring system, the sensor output signal of vibration sensors for the measurement of ultrasonic signals is one. To obtain complex ultrasonic signals from sensor signals, the inverting amplifier circuit is necessary. By generation of reversal signal of sensor signals without power consumption for amplitude amplification, the amplitude of two signals to input poly-phase circuit is half of the sensor signal. As this result, signal to noise ratio will be down in comparison with the sensor signal.

We proposed the complex analog circuit, of which input signal is one and output signal is two [9], operated as Hilbert filter. The complex analog circuit design is shown in Fig. 3. The basic circuit structure is the same of the RC poly-phase filter. The difference of proposed circuit and RC poly-phase filter is the number of input signals and measurement method of output signals. In this circuit, the input signal is one; therefore, the inverting amplifier circuit is not necessary. Moreover, two differential voltages between $V_{output\ 1}$ and $V_{output\ 3}$ and between $V_{output\ 2}$ and $V_{output\ 4}$ as output signals is measured. Here, the phase difference between these two differential signals is 90 degrees.

The design method of frequency characteristics of this complex analog circuit is explained. The overview diagram of ideal gain characteristics of this complex analog circuit is shown in Fig. 4. The frequencies, of which the gain values of $V_{output\ 1}$ and $V_{output\ 2}$ is the same, f_i ($i = 1 \sim n$) is decided by

$$f_i = \frac{1}{2\pi C_i R_i} \tag{1}$$

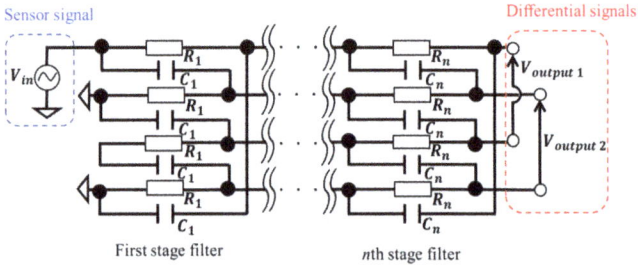

Fig. 3 Proposed multi-stage complex analog circuit with single input signal and two differential output signals.

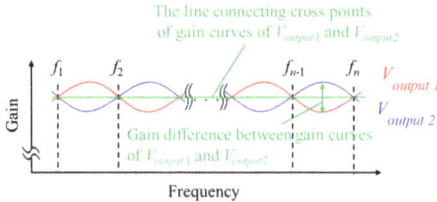

Fig. 4 Overview diagram of ideal gain characteristics of proposed complex analog circuit.

where, C_i and R_i are capacitance of capacitors and resistance of resistors in ith stage filter. The ideal frequency characteristics of this circuit satisfies two conditions. One of them is that line

Structural Health Monitoring: 10APWSHM
Materials Research Proceedings 50 (2025) 299-306

Materials Research Forum LLC
https://doi.org/10.21741/9781644903513-35

connecting cross points of gain curves of $V_{output\ 1}$ and $V_{output\ 2}$ is flat and its value is large. Another of them is that the gain difference between gain curves of $V_{output\ 1}$ and $V_{output\ 2}$ is small. The gain difference in the frequency range between f_i and f_{i+1} ($i = 1\sim n$-1) decreases with increase of stage number of this complex analog circuit; therefore, the stage number is larger, the approximation accuracy of Hilbert filter is improved. However, the absolute gain value decreases with increase of stage number of this complex analog circuit. As this result, the signal to noise ratio is low.

Numerical simulation of Complex Analog Circuit

In this section, the numerical simulations of this complex analog circuit to obtain the guidelines for the optimal design is explained. First, simulation conditions are described. Next, frequency characteristics from input signal to output signals are shown to explain about the Hilbert approximation of this circuit. Finally, the evaluation terms of this circuit to obtain the guidelines for the optimal design is discussed.

The purpose of this numerical simulation is to obtain the complex signals in the frequency range from 10 kHz to 100 kHz, which is usually used as guide waves in ultrasonic inspection; therefore, frequency f_1 and f_n are 10 kHz and 100 kHz. The frequency $f_2 \sim f_{n-1}$ were designed as equally divide the frequency range on logarithmic axis. The capacitance and resistance were designed to gain values at frequency f_i is the same in the case of a single-stage circuit. Designed frequencies and capacitances is shown in Table 1. As an initial investigation, Resistances of all resistor is 1 kΩ and capacitances are designed using Eq. 1.

Table 1 Specific parameters of designed frequencies and capacitances in this numerical simulation

Stage number	2	3	4	5
1st stage	$f_1 = 10$ kHz $C_1 = 15.915$ nF	$f_1 = 10$ kHz $C_1 = 15.915$ nF	$f_1 = 10$ kHz $C_1 = 15.915$ nF	$f_1 = 10$ kHz $C_1 = 15.915$ nF
2nd stage	$f_2 = 100$ kHz $C_2 = 1.5915$ nF	$f_2 = 31.62$ kHz $C_2 = 5.0329$ nF	$f_2 = 21.54$ kHz $C_2 = 7.3873$ nF	$f_2 = 17.79$ kHz $C_2 = 8.9499$ nF
3rd stage		$f_3 = 100$ kHz $C_3 = 1.5915$ nF	$f_3 = 46.42$ kHz $C_3 = 3.4289$ nF	$f_3 = 31.62$ kHz $C_3 = 5.0329$ nF
4th stage			$f_4 = 100$ kHz $C_4 = 1.5915$ nF	$f_4 = 56.25$ kHz $C_4 = 2.8302$ nF
5th stage				$f_5 = 100$ kHz $C_5 = 1.5915$ nF

Frequency characteristics in the case of 2-stage circuit and 5-stage circuit is shown in Fig. 5. It is confirmed that the phase difference between $V_{output\ 1}$ and $V_{output\ 2}$ is 90 degree at all frequency range in both cases. On the other hand, the gain curve at frequency range from 10 kHz to 100 kHz is not flat in cases of both circuits. Moreover, the gain difference between $V_{output\ 1}$ and $V_{output\ 2}$ exists. This gain difference in the case of 5-stage circuit is smaller than it in the case of 2-stage circuit; however, the mean curve of 5-stage circuit is convex function. Moreover, the absolute gain value in the case of 2-stage circuit is larger than in the case of 5-stage circuit. In terms of signal to noise ratio, it is better that the stage number is small. From these results, the optimal stage number of this circuit maybe exists.

To obtain evaluation terms for design of this circuit, details of gain curves are observed. Enlarged views of gain curves in the case of 2-stage and 5-stage circuits are shown in Fig. 6. The

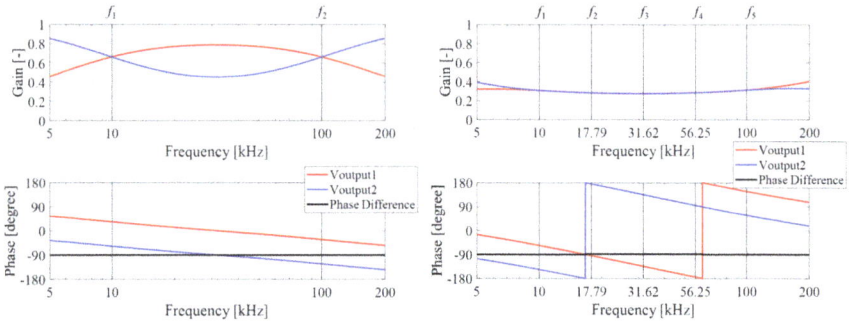

(a) In the case of 2-stage circuit

(b) In the case of 5-stage circuit

Fig. 5 Frequency characteristics of multi-stage complex analog circuit.

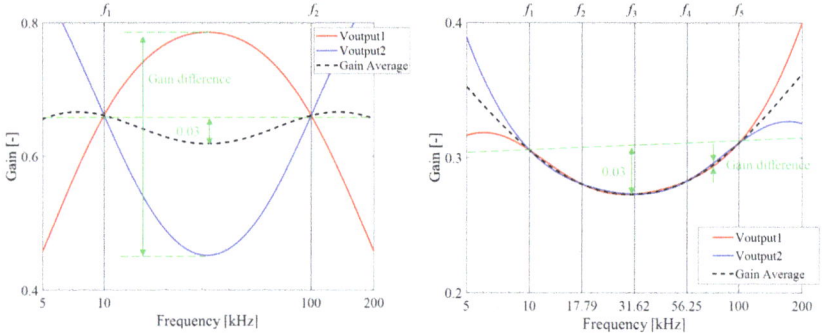

(a) In the case of 2-stage circuit

(b) In the case of 5-stage circuit

Fig. 6 Enlarged view of gain curves.

line connecting cross points of gain curves at frequency f_1 and f_2 in the case of 2-stage circuit or frequency f_1 and f_5 in the case of 5-stage circuit was described as green dashed lines. The average curve of between gain curves of $V_{output\,1}$ and $V_{output\,2}$ was described as black dashed lines. The value of the flat line connecting cross points g_1 is 0.661 in the case of 2-stage circuit and 0.306 in the case of 2-stage circuit in three significant digits. Moreover, the max value of gain difference between the flat line connecting cross points and the average curve g_2 is 0.03 in both cases. Here, the gain flat ratio GFR calculated by Eq. 2 is proposed.

$$GFR = \frac{g_2}{g_1} \tag{2}$$

GFR is the parameter of complex signal variation depending on the frequency variation of the input signal. The relationship between this GFR and stage number of analog circuits is shown in Table 2. The absolute value of gain flat line is decrease with increase of stage number. From this parameter, the signal to noise ratio worsens with increase of stage number. Next, the amplitude variation ratio of output complex signals depending frequency variation is under about 10 %. To reduce this variation, the optimal design of resistances and capacitances considered the influence of each stage circuits is necessary.

Structural Health Monitoring: 10APWSHM   Materials Research Forum LLC
Materials Research Proceedings 50 (2025) 299-306   https://doi.org/10.21741/9781644903513-35

Table 2 Relationship between GFR and stage number

Stage number	2	3	4	5
g_1	0.661	0.515	0.401	0.306
g_2	0.0422	0.0107	0.0293	0.0338
Gain flat ratio GFR	0.0638	0.0208	0.0731	0.111

Next, the gain difference between gain curves of $V_{output\,1}$ and $V_{output\,2}$ is discussed. Max values of the gain difference g_3 are shown in Table 3. The max value of gain difference ratio *MGDR* calculated by Eq. 3 is proposed.

$$MGDR = \frac{g_3}{g_1} \qquad (3)$$

MGDR is the parameter of distortion factor between the real part signal and the imaginary part signal. The relationship between this *MGDR* and stage number of analog circuits is shown in Table 3. The *MGDR* improves with the increase of stage number. The decrease rate of *MGDR* depending the increase of stage number becomes extremely large. From this parameter, the approximation accuracy of Hilbert transformation improves with increase of stage number.

Table 3 Relationship between MGDR and stage number

Stage number	2	3	4	5
g_3	0.334	0.0568	0.0114	0.0025
MGDR	0.505	0.110	0.0284	0.00820

From these results, it can be confirmed that the optimized stage number exists. These parameters evaluate the approximation accuracy of Hilbert transformation; therefore, the optimal design using combination of these parameters as an evaluation function may be effective.

Summary
In this paper, for optimal design of proposed complex analog circuit, the frequency characteristics of multi-stage circuit was investigated by numerical simulation. Three parameters, the absolute gain value of the flat line connecting cross points, the gain flat ratio and the max value of gain difference ratio, were proposed for optimal design guideline of proposed analog circuit. As the future work, the optimal design method using these parameters will be developed.

References
[1] A. Masuda, A. Akisada, T. Tanaka, A. Sone, Detection of adhesive failures via nonlinear piezoelectric impedance modulation, Proc. 12th Int. Conf. Mot. Vib. Cntrl. (2014) 2B21.

[2] S.E. Lee, H.J. Lim, S. Jin, H. Sohn, J.W. Hong, Micro-crack detection with nonlinear wave modulation technique and its application to loaded crack, NDT & E International 107 (2019) 102132. https://doi.org/10.1016/j.ndteint.2019.102132

[3] Y. He, Y. Xiao, Z. Su, Y, Pan, Z. Zhang, Contact acoustic nonlinearity effect on the vibro-acoustic modulation of delaminated composite structures. Mech. Sys. Signal Process. 163(15) (2022) 108161. https://doi.org/10.1016/j.ymssp.2021.108161

[4] T. Tanaka, T. Tamura, Y. Oura, Experimental verification of a translation model of linear transfer function of ultrasonic vibrations with nonlinear wave modulation for detection of contact-type failures. J. Vib. Contrl. OnlineFirst. https://doi.org/10.1177/10775463241237053

[5] T. Tanaka, A. Masuda, A. Sone, Integrity diagnosis method of bolted joint based on time fluctuation of reflection intensity caused by nonlinear wave modulation, Proc. ASME 2014 IDETC (2014) DETC2014-34879.

[6] T. Tanaka, A. Masuda, A. Sone, Estimation of damage location of beam subjected to impact hammer excitation based on nonlinear wave modulation phenomenon. Trans. JSME Series C 79(801) (2013) 1594–1601. https://doi.org/10.1299/kikaic.79.1594

[7] F. Haddad, L. Zaid, W. Rahajandraibe, O. Frioui, Polyphase filter design methodology for wireless communication application, in: S. A. Fares, F. Adachi (Eds.), Mobile and Wireless Communications Network Layer and Circuit Level Design, IntechOpen Ltd., London, 2010, 219-246.

[8] Y. Tamura, R. Sekiyama, S. Sasaki, K. Asami, H. Kobayashi, RC Polyphase Filter as Complex Analog Hilbert Filter, Appl. Mech. Mater. 888 (2019) 26-36. https://doi.org/10.1109/ICSICT.2016.7999091

[9] S. Nogami, T. Tanaka, Y. Oura, Detection of contact-type failure based on nonlinear wave modulation (Design of AM demodulation circuit using complex analog filter), Proc. Dyn. Des. Conf. 2024 (2024) D-OS6-1-02.

Structural Health Monitoring: 10APWSHM   Materials Research Forum LLC
Materials Research Proceedings 50 (2025) 307-314   https://doi.org/10.21741/9781644903513-36

Inversion of the spatially dependent mechanical field based on PIGNNs

Yuchen Bi[1,a], Hesheng Tang[1,b] *

[1]College of Civil Engineering, Tongji University, Shanghai, 200092, China

[a] 2410418@tongji.edu.cn, [b] thstj@tongji.edu.cn

Keywords: Spatially Dependent Mechanical Field, Parametric Inversion, Physics and Data-Driven, PIGNNs

Abstract. Deep learning (DL) is a promising approach to predicting physical phenomena, and has been widely researched in the inversion of the spatially dependent mechanical field due to their remarkable fitting abilities. The traditional physics-informed neural networks (PINNs) disregard the locality of physical evolution processes, which makes the model results lack representation ability and effectiveness. Physics-informed graph neural networks (PIGNNs) can adapt to various types of two-dimensional unstructured grids due to their flexible manipulative data, which enhances the interpretability of the model. Therefore, in this work, a physics and data-driven graph neural networks model is constructed for inversion of the spatially dependent parameter. This model combines traditional numerical methods with unstructured data, and embedded physical information into the graph networks model in the form of discrete differentiation, which made the model have better generalization and interpretability. To verify the feasibility of the model, parameter inversion is performed on plane structures with varying spatially dependent mechanical fields. The results demonstrate that PIGNNs can accurately identify the spatial stiffness variations, and its accuracy in inverting spatially dependent stiffness surpasses that of traditional PINNs. This indicates that the model has strong potential for solving inverse problems related to mechanical parameters.

1. Introduction

Solving mechanical inverse problems is a critical area of research in civil engineering. The mechanical inverse problems entail deducing the boundary conditions, initial conditions, constitutive relationships, boundary loads and governing equations of a structure based on partial structural responses. Common methods for addressing dynamic structural inverse problems include the selection method, the characteristic line method, and the augmented Lagrangian method[1]. Research on mechanical inverse problems is widely applied in fields such as engineering structural health monitoring, damage detection, and non-destructive testing. By analyzing and solving structural mechanics inverse problems, it is possible to detect and localize structural damage and heterogeneity. This is critical for ensuring the safety of structures, extending their service life, and guiding maintenance and repair efforts. However, from the mathematical perspective, the solution of mechanical inverse problems is inherently ill-posed. Additionally, when the model's mechanical parameters vary in a distributed manner, traditional solution methods often lack sufficient accuracy, presenting significant challenges in solving these inverse problems.

In recent years, with the maturation of neural network algorithms and significant advancements in computational power, research on replacing traditional numerical algorithms with neural network methods has proliferated, leading to substantial breakthroughs in deep learning across various fields. Physics-Informed Neural Networks (PINNs) have emerged as a promising framework that leverages observational data alongside physical equations by minimizing a loss function to constrain the outputs of deep neural networks to satisfy partial differential equations. PINNs map from the input domain to the solution through feedforward multilayer neural networks,

Structural Health Monitoring: 10APWSHM   Materials Research Forum LLC
Materials Research Proceedings 50 (2025) 307-314   https://doi.org/10.21741/9781644903513-36

where the partial derivatives can be easily computed using automatic differentiation[2]. PINNs have been widely applied to solve problems in fluid dynamics[3, 4], solid mechanics[5], and thermodynamics, among other engineering applications. In the context of inverse problem solving, PINNs offer the advantage of simultaneously iterating and optimizing parameters that need to be solved in the inverse problem while addressing the forward problem. There has been considerable research on solving inverse problems based on PINNs. For example, Guo[6] et al. successfully inverted parameters for a two-dimensional large-scale spatial distribution of permeability using known groundwater flow information. Haghighat employed the PINN framework for inversions in solid mechanics[5], identifying the volume and shear moduli in linear elastic problems, and demonstrated the performance of this method in nonlinear von Mises elastoplastic problems. These studies collectively validate the feasibility of using PINNs to solve inverse problems related to differential equations.

However, PINNs, by directly learning the functional mapping from spatial coordinates and time to the solution, neglect the locality and temporal nature of the physical evolution process, leading to a lack of representational capability in the models. As a result, methods utilizing numerical discretization to compute the derivative terms of the physical information loss have gained significant attention. Chen et al.[7] employed structured grids to discretize the computational domain and used the Finite Difference Method (FDM) for discretizing partial differential equations (PDEs). Moreover, to address parameterized partial differential equations on unstructured grids, recent works have focused on constructing generalized discrete loss functions based on the Finite Volume Method (FVM)[8] or the Finite Element Method (FEM)[9] and integrating them into physics-based neural network algorithms. Alongside numerical discretization, research on applying Convolutional Neural Networks (CNNs) and Graph Neural Networks (GNNs) for solving discrete PINNs is also evolving. However, classical CNN convolution operations face inherent limitations in handling irregular grid structures. In contrast, GNNs are better suited for processing unstructured data and can allow the network to learn spatial location evolution, enhancing the interpretability of the models. Currently, Jiang et al.[10] have proposed PhyGNNet, a network that employs a physics-informed graph neural network to solve spatiotemporal PDEs, specifically embedding physical information using FDM. Gao et al. [11] introduced a novel GNN-based discrete PINN framework, employing a mesh-free Galerkin method to construct the PDE residual. Furthermore, Xiang et al. [12]integrated these findings to propose a physics-informed framework based on GNNs and Radial Basis Function Finite Difference (RBF-FD) methods to tackle spatiotemporal PDE problems.

In this paper, based on the principles of the Finite Volume Method, we construct a graph neural network model for solving mechanical inverse problems with distributed parameters. This model is built upon MeshGraphNet [13], which is a graph neural network architecture characterized by an encoder-processor-decoder structure. To validate the feasibility of the model, a plane stress problem influenced by spatially distributed parameters is used as a case study, with three different conditions set. Furthermore, based on the plane strain data from numerical simulations, the inversion of spatially dependent mechanical parameters was achieved, demonstrating the effectiveness of the model.

2. Method

2.1 Linear Elastic Equations

The governing equations for linear elastic materials[5] typically consist of the momentum balance equations, geometric equations, and constitutive equations (Eq.1).

$$\sigma_{ij,j} + f_i = 0 \,.$$

Structural Health Monitoring: 10APWSHM
Materials Research Proceedings 50 (2025) 307-314

Materials Research Forum LLC
https://doi.org/10.21741/9781644903513-36

$$\sigma_{ij} = \lambda \delta_{ij} \varepsilon_{kk} + 2\mu \varepsilon_{ij} \,. \tag{1}$$

$$\varepsilon_{ij} = \frac{1}{2}(u_{i,j} + u_{j,i}) \,.$$

In the equations, σ_{ij} is the stress tensor, f_i is the body force, u_i is the structure displacement, ε_{ij} is the strain tensor and λ, μ are Lamé constants, which are related to Young's modulus and Poisson's ratio. In terms of boundary conditions, there are stress boundary conditions and strain boundary conditions.

In the following discussion of this paper, the two-dimensional plane stress problem will serve as an example.

2.2 Neural Network Model

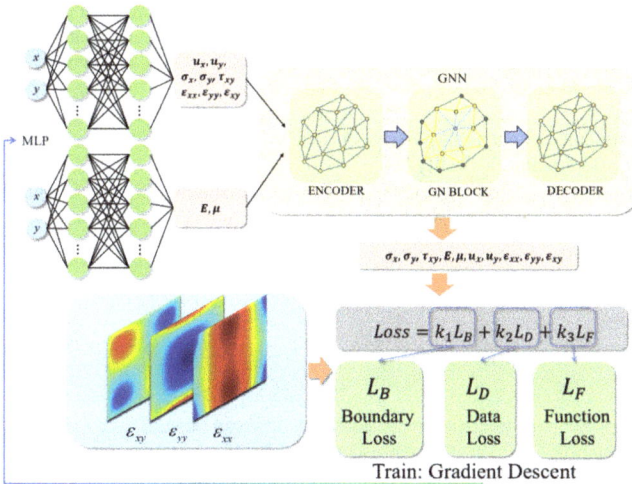

Fig. 1 Topology Diagram of PIGNN for Solving Linear Elasticity Problems

Fig.1 illustrates the overview for solving the inverse problem of spatially distributed mechanical parameters based on PIGNN. The process can be divided into two components: the neural network architecture and the formulation of the loss function. The neural network comprises two components: the first component is the fully connected layer, which takes the coordinates of the model's nodes as input and outputs the mechanical parameters of displacement, stress, strain, and spatial distribution. The second component is the graph network section, in which the MeshGraphNet[13] model is employed as the network architecture. This model is a graph neural network characterized by an Encoder-Processor-Decoder framework and has been applied in various applications. The encoder transforms the aforementioned node and edge features, while the processor utilizes a GN model to predict potential changes in node features. The processed information is then outputted through the decoder. The GN model can be described as follows:

$$m_i^k = \oplus_{j \in N(i)} \phi^k(v_i^k, v_j^k, e_{ij}) \,.$$

$$v_i^{k+1} = \gamma^k \left(v_i^k, m_i^k \right). \tag{2}$$

v_i^k denotes the feature of the i-th node in the k-th layer, m_i^k represents the aggregated information, e_{ij} represents the adjacency information., ϕ^k, γ^k represents a parameterized differentiable function within the deep neural network. $\oplus$ denotes the neighborhood aggregation function.

As shown in Fig.1, the loss function primarily consists of three components: the data-driven term, the boundary loss term, and the governing equation term. Among these, the governing equation loss term is based on Eq.1, and it involves the construction of nine residual equations. The data-driven term is constructed based on strain data obtained from numerical simulations, while the boundary loss term is developed using specific instances to formulate different displacement loss functions or stress loss functions.

2.3 Least Squares Gradient Calculation
In Physics-Informed Neural Networks (PINNs), automatic differentiation algorithms are commonly employed to construct the differential terms in the loss function. For graph neural networks, due to the inherent connections between nodes, discretization methods can often enhance the physical relevance of the structure. Graph structures are typically irregular mesh configurations, making the use of the finite volume method for gradient calculation more appropriate. And gradient computations based on the least squares method within the finite volume framework tend to offer greater flexibility [14].

For a point (P) and its neighboring nodes $(N(P))$, when considering the gradient solely within a two-dimensional plane, we can typically express it in the form of Eq.3.

$$\begin{bmatrix} \sum_{n=1}^{N(P)} \omega_n \Delta x_n \Delta x_n & \sum_{n=1}^{N(P)} \omega_n \Delta x_n \Delta y_n \\ \sum_{n=1}^{N(P)} \omega_n \Delta y_n \Delta x_n & \sum_{n=1}^{N(P)} \omega_n \Delta y_n \Delta y_n \end{bmatrix} \begin{bmatrix} \dfrac{\partial u_P}{\partial x} \\ \dfrac{\partial u_P}{\partial y} \end{bmatrix} = \begin{bmatrix} \sum_{n=1}^{N(P)} \omega_n \Delta x_n \Delta u_n \\ \sum_{n=1}^{N(P)} \omega_n \Delta x_n \Delta u_n \end{bmatrix} \tag{3}$$

This equation represents a system of linear equations that can be easily solved through matrix operations. In the equation, $\Delta x_n, \Delta y_n, \Delta u_n$ represents the differences in the x, y coordinates and the u values between point P and its neighbors. ω_n is the weight coefficient, defined by the following equation.

$$\omega_n = \frac{1}{\sqrt{\Delta x_n^2 + \Delta y_n^2}}. \tag{4}$$

Structural Health Monitoring: 10APWSHM   Materials Research Forum LLC
Materials Research Proceedings 50 (2025) 307-314   https://doi.org/10.21741/9781644903513-36

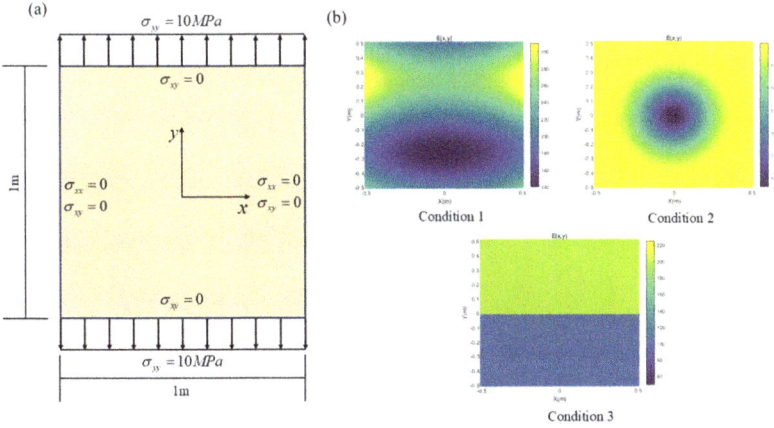

Fig.2 Problem Setting and Experimental Conditions

3. Experiment and Result

3.1 Problems Settings

To validate the feasibility of the model, we solve the plane stress model depicted in Figure 2(a), with the boundary conditions illustrated in the figure. The model does not account for body forces $(f_x = 0, f_y = 0)$.

Three different sets of spatial mechanical parameters are configured, as shown in Figure 2(b). The Poisson's ratio for all models is assumed to be 0.25 and all values of elastic modulus presented in this paper are expressed in MPa. In Condition 1, the Young's modulus (E) displays a sinusoidal distribution in the y-direction and a parabolic distribution in the x-direction (Eq.5). In Condition 2, there is a circular variation of Young's modulus at the center of the structure, and this values follow a Gaussian distribution (Eq.6). Condition 3 features a discontinuous Young's modulus, as described by the distribution function presented in Eq.7.

$$E_1(x, y) = 200 + 30\sin(2\pi y) + 100x^2 . \tag{5}$$

$$E_2(x, y) = 200 - \frac{200}{\pi} e^{-20(x^2+y^2)} . \tag{6}$$

$$E_3(x, y) = \begin{cases} 100, y < 0 \\ 200, y \geq 0 \end{cases} . \tag{7}$$

In terms of neural network configuration, the model is trained using the PyTorch and PyTorch Geometric platforms for graph neural networks, while also employing a PINN constructed with PyTorch for comparative experiments. The Tanh function is utilized as the activation function, and the Rprop algorithm is used as the optimizer, with a learning rate set to 0.001. The target strain data is derived from numerical simulation results. After 200,000 iterations of model training, the results of the model inversion are obtained.

3.2 Inversion of Spatially Continuous Mechanical Parameters

Conditions 1 and 2 involve spatially continuous distributions of mechanical parameters. The model solution results are presented in Fig.3 and Fig.4. Overall, with respect to the spatial identification of Young's modulus, the PIGNN model demonstrates a better performance compared to the PINN model.

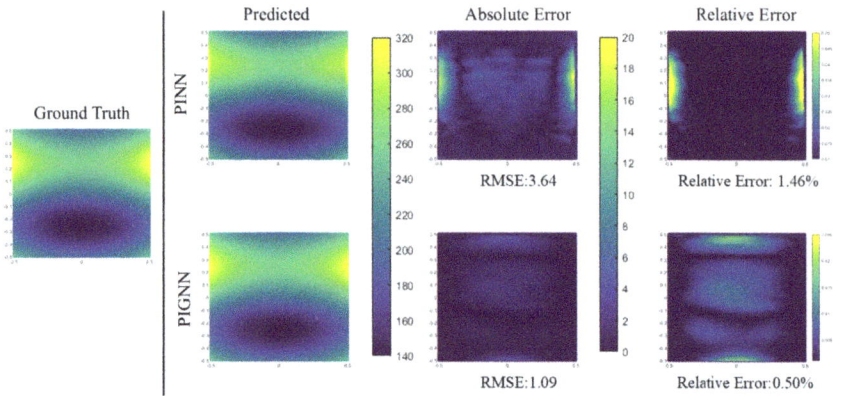

Fig.3 Condition 1 Inversion Results of Young's Modulus

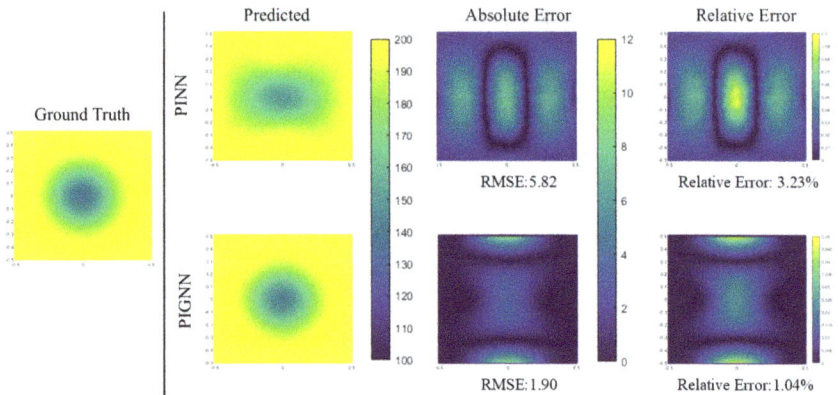

Fig.4 Condition 2 Inversion Results of Young's Modulus

In Condition 1, the PINN model exhibits lower accuracy at the boundaries, with an overall average relative error of 1.46% and a root mean square error (RMSE) of 3.64 MPa. In comparison, our PIGNN model achieves a maximum error of approximately 2%, with an overall average relative error of 0.50% and an RMSE of 1.09 MPa.

Similarly, in Condition 2, the PINN model shows lower accuracy at the center, with an overall average relative error of 3.23% and an RMSE of 5.82 MPa. The overall error of the PIGNN model

Structural Health Monitoring: 10APWSHM   Materials Research Forum LLC
Materials Research Proceedings 50 (2025) 307-314   https://doi.org/10.21741/9781644903513-36

is roughly one-third of that of the PINN model. Thus, it can be concluded that our model demonstrates a better performance in identifying spatially continuous mechanical parameters.

3.3 Inversion of Spatially Discontinuous Mechanical Parameters

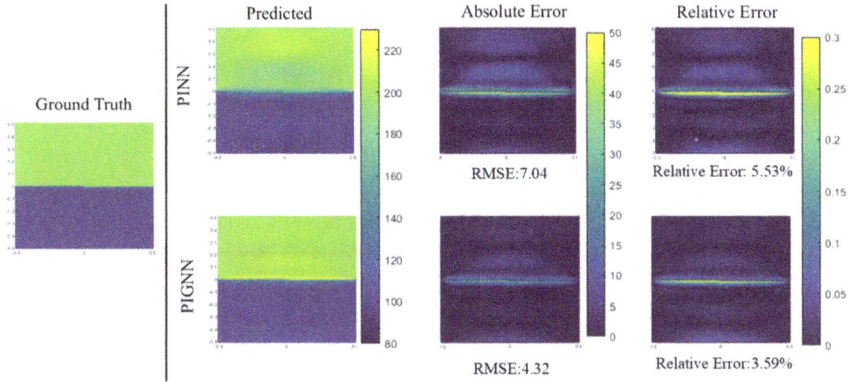

Fig.5 Condition 3 Inversion Results of Young's Modulus

When solving for mechanical parameters with spatial discontinuities in Condition 3, as shown in Fig.5, it is evident that the model errors are primarily concentrated at the points of discontinuity. Both the PINN and PIGNN models exhibit significant errors at these discontinuity points, while the errors in other regions remain relatively small. The overall RMSE for the PINN model is 7.04 MPa, with a relative error of 5.53%, whereas the RMSE for the PIGNN model is 4.32 MPa, with a relative error of 3.59%. This indicates that the PIGNN model retains a distinct advantage when addressing problems involving discontinuous mechanical parameters.

4. Summary

By comparing multiple sets of experimental results, it can be concluded that the PIGNN model demonstrates a certain level of fitting accuracy when solving for spatially continuous variations of mechanical parameters. However, when addressing spatially discontinuous variations, the model often exhibits insufficient accuracy at the discontinuity points. The results may be closely related to factors such as grid density, the number of model iterations, and the weights of the loss function. Further research and experimentation can be conducted to explore this issue in more detail.

Acknowledgments

The authors would like to acknowledge the funds from National Natural Science Foundation of China (No.52378184).

References

[1] Zhao X-Y, Sun D, Toh K-C. A Newton-CG Augmented Lagrangian Method for Semidefinite Programming. SIAM J Optim 2010; 20: 1737-65. https://doi.org/10.1137/080718206

[2] Baydin AG, Pearlmutter BA, Radul AA, Siskind JM. Automatic Differentiation in Machine Learning: a Survey. Journal of Machine Learning Research 2018;18:1-43.

[3] Mao Z, Jagtap AD, Karniadakis GE. Physics-informed neural networks for high-speed flows. Computer Methods in Applied Mechanics and Engineering 2020; 360: 112789. https://doi.org/10.1016/j.cma.2019.112789

Structural Health Monitoring: 10APWSHM   Materials Research Forum LLC
Materials Research Proceedings 50 (2025) 307-314   https://doi.org/10.21741/9781644903513-36

[4] Sun L, Wang J-X. Physics-constrained bayesian neural network for fluid flow reconstruction with sparse and noisy data. Theoretical and Applied Mechanics Letters 2020; 10: 161-9. https://doi.org/10.1016/j.taml.2020.01.031

[5] Haghighat E, Raissi M, Moure A, Gomez H, Juanes R. A physics-informed deep learning framework for inversion and surrogate modeling in solid mechanics. Computer Methods in Applied Mechanics and Engineering 2021; 379: 113741. https://doi.org/10.1016/j.cma.2021.113741

[6] Guo Q, Zhao Y, Lu C, Luo J. High-dimensional inverse modeling of hydraulic tomography by physics informed neural network (HT-PINN). Journal of Hydrology 2023; 616: 128828. https://doi.org/10.1016/j.jhydrol.2022.128828

[7] Chen Y, Huang D, Zhang D, Zeng J, Wang N, Zhang H, et al. Theory-guided hard constraint projection (HCP): A knowledge-based data-driven scientific machine learning method. Journal of Computational Physics 2021; 445: 110624. https://doi.org/10.1016/j.jcp.2021.110624

[8] Rezaei S, Harandi A, Moeineddin A, Xu B-X, Reese S. A mixed formulation for physics-informed neural networks as a potential solver for engineering problems in heterogeneous domains: Comparison with finite element method. Computer Methods in Applied Mechanics and Engineering 2022;401: 115616. https://doi.org/10.1016/j.cma.2022.115616

[9] Aulakh DJS, Beale SB, Pharoah JG. A generalized framework for unsupervised learning and data recovery in computational fluid dynamics using discretized loss functions. Physics of Fluids 2022; 34: 077111. https://doi.org/10.1063/5.0097480

[10] Jiang L, Wang L, Chu X, Xiao Y, Zhang H. PhyGNNet: Solving spatiotemporal PDEs with Physics-informed Graph Neural Network. Proceedings of the 2023 2nd Asia Conference on Algorithms, Computing and Machine Learning, New York, NY, USA: Association for Computing Machinery; 2023, p. 143-7. https://doi.org/10.1145/3590003.3590029

[11] Gao H, Zahr MJ, Wang J-X. Physics-informed graph neural Galerkin networks: A unified framework for solving PDE-governed forward and inverse problems. Computer Methods in Applied Mechanics and Engineering 2022; 390: 114502. https://doi.org/10.1016/j.cma.2021.114502

[12] Xiang Z, Peng W, Yao W. RBF-MGN: Solving spatiotemporal PDEs with Physics-informed Graph Neural Network 2022. https://doi.org/10.1145/3590003.3590029

[13] Pfaff T, Fortunato M, Sanchez-Gonzalez A, Battaglia PW. Learning Mesh-Based Simulation with Graph Networks 2021.

[14] Mavriplis D. Revisiting the Least-Squares Procedure for Gradient Reconstruction on Unstructured Meshes. 16th AIAA Computational Fluid Dynamics Conference, American Institute of Aeronautics and Astronautics; n.d.

Structural Health Monitoring: 10APWSHM
Materials Research Proceedings 50 (2025) 315-322

Materials Research Forum LLC
https://doi.org/10.21741/9781644903513-37

Load and temperature influence on a GW-SHM system for a composite fuselage

Maria Moix-Bonet[1,a] *, Daniel Schmidt[1,b], Benjamin Eckstein[2,c] and Peter Wierach[1,d]

[1] German Aerospace Center e.V., Institute of Lightweight Systems, Lilienthalplatz 7, 38108 Braunschweig, Germany

[2] Airbus Operations GmbH, Airbus-Allee 1, 28199 Bremen, Germany

[a] maria.moix-bonet@dlr.de, [b] daniel.schmidt@dlr.de,
[c] benjamin.eckstein@airbus.com, [d] peter.wierach@dlr.de

Keywords: Guided Waves, Environmental and Operational Conditions, Composite Structures, Aeronautic Structures

Abstract. A full-scale composite door surrounding aircraft structure was instrumented with a GW-SHM system and subjected to three representative quasi-static load cases using a hydraulic test rig. The test was performed in a hangar under uncontrolled temperature environment, resulting in broad temperature variations throughout the experiment. This work focuses on differentiating between benign environmental and operational conditions and barely visible impact damage. A data-driven approach based on Gaussian Processes is used to detect barely visible impact damage introduced during the test campaign, differentiating between benign environmental/operational conditions and barely visible impact damage.

Introduction

The design of aircraft structures has evolved over the years, with the introduction of composite materials leading to new requirements for damage tolerance and structural integrity ([1]). The integration of composite materials in aeronautics offers numerous benefits, but also introduces unique challenges related to damage and failure modes. Currently, the safety of composite structures is guaranteed through the use of damage-tolerant designs and traditional non-destructive inspection methods.

Guided Wave-based Structural Health Monitoring (GW-SHM) has emerged as a promising technique for identifying defects in composite aircraft structures. This method employs a network of permanently installed sensors to generate and receive ultrasonic Guided Waves, allowing for the detection of subtle changes in the structure's integrity ([2],[3]). However, one of the primary challenges facing GW-SHM is the impact of environmental and operational conditions on its performance.

Aircraft structures are subject to a wide range of variations in temperature, pressure, and other factors that can affect the propagation of Guided Waves. This can compromise the effectiveness of GW-SHM systems, which often rely on baseline data to detect anomalies. Compensating for these changes remains an active area of research, with numerous techniques having been proposed ([4],[5]). However, applying these methods in real-world scenarios is frequently challenged by the complex geometry of monitored structures and the simultaneous presence of multiple environmental and operational factors. Thus, this study focuses on using a GW-SHM system to monitor damage in a full-scale composite aircraft fuselage under varying load and temperature.

Experimental Setup: An Aircraft Composite Panel

A composite fuselage structure was employed in this study, which is representative of a passenger door region found on wide-body aircraft, such as the Airbus A350. The test structure has

Structural Health Monitoring: 10APWSHM   Materials Research Forum LLC
Materials Research Proceedings 50 (2025) 315-322   https://doi.org/10.21741/9781644903513-37

dimensions of 4100 mm x 5700 mm and comprises multiple frames, stringers, windows, and a surrounding door structure ([6]).

A custom-designed Guided Wave-based Structural Health Monitoring (GW-SHM) system was integrated onto the fuselage panel. The GW-SHM system's key features include a robust transducer network, reliable transducer connections, and a modular architecture ([7]). This design focuses on sensor reliability and practical application in an industrial setting.

The test rig described in [8] and illustrated in Fig. 1 was used to apply representative loads to the fuselage panel, simulating conditions encountered during flight. Three basic load cases were selected for this study: uniaxial tension, lateral bending left, and vertical bending down. Fig. 1 provides a visual representation of these selected load cases, including their orientation on a barrel level.

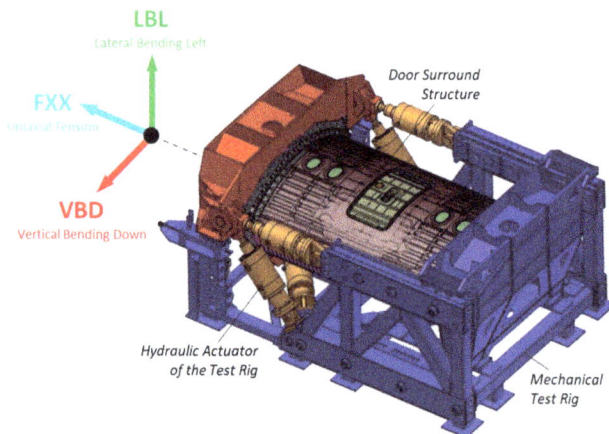

Fig. 1: Composite Fuselage Panel in a Test Rig ([8])

To evaluate the effectiveness of the GW-SHM system, three load cases were applied in a quasi-static manner until a load equivalent to 75% of the structure's Limit Load was reached. The Limit Load (LL) represents the maximum permissible load during normal operation. The quasi-static test paused at multiple steps between the unloaded condition and 75% LL to allow for measurements at various static load levels.

To introduce damage into the fuselage structure, impacts were applied using a gas gun, according the standardized method described in [9]. The impacts created Barely Visible Impact Damage, including delamination and debonding with sizes ranging from 364 mm² to 3720 mm². non-destructive inspection by phased array ultrasonics was performed at the impact locations to gather reference data.

The fuselage structure was equipped with a multi-sensor network, consisting of 148 piezoelectric transducers, 14 strain gauges, and 31 temperature sensors.

Guided Wave-based SHM and changing Environmental and Operational Conditions
There is extensive literature available regarding the sensitivity of GW to environmental and operational conditions and its consequences for the damage identification using GW-SHM systems.

Temperature is a main factor influencing GW propagation, with numerous studies showing its impact on wave velocity and amplitude in composite materials ([10],[11]). Temperature variations affect the response of such SHM systems through multiple mechanisms, including the thermal

Structural Health Monitoring: 10APWSHM   Materials Research Forum LLC
Materials Research Proceedings 50 (2025) 315-322   https://doi.org/10.21741/9781644903513-37

dependence of the piezoelectric effect, the material properties of the monitored structure and the coupling between transducer and structure.

In contrast to temperature's extensive literature, the behavior of guided waves under stress is less well understood. Changes in phase velocity ([12]) and amplitude ([13]) due to tensile load have been reported in correlation with tensile loads, while [14] observed anisotropic behavior in GW propagation, resulting in increased phase velocity in the direction of tension and decreased phase velocity perpendicular to it. However, research is mostly limited to specific cases involving isotropic materials and uniaxial loading conditions.

The current use case involves a combination of multi-axial loading and temperature variation. To investigate the effects of tensile loading and temperature on GW propagation, the signal response from actuator 1 and sensor 3, which corresponds to a highly loaded area, has been selected. To monitor strain, the measurements of an adjacent strain gauge installed in the actuator-sensor path direction have been used. The maximum strain in this direction during the tensile load case reached 1675με.

The signal response for an actuation frequency of 70 kHz is plotted at several load levels in Fig. 2(a) and within a temperature range of 12.8°C to 24.2°C in Fig. 2 (right). The signal has been cropped to focus on the main GW-mode propagating at 70kHz: the anti-symmetric (A$_0$) fundamental mode. In Fig. 2(a), the amplitude increases and the time-of-flight (ToF) decreases as the load is increased. In contrast, Fig. 2(b) shows a decrease in amplitude and an increase in ToF with increasing temperature.

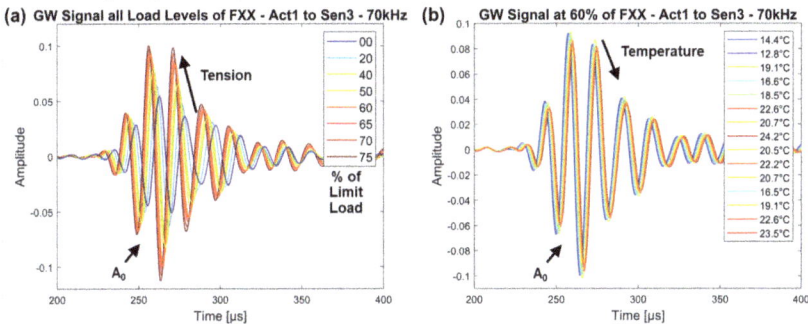

Fig. 2: Guided Wave Signal from Actuator 1 to Sensor 3 – Load (a) and Temperature (b) Effects

To quantify the effects of Environmental and Operational Conditions (EOC), two features have been derived from the GW signal: amplitude and ToF. The amplitude is defined as the maximum amplitude of the Hilbert envelope over the selected wave packet. The reference ToF is calculated using cross-correlation between the actuation pulse and received signal. Subsequent changes in ToF due to load variations are determined through local temporal coherence of the Cross Wavelet Transform, as shown in [15].

Fig. 3 displays all load-temperature combinations acquired during the test campaign in a load level vs. temperature plot. Each point represents the EOC at the time of a GW measurement. The experiment spanned 12 load-unload cycles over six weeks, during which the structure's temperature fluctuated due to the non-climatized environment.

The color range of the scatter plots in Fig. 3 indicates the amplitude and ToF of the GW measurement acquired at each plotted load-temperature combination. For this path, the analysis shows that the dominant effect on the variation of GW-propagation is the load level across both signal features, with a significant increase in amplitude and decrease in ToF.

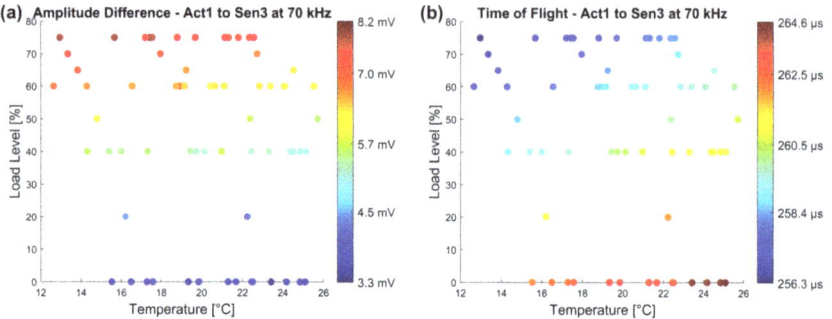

Fig. 3: Amplitude (a) and Time-of-Flight (b) for Temperature and Load Level

It is essential to note that both temperature and load fields are not constant throughout the entire structure, nor do they change uniformly across its geometry. The complex geometry of the structure also means that GW-signals often present interferences, which can lead to non-linear or non-monotonic signal changes in response to temperature and load variations. This is in contrast to the monotonic and even linear relationships between the GW-features and the load and temperature changes observed in Fig. 1 and Fig. 2. A more detailed description of the combined effects of load and temperature during this test campaign can be found in [16].

We now face the challenge of developing a statistical model that can capture the complexity of GW-signals and EOC. Our goal is to identify a model that can distinguish between benign load and temperature effects and damage effects, allowing us to offer reliable SHM capability under such challenging conditions.

Gaussian Process for Damage Detection in changing Environmental and Operational Conditions

Gaussian Processes (GPs) have been increasingly applied in the field of SHM to localize impacts using acoustic emission ([17]), to model of time-series in vibration analysis ([18]) or to quantify damage be means of guided waves ([19]), just to mention a few examples.

For detailed guidance on applying GPs to machine learning tasks, [20] provides an outstanding framework in their work, covering the mathematical formulation of GPs in depth.

This work uses Gaussian Process Regression (GPR) to perform an unsupervised damage detection by means of GW-SHM. For that, GPR is employed to connect a set of input operational variables with the resulting information from the GW-SHM system, using temperature and load as input features and a selected signal feature from the GW-signals as output variable.

The proposed method involves the following key steps:

1. Data Acquisition: Collect GW-data with the monitored structure in its pristine state, under EOC that are representative of the current application scenario. Simultaneously acquire relevant data to the EOC, such as load and temperature.
2. Training a GPR model: Use collected pristine-state data to train a GP model, where load and temperature serve as independent variables and ToF of the GW-data as the dependent variable. The chosen covariance kernel for the GPR model is the squared exponential kernel.
3. Likelihood of new data points: Upon acquiring new GW data and corresponding EOC information in an unknown structural state, apply the fitted GP model for classic anomaly detection. Calculate the likelihood of new data belonging to the pristine state using the GP distribution at specific load-temperature combinations.

4. <u>Damage Detection</u>: Identify potential damage when likelihood exceeds a predefined threshold and covariance uncertainty falls below a predetermined value. If the GP model's uncertainty is too high due to limited information in the region, mark the result as inconclusive.

Some advantages of GPs over other machine learning methods for damage detection with GW-SHM are ([20]):

1. <u>Handling uncertainty</u>: GP models inherently incorporate uncertainty into their predictions, manifesting increased covariance values in feature spaces where data points are scarce or absent. This is particularly useful in monitoring scenarios where data is often noisy and limited.
2. <u>Flexibility</u>: Ability to handle non-linear relationships between inputs and outputs. Complex geometries can lead to interferences in GW-signals, causing non-linear or non-monotonic changes in the signal features in response to temperature and load variations.
3. <u>Scalability</u>: GPs can handle large datasets and high-dimensional feature spaces. GPs can be extended to accommodate multiple input variables, such as complex multi-axial load states as independent variables or high dimensional features to describe GW-signals.

Results

This section aims to understand and assess GPR as a method method to detect damage using a GW-SHM system within the current application scenario: a composite fuselage structure under representative flight loads and temperature variations typical of a non-climatized hangar. For that the GP-based method previously described will be applied exemplarily on an actuator-sensor path affected by introduced impact damage.

The first step involves performing a GPR on the GW-signals acquired from the structure in its pristine state. Fig. 4(a) provides an illustration of how GPR fits to the observed data and demonstrates how data availability affects the GPR fit.

Fig. 4(a) displays the observed time shift values over the two independent variables (load level and temperature) fitted in a GPR model along with the 95% confidence bounds. Three additional views of the fitted GPR model are plotted in Fig. 4: plots (b) to (d). These views visualize the results in two dimensions and highlighting selected data points. Plot (b) shows a 2D cut at 0% load, where numerous GW data points were acquired in pristine condition, resulting in low uncertainty and narrow prediction intervals. In contrast, plot (c) illustrates a 2D cut at 30% load, where uncertainty is higher due to limited available data. Plot (d) presents a 2D cut at 18°C. While temperature is an uncontrolled parameter with randomly scattered data points across its range, the load level is a controlled parameter, resulting in consistent measurements only at predefined loads. This visualization highlights the impact of limited data at certain load levels.

To examine the influence of damage combined with load, an actuator-sensor path featuring damage presence have been selected. This particular path was affected by an impact event towards the end of the test campaign, which caused a delamination of approximately 20×29 mm² in size. This visual representation provides an insight into the interaction between operational factors and damage, and how GPs can be used to discriminate between the two.

Fig. 5 presents unsupervised damage detection with a GPR model fitted on GW-data acquired in a pristine state. The GPR fit in 3D, shown in Fig. 4(a), is displayed again but with an alternative actuator-sensor path that runs transversely to the applied load direction. This results in the Guided Waves experiencing compression instead of tension. Thus, the GPR fit shows a monotonic increase in time shift for both higher temperatures and higher load levels.

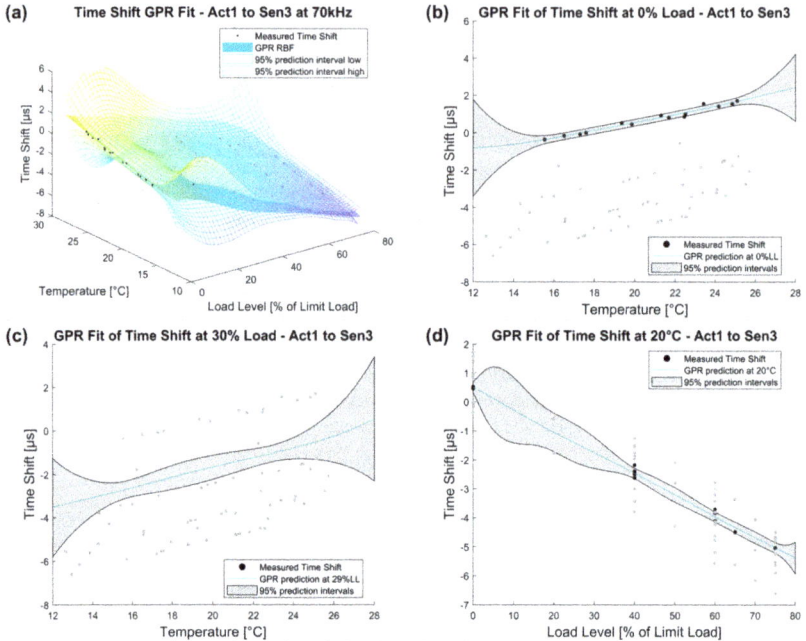

Fig. 4: (a) GPR fit of a single path with load level and temperature as independent variables and time shift as GP-output. 2D cut of GP model at 0% load (b), 30% load (c) and 20°C (d)

Fig. 5(b) provides a 2D cut of the 3D GPR fit, representing the model at a constant load level of 0% or unloaded state. The plot distinguishes between points used for training, which are colored in black for those at 0% load and grey for others. The colored points display the observed time shift values acquired during the unsupervised detection phase, where the color scale corresponds to the calculated likelihood or *how well the fitted GPR model explains the observed new data.*

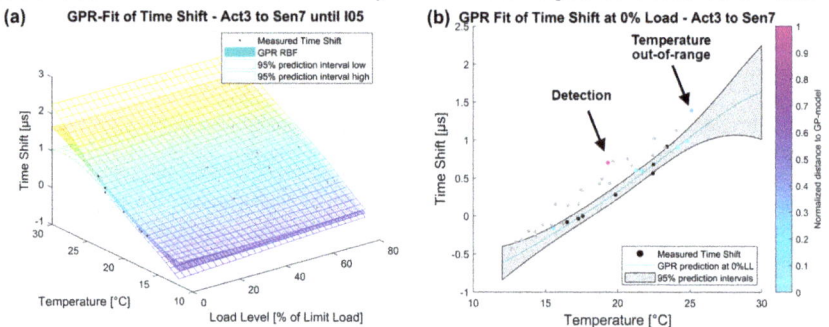

*Fig. 5: (a) Gaussian Process Regression fitted on Time Shift for Actuator 3 to Sensor 7 Path
(b) GPR fit at 0% Load with Training and Test Data Points*

The two marked data points, acquired after the impact event, clearly deviate from the 95% confidence bounds. One data point falls outside the temperature range used to train the GPR model due to high covariance in the affected area, rendering it difficult to clearly identify the damage. In contrast, the second point is within the temperature and load ranges used for training, and its deviation from the fitted GPR model indicates an anomalous observation.

The results show that excluding load information would not allow differentiation between operational conditions and damage, highlighting the importance of considering load effects in the architecture and signal processing design of GW-SHM-systems. The similarity in time shift between these anomalous points and those observed at other load levels (grey points) underscores the importance of accounting for load effects when designing GP models.

Summary

A Guided Wave-based Structural Health Monitoring (GW-SHM) system has been installed on a representative composite fuselage structure. A full-scale test, including multi-axial mechanical loading and the induction of Barely Visible Impact Damage, has generated a comprehensive dataset for the application of GW-SHM in realistic scenarios.

The dataset reveals the effects in GW propagation due to variations in both mechanical loads and temperature. An analysis highlights the relationship between two GW-signal features, time-of-flight (ToF) and amplitude, and the measured environmental and operational conditions.

This dataset has been used to evaluate a novel damage detection algorithm based on Gaussian processes (GP). The approach employs anomaly detection, where GW-data collected in a pristine state are used to train a GP model with load and temperature as input variables and a selected signal feature as output. When new data points arrive, the likelihood of each point belonging to the trained GP model is calculated. The results show that a GP-based detection algorithm is effective in distinguishing between Environmental and Operational Conditions (EOC) and damage for a successful structure monitoring in the presented application scenario.

Acknowledgements

We would like to express our gratitude for the support provided by the Federal Ministry of Economics and Climate Protection (BMWK) through their funding of this project under the LuFo VI framework concept, with grant agreement number 20D2105G. The success of this research initiative was made possible through the collaborative work with our partners.

References

[1] European Union Aviation Safety Agency. (2018). Easy Access Rules for Large Aeroplanes (CS-25) (Initial issue)

[2] Ricci, F., Monaco, E., Maio, L., Boffa, N. D., & Mal, A. K. (2016). Guided waves in a stiffened composite laminate with a delamination. Structural Health Monitoring, 15(3), 351-358. https://doi.org/10.1177/1475921716636335

[3] Moix-Bonet, M., Eckstein, B., Loendersloot, R., & Wierach, P. (2015). Identification of Barely Visible Impact Damages on a Stiffened Composite Panel with a Probability-based Approach. 10th International Workshop on SHM. https://doi.org/10.12783/SHM2015/290

[4] Douglass, A. C. S., & Harley, J. B. (2018). Dynamic Time Warping Temperature Compensation for Guided Wave Structural Health Monitoring. IEEE Transactions on Ultrasonics, Ferroelectrics, and Frequency Control, 65(5), 851-861. https://doi.org/10.1109/TUFFC.2018.2813278

[5] Worden, K., Cross, E. J., Antoniadou, I., & Kyprianou, A. (2014). A multiresolution approach to cointegration for enhanced SHM of structures under varying conditions - An exploratory study. Mechanical Systems and Signal Processing, 47(1-2), 243-262. https://doi.org/10.1016/j.ymssp.2013.10.012

Structural Health Monitoring: 10APWSHM   Materials Research Forum LLC
Materials Research Proceedings 50 (2025) 315-322   https://doi.org/10.21741/9781644903513-37

[6] Kruse, F., Kühn, M., Krombholz, C., Ucan, H., & Torstrick, S. (2015, September 2). Production of a Full Scale Demonstrator-Structure within the FP7 Project "Maaximus." 5th International Workshops on Aerostructures.

[7] Schmidt, D., Moix-Bonet, M., Galiana, S., & Wierach, P. (2023, September). Integration von angepassten und dezentralen Structural Health Monitoring Systemen in Faserverbundstrukturen. Deutscher Luft- Und Raumfahrtkongress (DLRK).

[8] Sachse, M., Götze, M., Nebel, S., Berssin, S., & Göpel, C. (2020). Testing approach for over wing doors using curved fuselage panel testing technology. Lecture Notes in Mechanical Engineering, 831-837. https://doi.org/10.1007/978-3-030-21503-3_65

[9] Federal Aviation Administration. (2018). Certification and Compliance Considerations for Aircraft Products with Composite Materials.

[10] Lanza di Scalea, F., & Salomone, S. (2008). Temperature effects in ultrasonic Lamb wave structural health monitoring systems. The Journal of the Acoustical Society of America, 124(1), 161-174. https://doi.org/10.1121/1.2932071

[11] Yinghui Lu, & Michaels, J. E. (2009). Feature Extraction and Sensor Fusion for Ultrasonic Structural Health Monitoring Under Changing Environmental Conditions. IEEE Sensors Journal, 9(11), 1462-1471. https://doi.org/10.1109/JSEN.2009.2019339

[12] F. Chen, & P.D. Wilcox. (2007). The effect of load on guided wave propagation. Ultrasonics, 47(1-4), 111-122. https://doi.org/10.1016/j.ultras.2007.08.003

[13] Roy, S., Lonkar, K., Janapati, V., & Chang, F.-K. (2014). A novel physics-based temperature compensation model for structural health monitoring using ultrasonic guided waves. Structural Health Monitoring, 13(3), 321-342. https://doi.org/10.1177/1475921714522846

[14] Gandhi, N., Michaels, J. E., & Lee, S. J. (2012). Acoustoelastic Lamb wave propagation in biaxially stressed plates. The Journal of the Acoustical Society of America, 132(3), 1284-1293. https://doi.org/10.1121/1.4740491

[15] Eckstein, B., Moix-Bonet, M., & Bach, M. (2014). Analysis of Environmental and Operational Condition Effects on Guided Ultrasonic Waves in Stiffened CFRP Structures. 7th European Workshop on SHM

[16] Moix-Bonet, M., Schmidt, D., Eckstein, B., & Wierach, P. (2024). A Composite Fuselage under Mechanical Load: a case study for Guided Wave-based SHM. 11th European Workshop on SHM. https://doi.org/10.58286/29842

[17] Hensman, J., Mills, R., Pierce, S.G., Worden, K., & Eaton, M. (2010). Locating acoustic emission sources in complex structures using Gaussian processes. Mechanical Systems and Signal Processing, 24(1), 211-223. https://doi.org/10.1016/j.ymssp.2009.05.018

[18] Avendaño-Valencia, L. D., Chatzi, E. N., & Tcherniak, D. (2020). Gaussian process models for mitigation of operational variability in the structural health monitoring of wind turbines. Mechanical Systems and Signal Processing, 142, 106686. https://doi.org/10.1016/j.ymssp.2020.106686

[19] Amer, A., & Kopsaftopoulos, F. (2023). Gaussian process regression for active sensing probabilistic structural health monitoring: experimental assessment across multiple damage and loading scenarios. Structural Health Monitoring, 22(2), 1105-1139. https://doi.org/10.1177/14759217221098715

[20] Rasmussen C.E., Williams C.B. (2006) Gaussian processes for machine learning. MIT Press. https://doi.org/10.7551/mitpress/3206.001.0001

Structural Health Monitoring: 10APWSHM
Materials Research Proceedings 50 (2025) 323-330

Materials Research Forum LLC
https://doi.org/10.21741/9781644903513-38

Identification of vortex induced vibration of long-span bridges based on transfer learning

Puyu Li[1,2,a], Yinan Luo[1,2,b], Jiale Hou[3,c], Chunfeng Wan[1,2,d,*], Changqing Miao[1,2,e], Songtao Xue[4,5,f,*]

[1]School of Civil Engineering, Southeast University, Nanjing, China

[2]Advanced Ocean Institute of Southeast University, Nantong, China

[3]School of Civil Engineering, Tsinghua University, Beijing, China

[4]Department of Disaster Mitigation for Structures, Tongji University, Shanghai, China

[5]Department of Architecture, Tohoku Institute of Technology, Sendai, Japan

[a]220231535@seu.edu.cn, [b]luo1nan@163.com, [c]hjl23@mails.tsinghua.edu.cn, [d]wan@seu.edu.cn, [e]chqmiao@163.com, [f]xue@tongji.edu.cn

Keywords: Transfer Learning, Vortex Induced Vibration Identification, Machine Learning, Long-Span Bridge

Abstract: Bridge vortex induced vibration (VIV) is a resonance phenomenon caused by the periodic shedding of vortices generated by natural wind passing through bridges. Bridge VIVs will not only cause fatigue damage to the structures but also affect driving safety for the passing vehicles. With the popularization of structural health monitoring (SHM) systems, machine learning technology is widely used in the field of vortex induced vibration identification for long-span bridges, due to its intelligence, real-time performance, and sensitivity to data. However, although traditional machine learning algorithms can identify the VIV based on the response data history of long-span bridges collected by SHM systems, they are difficult to apply to bridges which do not have historical vortex induced vibration data. Therefore, this paper proposes an adaptive transfer learning method for identifying VIV in the main girder of long-span suspension bridges. The proposed method can identify VIV without VIV history data of the target bridge. Results show that it can well identify VIVs at the earlier stage based on the SHM datasets of two long-span suspension bridges, verifying its effectiveness and generalization ability.

Introduction

Vortex-induced vibration (VIV) of bridges is a resonance phenomenon of bridges caused by the periodic vortex shedding produced by natural wind passing over the bridges. In recent years, VIV has been observed in many long-span bridges, such as the Tokyo Bay Bridge in Japan [1], the Great Belt East Bridge in Denmark [2], and the Xihoumen Bridge in China [3]. VIV has therefore been widely studied recently. The traditional research on VIV is mainly based on wind tunnel test and numerical simulation, but these methods can not completely simulate the real environment. Therefore, researchers have introduced the structural health monitoring (SHM) techniques [4] to identify VIV.

Annamdas et al. [5] proposed an automatic VIV extraction method based on unsupervised learning of clustering. Lim et al. [6] first used a semi-supervised method to label the historical data, and used the labeled dataset and the deep learning method to construct the VIV recognition algorithm for long-span bridges, and used the monitoring dataset to carry out several trainings in order to determine the optimal parameter range of the recognition algorithm. Hua Xugang et al. [7] proposed a novelty detection technique based on BP neural network to realize the automatic identification of VIV, and verified the effectiveness of the method by the massive monitoring data of Xihoumen Bridge. Huang et al. [8] introduced a random subtraction technique to preprocess the acceleration data and detected the VIV by setting the threshold value to the peak variation coefficient of the preprocessed dataset, and finally by Numerical simulation obtained a dataset of stochastic and VIV of a three-degree-of-freedom mass-spring-damper system, and the effectiveness of the proposed method was verified by running on this dataset. However, it should be noted that all these methods are inseparable from the VIV data in the bridge monitoring data set, and if there are no VIV samples in the monitoring data set of the target bridge, it is impossible to apply the above methods to construct the VIV identification algorithm.

However, it is difficult to collect the measured data VIV for newly built bridges or bridges with late installation of structural health monitoring systems. Therefore, new identification methods for bridges without VIV historical data are required to be developed. Transfer learning (TL) method provides the possibility to realize this need. TL can apply an already trained model to a new, similar task [9].

Inspired by the TL approach, this paper proposes an adaptive VIV identification method for main girders of long-span suspension bridges based on Transfer Component Analysis (TCA), which uses a bridge dataset with VIV monitoring data as the source domain to identify VIV of the target bridge and does not require that the target bridge contains VIV data in the existing dataset. The proposed VIV identification method for long-span bridges can not only identify the VIV of the target bridge in the historical data set, but also realize the real-time identification of VIV online, which can help to identify the VIV at the early stage of VIV automatically, and thus provide timely warning and control means. The effectiveness and strong generalization ability of the proposed method are also verified on SHM datasets of two long-span suspension bridges.

Methodology

Transfer Component Analysis

Transfer Component Analysis (TCA) is a method of transferring feature transformations that maps both the source and target domains into a new feature space, where a classifier can be used in source domain and target domain is constructed. TCA has adopted the following definitions: the source domain is $X_s = \{x_{s1}, x_{s2}, x_{s3}, \ldots, x_{sn_s}\}, x_i \in \mathbb{R}^d$, the source domain label is x_i, where $Y_s = \{y_1, y_2, y_3, \ldots, y_{n_s}\}$ and y_i are the source domain samples and their corresponding labels, and n_s is the number of source domain samples; The target domain is $X_t = \{x_{t1}, x_{t2}, x_{t3}, \ldots, x_{tn_t}\}, x_i \in \mathbb{R}^d$, and the target domain has n_t samples and no labels. TCA begins by assuming the existence of a mapping function ϕ that transfers the source domain X_s and the target domain X_t into a new feature space. This mapping aims to ensure a similar edge distribution distance between the source and target domains within the space $P(\phi(X_s)) \approx P(\phi(X_t))$, thereby achieving approximate equality in the conditional distribution of both domains $P(Y_s|\phi(X_s)) \approx P(Y_t|\phi(X_t))$. Subsequently, a classification algorithm is trained in the new feature space and then applied to the target domain. the pseudocode for TCA is presented in Algorithm 1 described below.

Structural Health Monitoring: 10APWSHM   Materials Research Forum LLC
Materials Research Proceedings 50 (2025) 323-330   https://doi.org/10.21741/9781644903513-38

Algorithm 1 : TCA
Input: Source domain data $X_s = \{x_1, x_2, x_3, \ldots, x_{n_s}\}, x_i \in \mathbb{R}^d$, **source domain label** $Y_s = \{y_1, y_2, y_3, \ldots, y_{n_s}\}$, **destination domain data** $X_t = \{u_1, u_2, u_3, \ldots, u_{n_t}\}, u_i \in \mathbb{R}^d$
Output: W
1.   Construct the kernel matrix K, matrix L, and central matrix H
2.   For Unsupervised TCA : Solve for the feature vector of the first m-order of $(KLK + \mu I)^{-1}KHK$ to obtain **W**
3.   For semi-supervised TCA: Solve for the first m-order eigenvector construction of W based on the labels $(K(L + \lambda)\mathscr{L}K + \mu I)^{-1}KH\widetilde{K}_{yy}$
4.   Output : **W**

Feature extraction

In this study, five parameter features, Root-mean-square, Variance, Kurtosis, Maximum PSD amplitude, PSD ratios, as shown in Table 1, are extracted from the vibration monitoring data as the index to identify VIV, where n is the number of signal sampling points, y_i is the acceleration or deflection data, and $\bar{y}$ is the mean value of the vibration signal. In particular, in PSD ratios, P and P_S are the first and second largest PSD amplitude, and R is defined as the ratio between the P and the P_S. The feature vector x_i are extracted every minute from 10-minute-long acceleration data and displacement data, where $x_i = \{R_{ms,i}, V_i, P_i, K_i, R_i\}, x_i \in R^d$.

Table 1. Features extracted from monitoring data

Features	Abbreviation	Formula
Root-mean-square	R_{ms}	$R_{ms} = \sqrt{\dfrac{1}{n}\sum_{i=1}^{n} y_i^2}$
Variance	V	$V = \dfrac{1}{n}\sum_{i=1}^{n}(y_i - \bar{y})^2$
Kurtosis	K	$K = \dfrac{\sum_{i=1}^{m} l_i}{\sum_{i=1}^{m} x_i + \sum_{i=1}^{m} l_i}$
Maximum PSD amplitude	P	---
PSD ratios	R	$R = \dfrac{P_S}{P}$

Datasets

Reference Bridge

The reference bridge is a two-span asymmetric steel box girder suspension bridge located in the East China Sea. Figure 1 shows the arrangement of the health monitoring system of the bridge. There are three sensors for measuring the vibration of the main girder, which are installed at 3/8, 1/2 and 5/8 spans, and only one acceleration sensor for each span is arranged at the guardrail on the downstream side of the bridge, with a sampling frequency of 50 Hz.

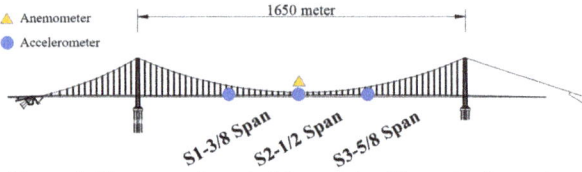

Figure 1. The sensor layout of the new health monitoring system

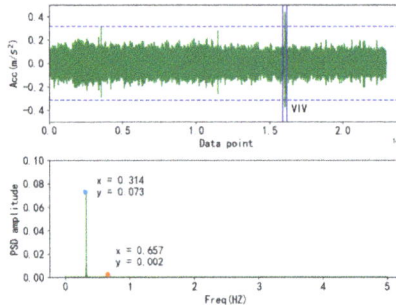

Figure 2. VIV event and the magnitude map of Power Spectral Density (PSD) for Reference Bridge occurred in January of 2021.

The monitoring dataset for the reference bridge contains acceleration data of the three accelerometers from January 2021. Manual analysis of this data revealed that a particular anomalous vibration event occurred at 11:09 p.m. on January 23 and ended at approximately 1:50 a.m. on January 24, for a total duration of nearly three hours. By analyzing the three acceleration data in this time period, it can be concluded that the main frequency of the bridge vibration is 0.314Hz, and the vibration mode occupies an absolutely dominant position, with a PSD amplitude of about 0.073, and the second main frequency is 0.657Hz, with a PSD amplitude of only about 0.002, so that the vibration event in the energy concentration coefficient is obviously lower than 0.1, and the acceleration amplitude is larger, so it is decided that the abnormal vibration is VIV. The acceleration time-range signal of the vortex-excited vibration of the reference bridge and its power spectral density are shown in Figure 2.

Target Bridge
The target bridge is a large-span, three-tower, four-span steel-hybrid combination suspension bridge which is located in the inland region of China where vortex induced vibrations are relatively rare. Figure 3 illustrates the layout of some sensors used in this study. The longitudinal accelerometers are mounted on the top of the intermediate towers, and the transverse and vertical accelerometers of the main girders are set at the bottom centers of the I, K, L, and Q sections as shown in the figure, and the accelerometers are sampled at a frequency of 20 Hz.

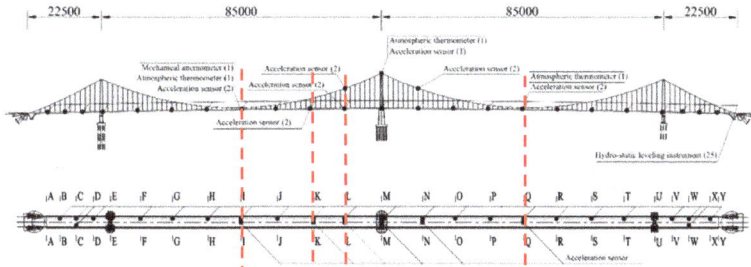

Figure 3. Target Bridge Layout diagram of monitoring sensors (unit: cm)

Verification

The VIV identification method based on TCA algorithm is applied to analyze its advantages in identifying VIVs. Mark the Reference Bridge as Bridge S and the Target Bridge as Bridge T. The proposed algorithm utilizes the monitoring dataset from Bridge S as the source domain and applies it to Bridge T to identify VIV events specific to Bridge T. It is important to note that the VIV algorithm is constructed independently for each sensor's monitoring data.

To assess the advantages of the proposed method algorithm over conventional algorithms, four evaluation indicators are employed, i.e., accuracy, recall, precision, and F1 score, which play a crucial role in evaluating the quality of VIV recognition results (as Figure 4 described). Each of these indicators is measured on a scale from 0 to 1, where a higher value signifies a superior algorithm model.

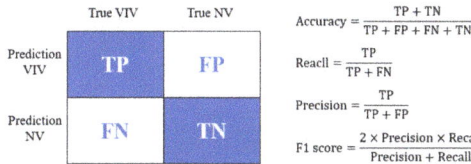

$$Accuracy = \frac{TP + TN}{TP + FP + FN + TN}$$

$$Reacll = \frac{TP}{TP + FN}$$

$$Precision = \frac{TP}{TP + FP}$$

$$F1\ score = \frac{2 \times Precision \times Recall}{Precision + Recall}$$

Figure 4. Definition of evaluation indicators for the evaluation of VIV tasks

The algorithm used in this case involves VIV identification with the S1(Source bridge, section 1) dataset as the source domain and the TL (Target bridge, section L) dataset as the target domain. The results obtained from this analysis are presented in Table 1. The "noDA" algorithm refers to a VIV recognition algorithm that is directly trained on the source data and applied to the target dataset without employing any transfer learning methods. To further enhance the VIV identification process, the subsequent step involves utilizing the TCA algorithm. The outcomes of this approach are presented in Table 2, highlighting the improved accuracy, recall, precision, and F1 score in comparison to the noDA benchmark model.

Table 1. S1 is the source domain and the results of VIV identification in TL

	Method	TN	FP	FN	TP	Accuracy	Recall	Precision	F1 score
S1-TL	noDA	931	0	71	7	0.930	0.090	1	0.165
	TCA	931	0	22	56	0.978	0.718	1	0.836

Compare and analyze the tag recognition results of TCA and noDA by plotting the tag into a figure (Figure 5), where the tag represents whether VIV has occurred, 1 represents VIV, and 0 represents normal vibration. From Figure 5, it can be seen that TCA has thicker bars compared to noDA, indicating the recognition of more VIV data. At the same time, analysis results of the VIV acceleration data recognized by TCA and noDA are shown in Figure 6. It can be intuitively seen that TCA recognizes more VIV data compared to noDA, and more of it is the formation and dissipation process of VIV. From the above analysis, it can be concluded that TCA can achieve better results in identifying VIV data compared to noDA. Moreover, TCA can detect the VIV at its earlier stage which provides great advantages for the following VIV control.

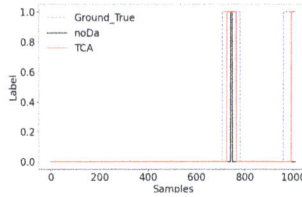

Figure 5. VIV identification results with noDA and TCA models

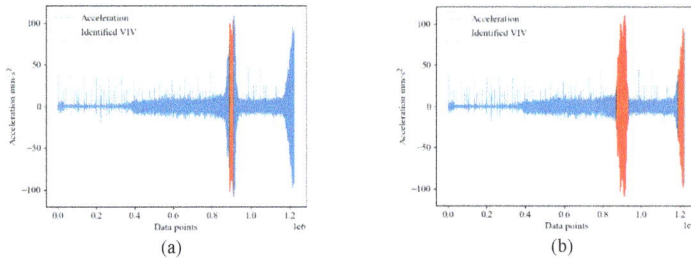

(a)   (b)

Figure 6. VIV data identification from the acceleration signal (a. with noDA, b. with TCA)

To verify the universality of the above conclusions, further verification has also been conducted for more vortex induced vibration identification tasks. The VIV data were identified using S1, S2 and S3 as the source domain and TQ and TL as the target domain, respectively, and the results are shown in Table 2, which shows obvious improvement in accuracy, recall, precision, and F1 score, compared to the results of the noDA algorithm.

Structural Health Monitoring: 10APWSHM — Materials Research Forum LLC
Materials Research Proceedings 50 (2025) 323-330 — https://doi.org/10.21741/9781644903513-38

Table 2 VIV identification with different dataset

	Method	TN	FP	FN	TP	Accuracy	Recall	Precision	F1 score
S1-TQ	noDA	886	0	33	90	0.967	0.732	1	0.845
	TCA	886	0	23	100	0.977	0.813	1	**0.897**
S2-TQ	noDA	886	0	30	93	0.970	0.756	1	0.861
	TCA	886	0	22	101	0.978	0.821	1	**0.902**
S3-TQ	noDA	886	0	33	90	0.967	0.732	1	0.845
	TCA	886	0	23	100	0.977	0.813	1	**0.897**
S2-TL	noDA	931	0	53	25	0.947	0.321	1	0.485
	TCA	931	0	22	56	0.978	0.718	1	**0.836**
S3-TL	noDA	931	0	77	1	0.924	0.013	1	0.025
	TCA	931	0	23	55	0.977	0.705	1	**0.827**

To demonstrate the robustness of the proposed method, we also evaluated its performance using the Bridge T monitoring dataset as the source domain and the Bridge S monitoring dataset as the target domain. From the data in Table 4, it can be seen that the TCA algorithm outperforms the noDA algorithm in all four evaluation indicators (accuracy, recall, precision, and F1 score), with an F1 score closer to 1.

Table 4 VIV identification results using Bridge T dataset as the source domain and the Bridge S dataset as the target domain

	Method	TN	FN	FP	TP	Accuracy	Recall	Precision	F1 score
TQ-S1	noDA	2710	0	46	114	0.984	0.713	1	0.832
	TCA	2710	0	19	141	0.993	0.881	1	**0.937**
TQ-S2	noDA	2710	0	49	111	0.983	0.694	1	0.819
	TCA	2710	0	20	140	0.993	0.875	1	**0.933**
TQ-S3	noDA	2710	0	44	116	0.985	0.725	1	0.841
	TCA	2710	0	19	141	0.993	0.881	1	**0.937**
TL-S1	noDA	2710	0	14	146	0.995	0.912	1	0.954
	TCA	2710	0	10	150	0.997	0.938	1	**0.968**
TL-S2	noDA	2710	0	14	146	0.995	0.912	1	0.954
	TCA	2710	0	10	150	0.997	0.938	1	**0.968**
TL-S3	noDA	2710	0	12	148	0.996	0.925	1	0.961
	TCA	2710	0	10	150	0.997	0.938	1	**0.968**

Based on the above analysis, it can be concluded that the TCA algorithm, achieves superior results compared to the baseline model in both scenarios where bridge T is used as the source domain to identify vortex samples in bridge S and vice versa.

Summary

This paper introduces an adaptive recognition method for VIV in long-span suspension bridges using the TCA technique. The proposed method addresses the limitations of traditional algorithms, such as difficulties in obtaining VIV data, applicability to only a single bridge, and reliance on manual judgment. It enables real-time analysis to identify VIV in the main girder of the bridge.

To verify the effectiveness of the method, monitoring data from Bridge T and Bridge S were utilized. The VIV recognition algorithm based on TCA consistently outperforms the benchmark model, regardless of whether it identifies the VIV samples in Bridge T using Bridge S as the source domain or identifies the VIV samples in Bridge S using Bridge T as the source domain. The proposed method incorporates transfer learning algorithms and could provide better results, which show that the proposed method can correctly identify the VIVs in the earlier stage, thus can brings great benefits to the following early warning and vibration control.

Acknowledgments

This work was supported by the National Natural Science Foundation of China (52178119), National Key R&D Program of China (2021YFE0112200), Nanjing Major Project of Science and Technology, and Research Fund for Advanced Ocean Institute of Southeast University, Nantong (GP202409).

Reference

[1] Fujino Y, Yoshida Y. Wind-Induced Vibration and Control of Trans-Tokyo Bay Crossing Bridge. Journal of Structural Engineering, 2002, 128(8): 1012-1025. https://doi.org/10.1061/(ASCE)0733-9445(2002)128:8(1012)

[2] Frandsen J B. Simultaneous pressures and accelerations measured full-scale on the Great Belt East suspension bridge. Journal of Wind Engineering and Industrial Aerodynamics, 2001, 89(1): 95-129. https://doi.org/10.1016/S0167-6105(00)00059-3

[3] Li H, Laima S, Ou J, et al. Investigation of vortex-induced vibration of a suspension bridge with two separated steel box girders based on field measurements. Engineering Structures, 2011, 33(6): 1894-1907. https://doi.org/10.1016/j.engstruct.2011.02.017

[4] Annamdas V G M, Bhalla S, Soh C K. Applications of structural health monitoring technology in Asia. Structural Health Monitoring, 2017, 16(3): 324-346. https://doi.org/10.1177/1475921716653278

[5] Arul M, Kareem A, Kwon D K. Identification of Vortex-Induced Vibration of Tall Building Pinnacle Using Cluster Analysis for Fatigue Evaluation: Application to Burj Khalifa. Journal of Structural Engineering, 2020, 146(11): 04020234. https://doi.org/10.1061/(ASCE)ST.1943-541X.0002799

[6] Lim J, Kim S, Kim H K. Using supervised learning techniques to automatically classify vortex-induced vibration in long-span bridges. Journal of Wind Engineering and Industrial Aerodynamics, 2022, 221: 104904. https://doi.org/10.1016/j.jweia.2022.104904

[7] HUA Xugang, SUN Ruifeng, WEN Qing, et al. Automatic detection of vortex-induced resonance events in bridges using novelty detection. Journal of Vibration Engineering, 2018,31(6):948-956.

[8] Huang Z, Li Y, Hua X, et al. Automatic Identification of Bridge Vortex-Induced Vibration Using Random Decrement Method. Applied Sciences, 2019, 9(10): 2049. https://doi.org/10.3390/app9102049

[9] Lu J, Behbood V, Hao P, et al. Transfer learning using computational intelligence: A survey. Knowledge-Based Systems, 2015, 80: 14-23. https://doi.org/10.1016/j.knosys.2015.01.010

Structural Health Monitoring: 10APWSHM   Materials Research Forum LLC
Materials Research Proceedings 50 (2025) 331-337   https://doi.org/10.21741/9781644903513-39

Machine learning-based estimation method of seismic response of building's unobserved floor

Daiki Kakehashi[1,a] *, Takenori Hida[1,b]

[1]Ibaraki University, 4-12-1, Nakanarusawa-cho, Hitachi-shi, Ibaraki, Japan

[a]24nm811y@vc.ibaraki.ac.jp, [b]takenori.hida.mn75@vc.ibaraki.ac.jp

Keywords: Seismic Response, Unobserved Floor, Machine Learning, Time Series Prediction, LSTM

Abstract. There is a need for technology that can automatically and immediately identify and evaluate the structural integrity of a building. This study aims to propose a method that utilizes machine learning to estimate the building responses of unobserved floors based on strong motion records. Numerical experiments were performed using a seismic response analysis model of RC super high-rise buildings to assess the estimation accuracy. As a result, the estimation accuracy was improved by increasing the amount of data used for training.

Introduction

In the 2011 off the Pacific coast of Tohoku Earthquake occurred in Japan, structures with long natural periods, such as RC super high-rise buildings, were resonated with long-period ground motion. Consequently, those buildings suffered damage and the stiffness of the building degraded [1]. Additionally, the first natural frequency of RC super high-rise buildings decreased during the seismic shaking [1].

In order to identify the damaged floors of such a damaged building, it is necessary to obtain the inter-story drift angles for all floors. Hence, the strong motion of all floors should be observed. However, due to constraints related to costs and technology, it is common to install accelerometers on a limited number of floors.

Considering this background, this study proposes a method to estimate the building response of unobserved floors using machine learning. In this method, the responses of floors where sensors are not installed are estimated based on the strong motion records observed at the limited floors of the building. To investigate the estimation accuracy of the building responses for each floor, we perform numerical experiments using data obtained from the seismic response analysis of RC super high-rise buildings.

Overview of LSTM (Long Short-Term Memory)

In this study, we utilize Long Short-Term Memory (LSTM), a type of Recurrent Neural Network (RNN) specifically designed for time series prediction. Figure 1 presents a conceptual diagram of LSTM. In Figure 1, where x_t denotes the input data at time t, h_t is the hidden state, and c_t is the memory cell. In addition, the functions σ and tanh correspond to the sigmoid and hyperbolic tangent functions, respectively. The memory cells control information forgetting, input, and output, as well as information retention and updating. Three gate mechanisms (Input Gate, Forgetting Gate, and Output Gate) manage the flow of information. This approach helps to address the vanishing gradient problem faced by traditional RNNs, allowing LSTM to learn long-term dependencies effectively.

Figure 2 demonstrates the method for one-step-ahead prediction using LSTM. A series of m data steps, ranging from the $n-m$ step to the n step, is input, and the $n+1$ step value is predicted as the output. By repeating this process sequentially, time series predictions are achieved. The error between the predicted data and its corresponding true values (target data) is evaluated at each step

Structural Health Monitoring: 10APWSHM   Materials Research Forum LLC
Materials Research Proceedings 50 (2025) 331-337   https://doi.org/10.21741/9781644903513-39

using Mean Squared Error (MSE). The back propagation through time (BPTT) algorithm is employed to minimize the average loss across all steps (hereafter, the dataset consisting of input data and their corresponding target data is referred to as training data). Finally, the trained LSTM model (hereafter referred to as the Trained LSTM) is used with untrained data to evaluate the estimation accuracy.

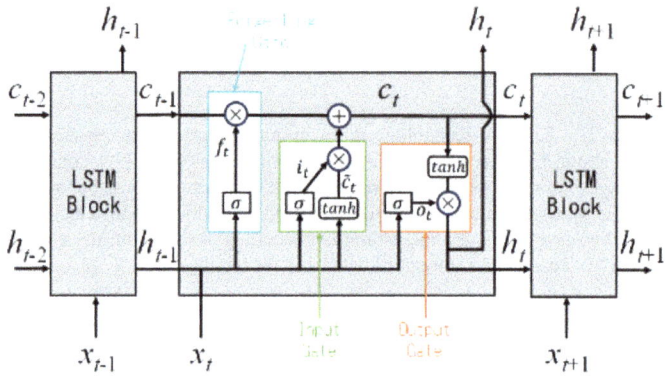

Figure 1 Conceptual Diagram of LSTM

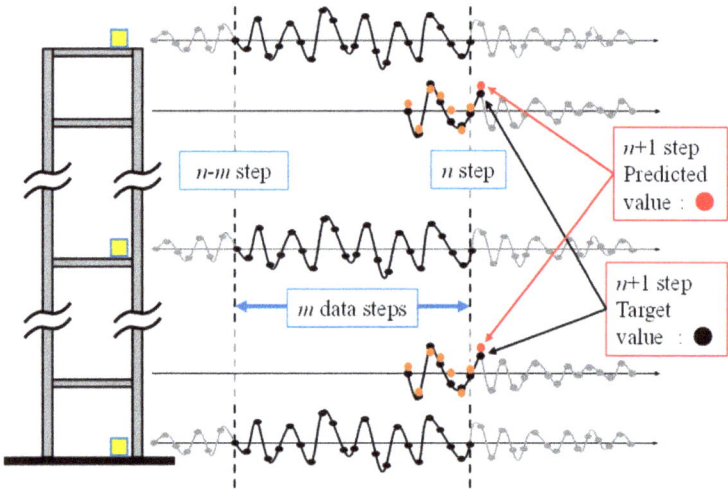

Figure 2 One-step-ahead Prediction Using LSTM

Structural Health Monitoring: 10APWSHM   Materials Research Forum LLC
Materials Research Proceedings 50 (2025) 331-337   https://doi.org/10.21741/9781644903513-39

Case study based on seismic analysis model of RC super high-rise building

Next, to investigate the estimation accuracy of the building responses for each floor, we perform numerical experiments using data obtained from the seismic response analysis of RC super high-rise buildings. The analytical model of the building is illustrated in Figure 3. This model is based on seismic response analysis models of 39 reinforced concrete (RC) super high-rise buildings with seismic-resistant structures, designed to represent the average response of these buildings. The analytical model is a shear-type multi-degree-of-freedom system, with restoring force characteristics of each story defined by a degrading trilinear model. Damping is set to be tangent stiffness-proportional, with a damping ratio of $h = 0.03$ for the first natural frequency. The standard floor area is set to 900 [m²], and the building has 20 stories.

Hereafter, the analytical model with the skeleton curves shown in Figure 4 is referred to as Model 1. Furthermore, the model with an initial stiffness of 0.6 times the initial stiffness K_1 of Model 1 is referred to as Model 2. The natural periods and frequencies of Models 1 and 2 are summarized in Table 1. Seismic response analysis is performed by SNAP Ver.8, which enables elasto-plastic seismic response analysis of building [2].

Table 2 lists the input motions used for the analysis, and Figures 5 and 6 show the acceleration time history of input motions and Fourier amplitude spectra (smoothed with a Parzen window with a bandwidth of 0.1 [Hz]) [3].

Table 1 Analytical Model Parameters

Analy. model name	K_1 Ratio	natural period [s]			natural frequency [Hz]		
		mode 1	mode 2	mode 3	mode 1	mode 2	mode 3
Model 1	1 times	1.200	0.468	0.295	0.833	2.135	3.388
Model 2	0.6 times	1.549	0.605	0.381	0.646	1.653	2.624

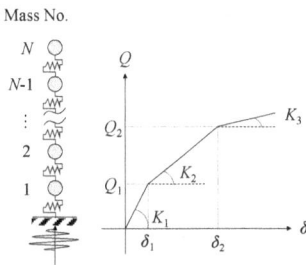

Figure 3 Analysis model of RC super high-rise building

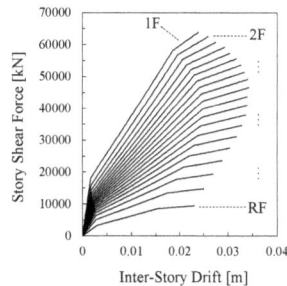

Figure 4 Restoring Force of each story

Table 2 list of input motions

wave name	Observation Date	Observation Location	component	Recording Time [s]
wave 1	2011.03.11.14.46	Ibaraki Prefecture (IBRH11)	EW	299.99
wave 2	2018.09.06.03.07	Hokkaido (HKD127)	EW	283.99
wave 3	2021.02.13.23.07	Miyagi prefecture (MYGH10)	NS	299.99
wave 4	2024.01.01.16.10	Ishikawa prefecture (ISK006)	EW	299.99
wave 5	2024.01.01.16.10	Ishikawa prefecture (ISK007)	NS	299.99

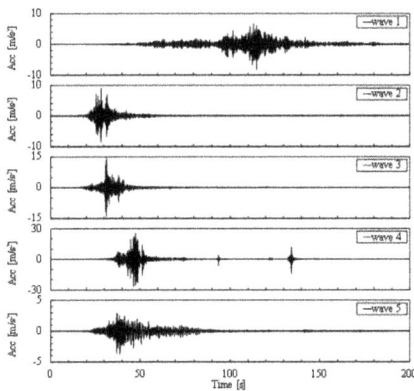

Figure 5 Acceleration Time History of Input Motions

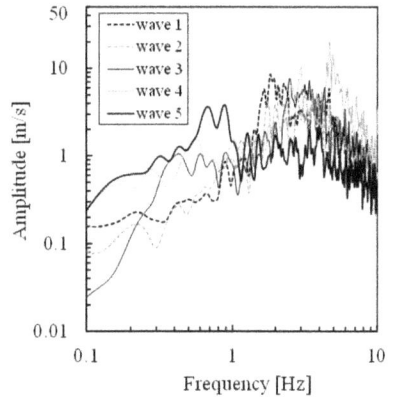

Figure 6 Fourier Amplitude Spectra

Summary of numerical experiments

In this study, we assume that strong-motion observations are performed only on the first floor, the intermediate floor (11th floor), and the roof floor of the building. Based on the strong-motion records obtained there, we estimate the response at the unobserved floors. The input data for training of LSTM consists of absolute acceleration time histories observed at the first floor, the intermediate floor and roof floor. The target data for training is the absolute acceleration time histories of the unobserved floors. The training dataset is generated by performing seismic response analysis of Model 1.

For training, we use the data obtained from the analytical results of Model 1 using wave 1 to wave 4 shown in Table 2. For accuracy evaluation, we use the data obtained from the analytical results of Model 1 and Model 2 (untrained) using wave 5 shown in Table 2.

In the following experiments, we use the Adam optimizer as the optimization method during training, with the mean squared error (MSE) as the loss function. The input data steps m is set to 400, batch size to 32, number of hidden layers to 2, number of LSTM blocks to 30, and the number of training epochs to 50.

Result

The loss curve during training is shown in Figure 7. It can be observed that the loss converges to a small value. Hereafter, we refer to this as Trained LSTM 1. We estimated the response of unobserved floors, inputting wave 5 into Models 1 and 2, by using Trained LSTM 1.

Structural Health Monitoring: 10APWSHM   Materials Research Forum LLC
Materials Research Proceedings 50 (2025) 331-337   https://doi.org/10.21741/9781644903513-39

Figure 8 presents the root mean squared error (RMSE), shown as a bar chart, calculated based on the difference between the target and predicted acceleration amplitude at each floor. The RMSE for each floor was calculated by normalizing the target and predicted acceleration amplitude. The RMSE values of Model 2 are generally larger than those of Model 1. Furthermore, RMSE tends to increase for floors further away from the observed floors.

As an example of the results predicted by Trained LSTM 1, the acceleration time history waveforms of the 17-floor of Model 1 and Model 2 are shown in Figures 9 and 10, respectively. In the legends, "T" denotes target, "P" denotes predicted, and "FL" denotes Floor Level. The estimation result of Model 1 showed high accuracy. On the other hand, the estimation result of Model 2 was relatively low. This is because the natural period of Model 2 is longer than Model 1.

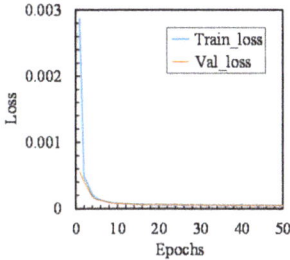

Figure 7 Loss Curve   Figure 8 RMSE

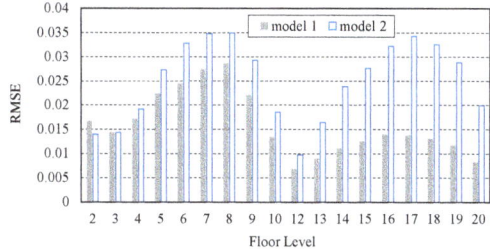

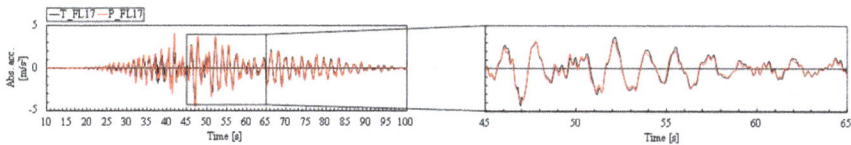

Figure 9 Time History of Floor Acceleration, Model 1

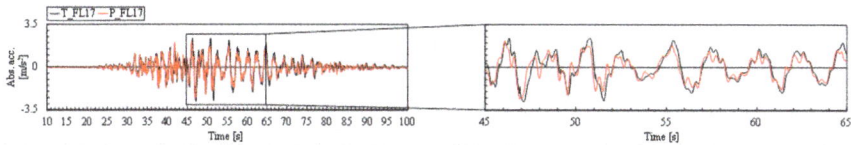

Figure 10 Time History of Floor Acceleration, Model 2

Response estimation of untrained building models

To improve the accuracy of response estimation for unobserved floors in an untrained analysis model (Model 2), we add further analysis models to the training dataset. Table 3 presents the natural period and the natural frequency of the additional analysis models. The initial stiffness of each model was set by multiplying coefficient to the initial stiffness K_1 of Model 1. For training, we use the data obtained from the analytical results of Model 1 and Model 3 to 6 using wave 1 to wave 4 shown in Table 2. For accuracy evaluation, we use the data obtained from the analytical results of Model 2 using wave 5 shown in Table 2.

Table 3 Analytical Model Parameters

Analy. model name	K_1 Ratio	natural period [s]			natural frequency [Hz]		
		mode 1	mode 2	mode 3	mode 1	mode 2	mode 3
Model 1	1 times	1.200	0.468	0.295	0.833	2.135	3.388
Model 2	0.6 times	1.549	0.605	0.381	0.646	1.653	2.624
Model 3	0.75 times	1.386	0.541	0.341	0.722	1.849	2.934
Model 4	0.9 times	1.265	0.494	0.311	0.791	2.025	3.214
Model 5	1.3 times	1.052	0.411	0.259	0.950	2.434	3.863
Model 6	1.5 times	0.980	0.383	0.241	1.021	2.614	4.149

Result

The loss curve during training is shown in Figure 11. Hereafter, this model is referred to as Trained LSTM 2. Additionally, in the following sections, the estimation result of Model 2 calculated by Trained LSTM 1 is referred to as case 1. Whereas the estimation result of Model 2 calculated by Trained LSTM 2 is referred to as case 2.

Figure 12 shows a comparison of RMSE values for each floor between case 1 and case 2, shown as a bar chart. RMSE values of case 2 are lower than those of case 1 on all floors.

The acceleration time history waveform of the 17-floor of case 2 is shown in Figure 13. Compared to Figure 10 shown previously, the prediction accuracy was improved.

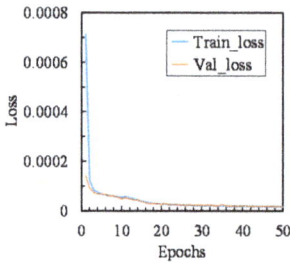

Figure 11 Loss Curve

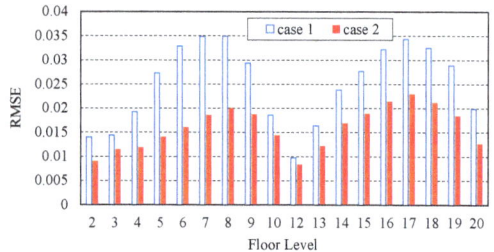

Figure 12 RMSE

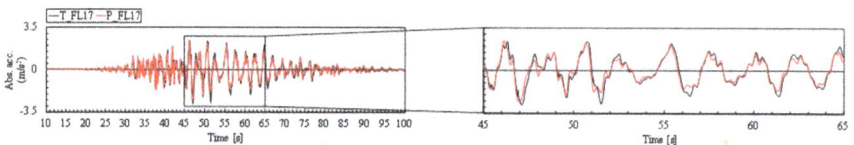

Figure 13 Time History of Floor Acceleration, case 2

Peak floor response and inter-story drift angle

The height distribution of Peak Floor Acceleration (PFA), Peak Floor Velocity (PFV), Peak Floor Displacement (PFD), and Peak Inter-Story Drift Angle (PIDA) of Model 2 are shown in Figure 14. Velocity and displacement were calculated by numerical integration of acceleration.

Focusing on each Target data, PFA tends to be large on both the lower and upper floors. The higher the floor, the larger the PFV and PFD. PIDA on all floors remain are less than 1/100 [rad], which is the design criterion in Japan. Focusing on the estimation results, some PIDA values exceed 1/100 [rad], but case 2 is closer to the Target value for more floors than case 1. PFD shows larger errors on the upper floors, whereas case 2 results are closer to the Target than those of case 1. PFA, PFV, and PFD are closer to the Target in case 2 than in case 1.

These results indicate that the estimation accuracy of untrained model response is improved by increasing the number of building model used for training of LSTM.

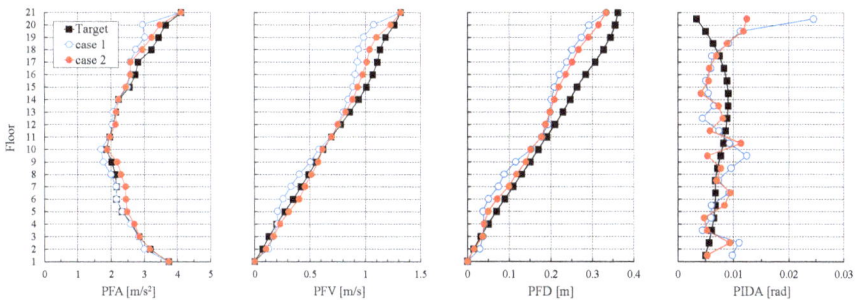

Figure 14 Peak floor response

Conclusion

In this study, based on the seismic response analysis of an RC high-rise building, we proposed a method for estimating the seismic response of unobserved floors using LSTM. The estimation result of trained building model showed good accuracy. The estimation accuracy of untrained building model was improved by increasing the number of building models used for training of LSTM.

References

[1] National Institute for Land and Infrastructure Management, Ministry of Land, Infrastructure, Transport and Tourism; Building Research Institute: Report on the Damage Investigation of the 2011 off the Pacific coast of Tohoku Earthquake, March 2012. (in Japanese)

[2] Kozo system: SNAP Ver. 8, https://www.kozo.co.jp/index.php (accessed 2024.9.9) (in Japanese)

[3] National Research Institute for Earth Science and Disaster Resilience: Strong-motion seismograph networks (K-NET, KiK-net), https://www.kyoshin.bosai.go.jp/

Structural Health Monitoring: 10APWSHM
Materials Research Proceedings 50 (2025) 338-346

Materials Research Forum LLC
https://doi.org/10.21741/9781644903513-40

Real-time seismic response prediction for electrical equipment using CNN-MLP model

Renpeng Liu[1,a], Zhihang Xue[2,b], Qiang Xie[1,c *]

[1]Tongji University, Shanghai 200092, China

[2]State Grid Sichuan Electric Power Research Institute, Sichuan Province 610041, China

[a]renpengliu@tongji.edu.cn, [b]xuezhihang2@163.com, [c]qxie@tongji.edu.cn

Keywords: Substation Equipment, Seismic Response Prediction, Temporal Prediction, CNN-MLP, Wall Bushing

Abstract. Monitoring the seismic response of substation equipment is critical for developing effective real-time emergency response strategies. However, traditional monitoring approaches face significant challenges due to the electromagnetic sensitivity of electrical equipment. This study proposes a CNN-MLP neural network architecture that aggregates both long-duration seismic data and local temporal window information to enable real-time prediction of equipment acceleration responses based on seismic accelerations. Additionally, a beyond-range training strategy is introduced to enhance model performance under high-intensity seismic conditions. The proposed model is evaluated using a damped ±800kV wall bushing, which has been validated through shaking table tests, and compared to an LSTM model. Results show that the CNN-MLP model significantly outperforms the LSTM model, with the beyond-range training strategy effectively improving prediction accuracy and stability. This method requires only ground motion signals as input, avoiding the need to install sensors on electromagnetically sensitive equipment surfaces, thereby providing reliable support for emergency decision-making in substations during earthquakes.

1. Introduction

Substations and converter stations serve as critical roles within modern power grids, housing a wide array of complex electrical equipment. Studies have shown electrical equipment is particularly vulnerable to seismic events[1]. During an earthquake, promptly monitoring the seismic response of numerous electrical devices within substations or converter stations is crucial for effective emergency response. By monitoring real-time seismic data, it is possible to issue early warnings or initiate power shutdowns for overstressed equipment, which can help prevent mechanical damage from escalating into severe electrical failures, such as fires or explosions.

However, due to the high voltage and strong current characteristics of electrical equipment, there are significant challenges. On one hand, the strong electromagnetic interference generated by such equipment severely affects the measurement accuracy of various sensors. On the other hand, ultra-high voltage equipment requires specific insulation clearances, and installing sensors on the equipment surface would compromise these insulation conditions, thereby affecting the normal electrical functions of the equipment. Therefore, monitoring the seismic response of electrical equipment using contact-based sensors is not a feasible approach.

For electrical equipment, developing finite element models and utilizing incremental dynamic analysis (IDA) for seismic performance assessment is a commonly used and well-established approach[2]. Additionally, for some structurally simple electrical equipment, such as post equipment, theoretical models can be established for mathematical solutions to compute their mechanical responses under seismic loads[3]. However, finite element analysis (FEA) is time-consuming and cannot meet the demands for rapid assessment during an earthquake. Although

Structural Health Monitoring: 10APWSHM
Materials Research Proceedings 50 (2025) 338-346

Materials Research Forum LLC
https://doi.org/10.21741/9781644903513-40

theoretical models are efficient for real-time solutions, they involve considerable structural simplifications and are applicable to a limited range of electrical equipment, making them inadequate for post-earthquake emergency repairs.

Some researchers have utilized machine learning methods to develop "point-to-point" prediction models that map seismic parameters to the peak response of electrical equipment. Zhu et al. constructed a surrogate model using machine learning to map monitoring parameters to the electrical contact condition of transformer bushings[4]. Furthermore, Zhu proposed a rapid assessment framework for substation loop systems based on seismic signals, using an optimized machine learning model to evaluate the post-earthquake response of electrical equipment[5]. However, these methods often require complete seismic data to make predictions, which means that they can only be applied after the seismic event has ended, making them unsuitable for real-time emergency decision-making during an earthquake.

With the development of deep learning, significant progress has been made in constructing seismic response time-series prediction models for buildings, bridges, and other structures using various deep neural networks (DNNs). Torky et al. utilized a convolutional long short-term memory network (ConvLSTM) to capture the nonlinear response of buildings under seismic loading[6]. For underground structures, Huang et al. employed a one-dimensional convolutional neural network (1D-CNN) and a long short-term memory network (LSTM) to model the time-series response of a two-story, three-span subway station, applying the findings to seismic response analysis[7]. Liao et al. combined the long-sequence prediction capabilities of LSTMs to propose a stacked residual LSTM network, which effectively predicted the response of complex bridge structures[8]. In summary, time-series prediction methods based onDNNs have been validated for their feasibility in predicting nonlinear system seismic responses, showing notable computational efficiency advantages during the prediction stage[9]. However, in the domain of electrical equipment, deep learning-based seismic response time-series prediction is still in its early research stages[10].

Inspired by existing research, this paper proposes a CNN-MLP neural network architecture for real-time prediction of the seismic response time-series of substation equipment. Additionally, a beyond-range training strategy is introduced to enhance the accuracy of time-series prediction. The model is trained using a damped ±800kV DC wall bushing, which has been validated through shaking table tests, demonstrating the effectiveness and practical value of the proposed model architecture and training strategy in predicting the seismic response of electrical equipment.

2. Methodology

2.1 Dataset Structure
Taking a single seismic event as an example, its tri-axial acceleration time-series data can be represented as $X = \{x_1, x_2, ..., x_T\}$, $Y = \{y_1, y_2, ..., y_T\}$, and $Z = \{z_1, z_2, ..., z_T\}$, where T is the number of time steps for the seismic event. Correspondingly, the key monitoring target of the equipment, such as acceleration response, can be represented as $S = \{s_1, s_2, ..., s_T\}$. During training, the CNN-MLP model uses the equipment response at time step $t(s_t$, where $1 \leq t \leq T)$ as the label, with the corresponding feature array $F_t \in \mathbb{R}^{4\times(w+l)}$ as shown in Eq.1.

$$F_t = \begin{bmatrix} x_{t-(w+l)+1} & \cdots & x_{t-w+1} & \cdots & x_t \\ y_{t-(w+l)+1} & \cdots & y_{t-w+1} & \cdots & y_t \\ z_{t-(w+l)+1} & \cdots & y_{t-w+1} & \cdots & z_t \\ s_{t-(w+l)} & \cdots & s_{t-w} & \cdots & s_{t-1} \end{bmatrix}. \tag{1}$$

In Eq.1, w and l represent the lengths of the short-term and long-term time windows, respectively. If $t - (w + l) < 0$, zero-padding is applied, indicating that no seismic motion occurred before time step $t = 1$, and the corresponding equipment response is zero.

Structural Health Monitoring: 10APWSHM   Materials Research Forum LLC
Materials Research Proceedings 50 (2025) 338-346   https://doi.org/10.21741/9781644903513-40

In summary, F_t and s_t correspond one-to-one, forming a training sample (F_t, s_t) for the CNN-MLP model. Assuming there are N seismic events, each with a time length of $T_i (1 \le i \le N)$, a total of $T_{all} = \sum_{i=1}^{N} T_i$ training samples can be obtained.

2.2 CNN-MLP Architecture

The proposed CNN-MLP model architecture is illustrated in Fig. 1. For each sample F, it can be divided into two parts: the long-term time window $F_l \in \mathbb{R}^{4 \times l}$ and the short-term time window $F_s \in \mathbb{R}^{4 \times w}$. On one hand, F_l is fed into the CNN module, where long-term time-series information is extracted and aggregated through 1D grouped convolution and adaptive average pooling. After flattening, it is concatenated with the short-term time window F_s and passed through multiple fully connected layers, yielding the prediction y_1 from the CNN module. On the other hand, F_s is flattened and fed separately into the MLP module, which also passes through multiple fully connected layers to extract the detailed information within the short-term window, yielding the prediction y_2 from the MLP module. In the final output stage of the model, the combined prediction is calculated as $y = \alpha \cdot y_1 + \beta \cdot y_2$, where the weight coefficients α and β are learned automatically during backpropagation to achieve the optimal combination of the CNN and MLP modules under different prediction scenarios.

The advantage of the proposed CNN-MLP model architecture lies in its hybrid design, which considers both the global variation trends within the long-term time window and retains the local fluctuations of seismic acceleration and equipment response within the short-term time window. Compared to other time-series prediction models, such as LSTM and Transformer, the CNN-MLP model has a simpler structure, achieving both accuracy and enhanced training and prediction efficiency, thereby better supporting the real-time post-earthquake assessment needs of substations.

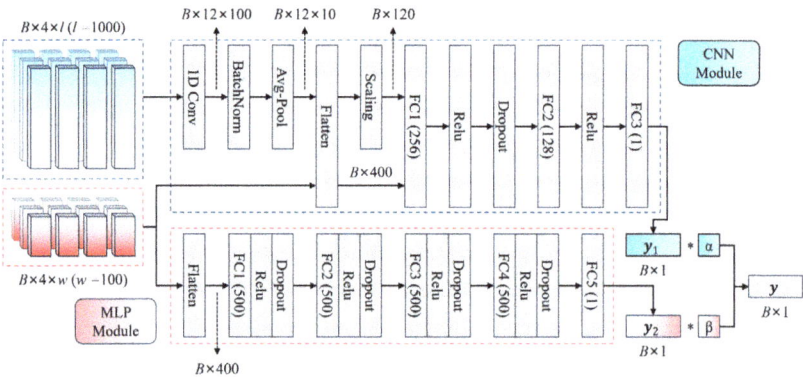

Fig.1 CNN-MLP Architecture

2.3 Beyond-Range Training Strategy

In previous studies on seismic response prediction, dataset construction is often aligned with the model's target peak ground acceleration (PGA) range. For instance, to train a model capable of providing good prediction accuracy within the range of [0.1g, 0.4g], the input seismic waves are scaled to the same acceleration range during dataset construction. This approach is suitable for "point-to-point" prediction models; however, it presents certain limitations when applied to seismic time-series prediction problems.

Specifically, seismic time-series data often exhibit a long-tailed distribution, meaning that a large portion of the data is concentrated in lower acceleration ranges, while only a few time points

Structural Health Monitoring: 10APWSHM   Materials Research Forum LLC
Materials Research Proceedings 50 (2025) 338-346   https://doi.org/10.21741/9781644903513-40

fall within higher acceleration ranges. When performing seismic response time-series prediction, the scarcity of seismic data in high-acceleration ranges may result in insufficient training samples for time-series models, thereby affecting their predictive performance under high-intensity seismic conditions.

To address this issue, this paper proposes a beyond-range training strategy. This strategy involves using seismic samples that exceed the target PGA range during dataset construction as excitation inputs. For instance, seismic samples in the range of [0.4g, 0.5g) are used to improve the model's predictive capability for the [0.1g, 0.4g) range. The feasibility of this approach lies in the continuity of structural response time-series and seismic excitation data; the response of a structure under a PGA of 0.5g inherently includes information about its response under lower accelerations. By deliberately using seismic waves with amplitudes beyond the target range, this approach enriches the response data for high-acceleration ranges, mitigating the issue of uneven sample distribution and improving the model's predictive performance. The effectiveness of this training strategy is validated in the case study in Section 3.

3. Case Study

3.1 ±800kV Wall Bushing

The case study selected is a ±800kV wall bushing installed in a converter station in China, serving as a critical link between the high-voltage valve hall and direct current (DC) field equipment. The overall structure of the wall bushing is shown in Fig. 2(a), which comprises an outdoor insulator, an indoor insulator, corona shields fixed at both ends of the bushing, and an intermediate flange. The flange is connected to eight ring dampers on both the top and bottom through connecting plates, with the other ends of the dampers securely attached to the steel structural frame of the valve hall, as depicted in Fig. 2(b). During the shaking table tests, the entire equipment was fixed to a rigid support, with the bushing axis forming a 10° angle with the horizontal direction. Previous studies have indicated that wall bushing equipment, due to its length, weight, and flexibility, exhibits a high degree of vulnerability during earthquakes. Therefore, monitoring its seismic response is essential.

(a) Wall bushing and shaking table setup   (b) Ring dampers
Fig.2 ±800kV DC Wall Bushing

Structural Health Monitoring: 10APWSHM   Materials Research Forum LLC
Materials Research Proceedings 50 (2025) 338-346   https://doi.org/10.21741/9781644903513-40

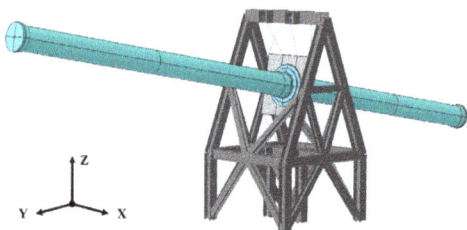

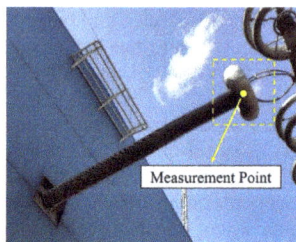

Fig.3 Finite Element Model of Wall Bushing   *Fig.4 Outdoor End of the Bushing*

Due to the limitations of the shaking table tests, only a limited number of scenarios could be simulated, resulting in a restricted amount of experimental data. Therefore, a finite element model was developed using ABAQUS, as shown in Fig. 3. In this model, the indoor and outdoor bushings, as well as the metal flange, were modeled using C3D8R solid elements. The corona rings at both ends of the bushing were simplified as concentrated masses, and the mechanical properties of the ring dampers were represented by variable friction springs. Detailed modeling information can be found in Reference [11].

The FEA results were compared with the test results, including the first four natural frequencies and the acceleration response at the top of the indoor bushing under artificial seismic excitation, as shown in Table 1. A positive sign indicates that the FEA result is greater than the experimental result, while a negative sign indicates the opposite. As shown in Table 1, the differences between the finite element model and shaking table test results, for both frequency and seismic response, are within 11%, which is within an acceptable range. Therefore, the finite element model can accurately simulate the dynamic response of the equipment, and its results are considered reliable for constructing a time-series dataset. Due to space limitations, details of the shaking table tests can be found in Reference [12].

Table 1 Comparison of experimental results and FEA results

	Frequency [Hz]				Acceleration response [m/s²]	
	First Order (Y)	Second Order (Z)	Third Order (Y)	Fourth Order (Z)	Y	Z
FEA	0.95	1.54	2.94	3.00	37.21	33.59
Tests	1.01	1.39	3.27	3.30	38.53	36.46
Error	-5.94%	10.79%	-10.09%	-9.09%	-3.43%	-7.87%

As shown in Fig. 4, the top end of the outdoor bushing in the converter station is connected to the DC field equipment using flexible conductors, which are prone to tensile forces between the equipment during seismic events. Therefore, the subsequent prediction model targets the Y-direction acceleration time-series at the end of the outdoor bushing (assumed to be the primary vibration direction under seismic loading) to simulate real-time monitoring of the seismic response at the outdoor bushing end during an earthquake.

3.2 Seismic Motion Selection
The converter station housing the aforementioned wall bushing is classified as a Category II site condition according to Chinese standards[13], with a corresponding shear wave velocity Vs30 ranging from 265m/s~550m/s. Based on this shear wave velocity, 180 seismic waves were randomly selected from the NGA-WEST2 database[14], which includes three-dimensional ground motion time-series data. To ensure the model's generalization performance, no additional parameter restrictions were applied. According to Chinese standards, the target seismic

Structural Health Monitoring: 10APWSHM | Materials Research Forum LLC
Materials Research Proceedings 50 (2025) 338-346 | https://doi.org/10.21741/9781644903513-40

fortification intensity for the equipment is set at 9 degrees[15], corresponding to a design basic acceleration value of 0.4g. Therefore, the prediction algorithm must ensure the accuracy of device response predictions for seismic actions of 0.4g and below.

According to the beyond-range training strategy proposed in Section 2.3, 160 seismic waves were randomly selected, with a PGA scaling range of [0.4g, 0.5g]. Among these, 128 seismic waves and their corresponding equipment response time-series were used as the training set, while 32 seismic waves and their response data were set aside as the validation set to identify the optimal model hyperparameters and implement early stopping. The ratio of the training set to the test set is 0.8:0.2. The remaining 20 seismic waves, with a scaling range of [0.1g, 0.4g], were used as the test set to evaluate the performance of the prediction model. The equipment response data were generated using ABAQUS, with the scaled seismic waves serving as excitation, and the IDA method was employed to obtain the data. Notably, to facilitate model training, all seismic waves were trimmed to retain only the time-series data for 10 seconds before and after the peak acceleration occurrence. Since the input features for the CNN-MLP model are time-series windows, this approach does not affect the training results.

3.3 Real-time Rolling Prediction

During an earthquake, to achieve real-time prediction of equipment responses, a recursive prediction strategy is adopted[16]. The prediction process is illustrated in Fig. 5. At time step $t = 1$, the seismic monitoring system within the substation captures three-dimensional acceleration signals and performs zero-padding, creating the input array I_1 alongside the initial equipment response (which is 0). This input array is fed into the trained CNN-MLP model to obtain the predicted equipment seismic response value $\hat{s}_1$. $\hat{s}_1$ is then used as part of the input array I_2 for time step $t = 2$, leading to the prediction of $\hat{s}_2$. This recursive prediction process continues in a similar manner, ultimately yielding the complete response time series of the equipment under seismic loading, $\hat{S} = \{\hat{s}_1, \hat{s}_2, ..., \hat{s}_T\}$. Through this recursive prediction strategy, predictions can replace real-time monitoring, supporting emergency decision-making during seismic events.

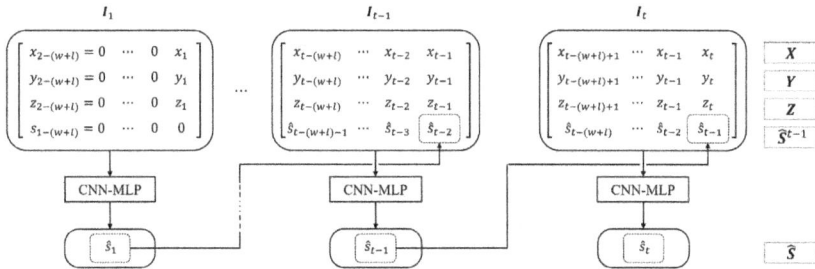

Fig.5 Recursive prediction strategy

3.4 Performance Evaluation

To validate the effectiveness of the proposed CNN-MLP architecture and the beyond-range training strategy, comparative models are set up as shown in Table 2. Models 1 and 2 use different training set PGA ranges to compare the prediction accuracy before and after applying the training strategy. Model 3 employs an LSTM time-series neural network, which is compared to Model 1 to assess the effectiveness of the proposed CNN-MLP model. The dataset divisions among the models are consistent, and hyperparameter optimization has been conducted to ensure that each model achieves optimal predictive performance.

Structural Health Monitoring: 10APWSHM   Materials Research Forum LLC
Materials Research Proceedings 50 (2025) 338-346   https://doi.org/10.21741/9781644903513-40

Table 2 Model Settings

No.	Model	PGA range of training set	PGA range of testing set
1	CNN-MLP-1	[0.4g,0.5g]	[0.1g,0.4g]
2	CNN-MLP-2	[0.1g,0.4g]	[0.1g,0.4g]
3	LSTM	[0.4g,0.5g]	[0.1g,0.4g]

For the prediction results of the selected 20 seismic waves, mean squared error (MSE) and peak relative error (PRE) are used as evaluation metrics. The former measures the overall level of error between the predicted and actual values, and for a single seismic wave sample, it is calculated as shown in Eq. 2. The latter evaluates the model's ability to predict the peak response of the equipment under seismic excitation, as shown in Eq. 3.

$$MSE = \frac{1}{T}\sum_{t=1}^{T}(s_t - \hat{s}_t)^2 \tag{2}$$

$$PRE = \left|\frac{\max(\boldsymbol{S}) - \max(\widehat{\boldsymbol{S}})}{\max(\boldsymbol{S})}\right| \times 100\% \tag{3}$$

In Eq. 2 and Eq. 3, s_t represents the true response of the equipment, $\hat{s}_t$ is the predicted value from the model, and T is the number of time steps for a given seismic event. $\boldsymbol{S}$ and $\widehat{\boldsymbol{S}}$ denote the actual and predicted response time-series, respectively. The smaller the MSE and PRE, the higher the prediction accuracy of the model.

Fig. 6(a) and Fig. 6(b) illustrate the changes in MSE and PRE for three different models (CNN-MLP-1, CNN-MLP-2, and LSTM) as a function of PGA. In terms of MSE performance shown in Fig. 6(a), the average MSE of the CNN-MLP-1 model is significantly lower than that of the CNN-MLP-2 and LSTM models. Specifically, the average MSE for CNN-MLP-1 is 0.0223, while for CNN-MLP-2 it is 0.0552, indicating that the beyond-range training strategy enhances the model's overall predictive capability for time-series data. Additionally, the MSE distribution for the LSTM model is more dispersed; under higher PGA values, the MSE of the LSTM model significantly increases, reaching an average MSE of 0.1322, which is considerably higher than that of both CNN-MLP-1 and CNN-MLP-2. This result demonstrates the effectiveness of the proposed CNN-MLP model architecture.

As shown in Fig. 6(b), for the PRE metric, the CNN-MLP-1 model also demonstrates the lowest error level, with an average PRE of 6.52%. In contrast, the average PRE for CNN-MLP-2 is 9.18%, while the LSTM model reaches as high as 15.38%. This indicates that CNN-MLP-1 has greater accuracy in predicting peak responses. More importantly, the CNN-MLP-1 model using the beyond-range training strategy maintains a maximum PRE of no more than 15% within the range of [0.1g, 0.4g]. This prediction stability is crucial for emergency response during earthquakes, as it effectively reduces the risk of misjudging the equipment's condition under high-intensity seismic actions, ensuring the accuracy of critical response decisions.

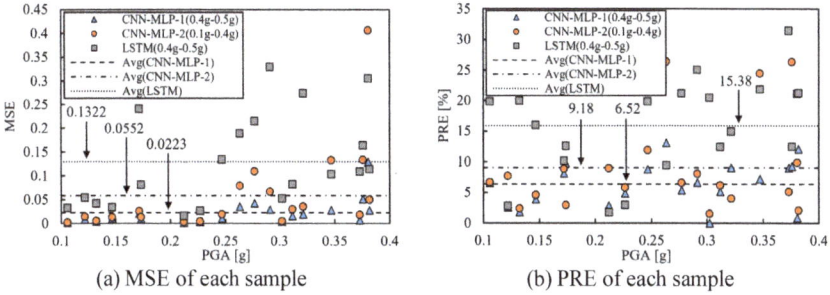

(a) MSE of each sample (b) PRE of each sample

Fig.6 Comparison of prediction errors for each model

In summary, the proposed CNN-MLP model architecture demonstrates excellent prediction accuracy and stability. With the introduction of the beyond-range training strategy, the CNN-MLP-1 model outperforms both the CNN-MLP-2 and LSTM models in terms of MSE and PRE metrics, while also exhibiting lower prediction variability under high-intensity seismic excitation. These results fully validate the effectiveness of the proposed CNN-MLP architecture and the beyond-range training strategy in improving prediction accuracy and handling severe seismic conditions, providing robust support for earthquake emergency decision-making.

Conclusion

This study proposes a CNN-MLP neural network architecture for predicting the seismic response of substation electrical equipment. By introducing a beyond-range training strategy, the model's predictive accuracy and stability are significantly improved. Using a damped ±800kV DC wall bushing as an example, the effectiveness of the proposed model architecture and training strategy is validated. The main conclusions are as follows:

1) The CNN-MLP model outperforms the traditional LSTM model in terms of both mean squared error and peak relative error, especially demonstrating higher stability under high-intensity seismic conditions.

2) The beyond-range training strategy enables the model to exhibit excellent generalization within the target PGA range, effectively reducing the average error and significantly enhancing peak prediction accuracy.

3) The CNN-MLP model trained with the beyond-range strategy shows low error variability, ensuring reliability and consistency in predictions under high-intensity seismic excitation. This is crucial for emergency response during earthquakes, as it helps prevent misjudgment of equipment status and ensures accurate decision-making.

References

[1] Xie Qiang, Zhu Ruiyuan. Earth, Wind, and Ice[J]. IEEE Power and Energy Magazine, 2011, 9(2): 28-36. https://doi.org/10.1109/MPE.2010.939947

[2] He Chang, He Ziwei, Zhu Wang. Seismic interconnecting effects of multi-span flexible conductor-post electrical equipment coupling system[J]. Journal of Constructional Steel Research, 2024, 212: 108209. https://doi.org/10.1016/j.jcsr.2023.108209

[3] Shi Gaoyang, Xie Qiang, Liu Yun, et al. A design method on seismic isolation and reduction of post equipment in substation [J]. Journal of Vibration and Shock, 2023, 42(24): 109-116+142.

[4] Zhu Wang, Wu Minger, Xie Qiang, et al. Post-earthquake rapid assessment method for electrical function of equipment in substations[J]. IEEE Transactions on Power Delivery, 2023: 1-9. https://doi.org/10.1109/TPWRD.2023.3270178

Structural Health Monitoring: 10APWSHM   Materials Research Forum LLC
Materials Research Proceedings 50 (2025) 338-346   https://doi.org/10.21741/9781644903513-40

[5] Zhu Wang, Xie Qiang. Post-earthquake rapid assessment for loop system in substation using ground motion signals[J]. Mechanical Systems and Signal Processing, 2024, 208: 111058. https://doi.org/10.1016/j.ymssp.2023.111058

[6] Torky Ahmed A, Ohno Susumu. Deep learning techniques for predicting nonlinear multi-component seismic responses of structural buildings[J]. Computers and Structures, 2021, 252: 106570. https://doi.org/10.1016/j.compstruc.2021.106570

[7] Huang Pengfei, Chen Zhiyi. Deep learning for nonlinear seismic responses prediction of subway station[J]. Engineering Structures, 2021, 244: 112735. https://doi.org/10.1016/j.engstruct.2021.112735

[8] Liao Yuchen, Zhang Ruiyang, Lin Rong, et al. A stacked residual lstm network for nonlinear seismic response prediction of bridges [J]. Engineering Mechanics, 2022: 1-12.

[9] Xu Zekun, Chen Jun, Shen Jiaxu, et al. Recursive long short-term memory network for predicting nonlinear structural seismic response[J]. Engineering Structures, 2022, null: null. https://doi.org/10.1016/j.engstruct.2021.113406

[10] Guo Yanyan, Chen Yafang, He Chang, et al. Seismic response prediction of electrical equipment interconnected system of traction station based on LSTM neural network [J]. Journal of Railway Science and Engineering, 2023: 1-11.

[11] Wang Xiaoyou, Xie Qiang, Luo Bing, et al. Seismic Responses and Vibration Control of ±800 kV Wall Bushing [J]. High Voltage Apparatus, 2018, 54(1): 16-22.

[12] Xie Qiang, Wang Xiaoyou, Hu Rong, et al. Shaking Table Tests on ±800 kV UHV DC Wall Bushing with Damper Devices [J]. High Voltage Engineering, 2018, 44(10): 3368-3374.

[13] GB 50011-2010, Code for seismic design of buildings, Beijing, 2010 (in Chinese).

[14] Ancheta Timothy D, Darragh Robert B, Stewart Jonathan P, et al. NGA-West2 database[J]. Earthquake Spectra, 2014, 30(3): 989-1005. https://doi.org/10.1193/070913EQS197M

[15] GB 50260-2013, Code for Seismic Design of Electrical Installations, Beijing, 2013 (in Chinese).

[16] Xu Zekun, Chen Jun. Neural network algorithm for nonlinear structural seismic response [J]. Engineering Mechanics, 2021, 38(9): 133-145.

Keyword Index

d

www.ingramcontent.com/pod-product-compliance
Lightning Source LLC
Chambersburg PA
CBHW071320210326

41597CB00015B/1291